"This is a brave book that collaboratively experiments with methods to look at land and water as essential resources during occurring and past crises. Operating with the notion of 'polycrisis' it criticizes using 'crisis' as spectacle; instead, the authors carefully apply long-term cartography and local experience to reflect on ways of dealing with insecurities caused by climate change, pandemics, famine and war, but also political decision making and international prescriptions. Thorough research and an eye and ear for the complexities around these essential topics make this book an essential read for everyone who wants to deepen their understanding of geographic, social and cultural ways of dealing with changes in the Sahel Region."

Andrea Behrends, *Professor of Anthropology, University of Leipzig, Germany*

"Based on solid, in-depth, multi-year field research and a fruitful interdisciplinary dialogue between anthropology, geography and other social sciences, as well as between European and Sahelian researchers, this book offers an up-to-date re-reading of the Sahel and the ecological, economic and social dynamics and tensions that run through it, offering critical cartographic, interpretations and innovative keys to keep the focus on a territory in which we seem to have lost our compass and need to find our way back into a post-colonial pluriverse. A 'reSahelient' read, to find new routes between water and land."

Egidio Dansero, *Professor of Political and Economic Geography, University of Turin, Italy*

"An ambitious and original book on the multiple crises of the Sahel, examined from the perspective of the water-land relationship and from a wide range of perspectives: from ecological rhythms to development, from the long-term to current health and security issues, from social representations to public modernisation policies. Drawing on 30 years of research and mixed methods (field surveys, remote sensing, GIS), the 17 chapters, ranging from general overviews to localised case studies, weave the complexity of the material into the unity of Sahelian issues. An impressive achievement."

Géraud Magrin, *Professor of Geography, University of Paris 1 Panthéon-Sorbonne, France*

"Water is an extremely important topic, looking at drylands and particularly the Sahel. This book is a must-read for those who are interested in the backgrounds to the polycrisis in the Sahel and in drylands in general. Based on decade-long engagement with the region, it provides us with in-depth insights from a variety of disciplinary perspectives on the role of water and its connections with the daily lives of the Sahelians, political economy and conflict. It shows that our 'solutions' failed and that we need to refresh our look at the region and critically examine our narratives and prejudices about this amazingly complex world."

Han van Dijk, *Professor, Sociology of Development and Change Group, Wageningen University, The Netherlands*

Water and Land in the Sahel

Drawing on over 30 years of research, this book investigates the intermingling of land and water in the Sahel, analysing landscapes defined by the ebb and flow of rainfall, flooding and population movements, as well as environmental, political and social crises.

Exploring both the seasonal flooding of rivers around the Nile, Chari-Logone, Niger and Senegal, as well as agricultural irrigation projects such as dams, canals and pumping stations, the book examines the different narratives related to water and land in the Sahel. It combines fieldwork research with remote sensing, big data and GIS mapping to outline the ever-changing interplay between land and water in the region. Beyond this, the book also reinterprets the colonial and post-colonial legacies of large-scale irrigation projects and the geopolitical interests which defined them.

Supported by an Open Access website with a WebGIS and further maps and analysis, this book is an essential read for policy makers and development practitioners in the region, as well as for researchers and scholars across the fields of geography, history, political science, sustainable development and African studies.

Andrea Pase is a full professor of geography at the University of Padua. In the Sahel, his research interests concern territorial processes linked to water seen both as common resource and as the basis for the modernization of agriculture.

Angela Kronenburg García holds postdoctoral fellowships at UCLouvain (FRS-FNRS) and the University of Padua. She is also affiliated with the University Eduardo Mondlane. As an anthropologist, she conducts research on religious and land-tenure change, and on pastoralism, milk and social change in Kenya, as well as on energy transition and graphite.

Mariasole Pepa is a postdoc researcher at the University of Padua and an affiliate researcher at CEDEJ-Khartoum. She is interested in the transformation in the Sahel through the lenses of water and land as well as exploring alternative methodologies to research approaches, ethics and practices.

Federico Gianoli is a geographer specialized in GIS and Geoinformatics, focusing on open-source data in spatial analysis. As a consultant for the Joint Research Centre (JRC-EC), he conducts research on land degradation and productivity dynamics. He's also pursuing a PhD on these topics at the University of Seville, Spain.

Marina Bertoncin is an honorary professor of geography at the University of Padua. In Sahelo-Sudanese Africa her research interests concern the geography of the hydraulic territorialization with a focus on the role of the irrigation projects for the local development (Lake Chad basin, Nile, Niger and Senegal River regions).

Carla Braga is an associate professor of Anthropology at the Faculty of Arts and Social Sciences of University Eduardo Mondlane in Mozambique. Her research interests cover decolonial readings of knowledge production about Africa, onto-epistemic dimensions of the current planetary crisis, nature governance and health.

Routledge Studies in African Geography

Transport Planning and Mobility in Urban East Africa
Edited by Nadine Appelhans, Wolfgang Scholz and Sabine Baumgart

Climate Change Epistemologies in Southern Africa
Social and Cultural Dimensions
Edited by Jörn Ahrens and Ernst Halbmayer

Water and Land in the Sahel
Mapping the Flow
Edited by Andrea Pase, Angela Kronenburg García, Mariasole Pepa, Federico Gianoli, Marina Bertoncin and Carla Braga

Water and Land in the Sahel

Mapping the Flow

Edited by Andrea Pase, Angela Kronenburg García, Mariasole Pepa, Federico Gianoli, Marina Bertoncin and Carla Braga

Routledge
Taylor & Francis Group
LONDON AND NEW YORK

Designed cover image: On the embankment between Lake Maga and the Logone River, near Pouss, Cameroon (Marina Bertoncin, 2011)

First published 2025
by Routledge
4 Park Square, Milton Park, Abingdon, Oxon OX14 4RN

and by Routledge
605 Third Avenue, New York, NY 10158

Routledge is an imprint of the Taylor & Francis Group, an informa business

British Library Cataloguing-in-Publication Data
A catalogue record for this book is available from the British Library

ISBN: 978-1-032-55122-7 (hbk)
ISBN: 978-1-032-55125-8 (pbk)
ISBN: 978-1-003-42911-1 (ebk)

DOI: 10.4324/9781003429111

Typeset in Times New Roman
by KnowledgeWorks Global Ltd.

Contents

Figures

Tables

Contributors

Mohammed Hamed Hassan Abubakr is a PhD student in Geographic Information Systems (GIS) and Remote Sensing (RS) at the University of Khartoum. He is a GIS Specialist with a Master's degree in GIS and RS, complemented by a bachelor's degree in Geoinformatics and a Diploma in Cartography.

Edoardo Baldaro is assistant professor of political science and international relations at the University of Palermo (Italy). He holds a PhD in political science from the Scuola Normale Superiore (Florence, Italy), and he previously served as a postdoc researcher at the Université Libre de Bruxelles, the Sant'Anna School of Advanced Studies (Pisa, Italy) and the 'L'Orientale' University (Naples, Italy).

Andreas Brink works at the Joint Research Centre of the European Commission, leading the global component of the Copernicus Land Monitoring Services. He specializes in the production and dissemination of global land cover and land productivity dynamics (LPD) data, contributing to the development of essential Earth Observation variables and services.

Martina Angela Caretta is an associate professor at the Department of Human Geography at Lund University, Sweden. She is a feminist geographer researching human-environment interactions with expertise in water, climate change adaptation, extractivism and gender. She uses feminist epistemology prioritizing situated knowledge, epistemic plurality, participatory methodologies, ethics of care and a decolonizing approach.

Michael Cherlet has focused on the Sahel region and land degradation, initially working with the FAO and later at the European Commission's Joint Research Centre (JRC). He spearheaded the third edition of the World Atlas of Desertification, introducing innovative approaches to studying global land degradation.

Ludovica Crocitto is a geologist specializing in GIS, remote sensing and geological surveying. Since 2017, she has worked as a freelancer, integrating advanced technologies like Google Earth Engine into her mapping and environmental analysis projects. Passionate about the Dolomites, she focuses her work on mountain environments and innovative use of geospatial data.

Luca de Felice is a geospatial scientist with experience in Earth Observation, hydrological early warning systems and data science. He contributed to projects like the Global Surface Water Explorer and presented at international conferences. As a former researcher at the Joint Research Centre of the European Commission, Luca combines technical expertise with a passion for scientific advancement, and an interest in history.

Evelien de Hoop has a background in STS and human geography. She works, amongst others, on the onto-epistemic politics of landscape change and transformation, bringing historical research in conversation with transformation-oriented knowledge co-production. Evelien is part of the 'Alternatives2Extractivism' initiative, which attends to diverse ways of being-doing-knowing and contributes to pluralizing development imaginaries.

Chiara Forner is a student in the Master's program in Historical Sciences at the University of Padua. Her bachelor's thesis research focused on African cartography, particularly in the Sahel region, during the Modern Era.

Chiara Gallanti is the curator of the Museum of Geography at the University of Padua. Since her PhD, she has been conducting research on its tangible and intangible heritage, as well as on the history of academic geography.

Eltaib Saeed Ganawa is a professor at the University of Khartoum. He has a PhD and an MSc in applications of GIS and remote sensing in soil, land and water. He has more than 28 years of experience practising and teaching GIS, GPS and remote sensing applications in natural resources and the environment and projects in various kinds of soil management.

Aboukar Mahamat is an environmentalist from Cameroon. His expertise focuses on wetlands in the Lake Chad Basin. He is interested on integrated water resource management, conflicts dynamics, climate change and community development. He works with institutions such as the LCBC and UNESCO.

Dobah Marsala is a Chadian lawyer and doctoral student in international environmental law at the University of Yaoundé II. His interests focus on decolonial studies and the critical analysis of development. He is co-founder of the CSO 'ESPOIR', which is active in the field of environmental education, and works as a consultant with the ACRA Foundation.

Esther Miedema has a background in educational and feminist theory. She focuses on gendered, racialized and classed underpinnings of (epistemic) privilege and marginalization, and narratives of health and belonging. Esther is part of the 'Alternatives2Extractivism' initiative, which attends to diverse ways of being-doing-knowing and contributes to pluralizing development imaginaries.

Abdelrahman Eltahir A. Musa, PhD, is a biologist with over 28 years of expertise in environmental management, biodiversity assessment and ecosystem management. He is a research fellow and lecturer at the University of Khartoum, Sudan. His work focuses on environmental management, cultural geography, participatory land-use planning, and climate change in Africa and the GCC states.

Moumouni Ouande holds a PhD in geography, with a specialization in regional planning, at the Joseph KI-ZERBO University, where he has been a teacher-researcher since April 2023. In his doctoral research he studied hydro-agricultural development projects and territorial transformations in Burkina Faso.

Luca Pes is an associate professor of comparative law at the University of Trento. He holds a double PhD in Private Law (Università di Torino) and Anthropology (London School of Economics). His research focuses on politico-legal relations in rural Mali, contributing to the anthropological tradition of studying law, Anthropology and the state in Africa.

Elvira Pietrobon is an architect who graduated from EPFL in Lausanne and is currently a PhD student in urbanism at IUAV in Venice. Since 2018, she has been living in Mali where she works as a member of the Mieruba collective, a cultural centre and publishing house in Ségou.

Ibrahima Poudiougou is a postdoctoral researcher at the Norwegian University of life sciences. He holds a double PhD (2023) in Anthropology and African Studies (Università di Torino and Universiteit Leiden). His research focuses on the articulations between land-use conflicts and armed violence in Mali and the central Sahel.

Francesco Staro is a researcher at the University of Padua. He holds a PhD in Anthropology at the University of Paris 8 and Milano-Bicocca. He carried out field research with mobile pastoralists in several countries in the Horn of Africa and Central Africa, focusing on the study of water management and access, and socio-environmental change in development contexts.

Edmond Totin is a social scientist with expertise in the management of agricultural innovation, climate adaptation and governance. He serves as a senior lecturer at the university and a scientist at the World Vegetable Center. He was one of the Coordinating Lead Authors on the AR6 of the IPCC 6th assessment.

Jeroen Warner is an associate professor of Crisis and Disaster Studies at Wageningen University. He led the European Horizon 2020 project on disasters and cultures, EDUCEN, and is currently co-ordinator of a Dutch-Indian Merian Fund project on early warning, LODESTAR. He is Editor-in-Chief of the *International Journal of Water Governance*.

Mélanie Weynants is a scientist at the Max Planck Institute for Biogeochemistry. Her research spans from soil hydrology and land degradation to ecosystem responses to extreme weather events, drawing on her skills in spatial data analysis and her understanding of the Earth system.

Lassane Yameogo is a professor of geography at the Université Joseph KI-ZERBO in Ouagadougou, where he has taught since 2008, after completing a doctorate at the University of Padua. His works focus on geopolitics of borders, spatial planning and hydraulic territorialization.

Acknowledgements

In all these years of research, the people we have met and those whose mark remains in this volume have been many. Some are living in troubled irrigation mega-schemes, others work in the fields, some fish, while others trade on the rivers and lakes of the Sahel. Still others are with livestock who knows where. Technicians, NGO operators and scholars have generously given us their time and expertise. Some we hope to meet again through this volume: their "voices", hopes and disappointments, have been carefully collected. Yet, even profound silences have not been forgotten. Many we spoke with do not read or write, but the wind carries their voices and silences, and the wordless expressions also speak. We hope we have not erred too much in translation—not linguistically, but semantically. Specific thanks, where needed, are given in the different chapters, but here we would like to thank all the people and all the faces, one by one, who welcomed us in their everyday lives and were open to listening to our questions.

We would like to thank Patricia Sours, Prabhu Chinnasamy, Francesco Ferrarese, Roberto Galati, Valentina Casari, Beatrice Bezzon, Michela Marini, Melania Brolis and Marco Orlandi for their continuous help and essential support for towards the successful development of this volume.

This volume has been published under the frame of the PRIN 2017 project "Contested wetlands in Sahelian drylands: Which kind of development and for whom?" (2017B4JBWN) and the frame of PRIN 2022 project, financed by the European Union – NextGenerationEU, "Socio-Environmental Crises and Local Responses in African Drylands: Lessons from the Global South (SECA)" (2022CZZXZS).

Introduction

Perspectives on water and land in the Sahel: facing polycrisis

Andrea Pase, Angela Kronenburg García, Mariasole Pepa, Federico Gianoli, Marina Bertoncin and Carla Braga

Doing research in the midst of troubled times

This volume is the result of two different research paces. The first, a slower one, began in the early 1990s by geographers at the University of Padua—Marina and Andrea—and has continued developing research for over 30 years within the Sahelian countries from the Atlantic Ocean to the Red Sea. The research focal points have been the Nile River Valley in Sudan, the Senegal River Valley between Senegal and Mauritania, the Sourou Valley in Burkina Faso, the Office du Niger zone in Mali and, with special intensity, the Lake Chad region between Cameroon, Nigeria, Niger and Chad (Bertoncin & Pase, 2012). Constantly returning to these places, several times and in different years and seasons, has made it possible to observe transformations that are sometimes silent and barely visible but also profound (Jullien, 2009).

The second one is faster and troubled. It was funded by the Italian Ministry of University and Research, thus making it possible to initiate new research efforts with an increase of the researchers involved.[1] This research turned into what we refer to as the "AtlaSahel Project" (https://atlasahel.it). The problem was the starting date of the availability of funds: March 2020. At that point, Italy had just entered lockdown, having been hit hard by the COVID-19 pandemic. The entire research framework had to be redesigned, which brought us to reflect on the how(s) of doing research during a global pandemic, as is discussed in Chapter 16. Since then, exceptional crises have accompanied the research group's journey. Researchers in Burkina Faso, in the Far North Province of Cameroon and Chad, have struggled and sometimes failed to reach some of the sites of interest due to security problems (Chapters 11 and 12). In March 2023, two of us, Mariasole and Andrea, were in Sudan. It was an intense fieldwork: many future projects had been set up with Sudanese colleagues for new lines of investigation. In mid-April 2023, a war broke out in the Sudanese capital of Khartoum, triggering, even as we are writing in the fall of 2024, one of the worst humanitarian crises globally. Our Sudanese colleagues were forced into exile in Egypt, the Emirates, the Netherlands or otherwise to leave Khartoum and seek temporary homes. Chapter 10 describes a situation that was about to be devastated—the irrigated spaces in Sudan—leaving a ruined world in

DOI: 10.4324/9781003429111-1

its wake. Some of us are Mozambican or are based in Mozambique. Since 2022, Angela and Carla, in parallel with their participation in this study, have co-organised in the capital city of Mozambique, Maputo, activities for a different research project ("Alternatives to Extractivism", or A2E), which places Mozambique at the centre of a global reflection network on "diverse ways of knowing, being and doing in relation to the (sub)surface". Researchers from all over the world contributed. Such a rich and promising network now has to contend with deep political instability, which in October of 2024—as we write this introduction—led to the assassination of two opposition leaders, one of them a participant in the A2E's project. Thus, the final drafting of Chapter 17 took place in the gloom of tense days in Maputo. One of the greatest challenges that we faced whilst researching and writing this volume was dealing with the exceptional nature of the situations in which we found ourselves living, fearing and reflecting.

Uneven edges of the crises: Dominant narratives, different temporalities and multiple layers

The crisis scenario provoked by the COVID-19 pandemic (see Chapter 16) and securitarian issues (see Chapter 4) is therefore essential to frame the process that produced this volume. Furthermore, the word "crisis" has for at least 50 years been closely linked to the study area that the research deals with: the Sahel. It is the droughts that repeatedly hit the region, starting in 1969 and with particular force in 1973 and 1984, that have led to talk of a "historic crisis" for the entire region (Turri, 1990, pp. 513–516). The devastating famine caused by the drought became one of the first highly mediatised crises: the photographs taken by Sebastião Salgado of the 1984–1985 famine are perhaps one of the well-known testimonies to this fact. There is "an uneasy proximity" (Morton, 2006, p. 176) in photographing (well) such a (horrific) drama. An ongoing debate questions Salgado's work and the risk of transforming tragedy into spectacle through photography, even with the intention of collecting funds for an organisation like *Médicins sans Frontières*. At the time of the release of the first English edition of Salgado's photography book *Sahel: The End of the Road* (Salgado, 2004), anthropologist Christopher Morton (2006) wondered, "What have the people of the Sahel made of these images? How do they interpret these images some twenty years afterwards, and is there a perception that they have helped, as Salgado intended? Or do they resent images that portray them in only one modality, that of physical suffering?" (p. 176). Alexander Gerstenhaber (2023), a columnist for *The Harvard Crimson*, has a different opinion about Salgado's work: "Throughout the project, he depicts countless people dying as they cling to their last moments of life. In this way, criticisms of Salgado's dilution of human suffering are unfounded because none of these images are easy to look at". Indeed, in a world where climate change increases the numbers of extreme weather events, such as droughts and floods that are ravaging millions of people, "photography like Salgado's is critical to transport the realities of the human experience beyond borders and across continents". Beyond the discussion of the appropriateness of the use of high-impact emotional images as the "biggest dilemma of photography" (Gerstenhaber, 2023), the nexus, simplified and rooted in Eurocentric beliefs, between population growth, land degradation and reduced, erratic rainfall has taken root in the dominant narratives on the Sahel crisis, which, in turn, have led to enduring outcomes on development policies, as described and critiqued in Chapter 2.

However, to blame nature (i.e. drought) and then climate change (i.e. overheating) in conjunction with population growth and resource overexploitation for the "desertification of the Sahel" (see Chapter 6), among other things, overshadows long-lasting processes that have led to the progressive marginalisation of the region. Historically, the Sahel has been, in the words of the great French historian Lucien Febvre (1949, p. 369), one of "those privileged places for the emergence of stable political groupings", as well as kingdoms, empires and state organisations because it is located "on the frontiers of large different natural formations (steppes and savannas, savannas and intertropical forests) and at the points of contact of these formations". Trans-Saharan and latitudinal connections have made the area an essential "hinge" of the spatial organisation between the Mediterranean and the Gulf of Guinea (Retaillé, 2018, pp. 64–68). The destabilisation of these political realities and economic potentials has largely been due first to the shifting of the centre of trade from the Sahel to the Atlantic coast in the era of mercantilism and later to the colonial appropriation and partitioning of the region, which permanently broke the Sahelo-Saharan unity (Retaillé & Walther, 2019, pp. 6–8). In more recent times, the narrative of the Sahel crisis has been enriched with a new declination, that of the environment–security nexus, which links "ecological disequilibrium" and climate change with the rise of conflictual violence (Chapter 2). This interpretive lens is simplistic and decidedly rough as well as slippery in terms of the political consequences that can be drawn from it: to break out of a one-dimensional view of the Sahelian crisis, it is necessary to examine it from multiple angles. A multifaceted view is what one of the latest papers by the French anthropologist Jean-Pierre Olivier de Sardan (2023, pp. 11–39) proposes. He identifies eight different intersecting crises in the Sahel: the agro-pastoral crisis, labour, political elites, public services, Islam, Western-centrism, security and national armies. In recent years, the term "polycrisis" has re-emerged in academic discourse and discussions within international agencies to describe situations characterised by multiple interconnected and compounding crises. Initially introduced by Morin and Kern (1999, p. 74), the concept of polycrisis highlights the complexity of interrelated challenges and emphasises that "there is no single vital problem, but many vital problems". Moreover, it is the intricate interconnections among these issues that constitute "one vital problem". In Chapter 16, this concept is further explored by drawing on the work of Henig and Knight (2023), who propose polycrisis as a framework to investigate the intertwined crises that we understand as affecting both communities and researchers. Thus, the Sahel can be viewed from this perspective as a laboratory of a world increasingly immersed in polycrisis, so much so that in his latest book Ulrich Beck proposes the term "metamorphosis" to denote a state of radical change that goes beyond "transformation" (Beck, 2016).

Among the many crises of a world in metamorphosis is the epistemological one, which is related to the difficulty of usefully mobilising the analytical concepts hitherto in use, hence the importance of recognising ways out of this "crisis of research". In our limited sphere of action, we have tried to identify one among the many threads in the tangled skein of the Sahelian polycrisis, following it while attempting to untie at least some of its knots. The thread we followed is that of land and water because in the Sahel almost everything revolves around water and the different uses of the land.

Focus on water and land: Re-drawing the "Sahelian condition"

In 1975, the French geographer Jean Gallais defined the cornerstones of the "Sahelian condition" as a set of possibilities, constraints and ways of living of the population in the region of Gourma (Mali) as an example of a widespread Sahelian situation. As Gallais noted, the different uses of water and land resources varied in the alternation of wet or dry seasons, and were changeable in occurrence with the multi-year drought or rainy cycles and location variability—because of the topo-hydrographic accidents—play a central role in determining the "sedentary-nomadic opposition" (Gallais, 1975, p. 219). This "sahélité" has been criticised by many, most recently by the anthropologist Jean-Loup Amselle (2022), as "geographic essentialism" easily combined with an "ethnic essentialism" (pp. 18–19), which are both of colonial derivation. Actually, Gallais's discourse is much more multifaceted: there are so many categories ("pastoralists", "herders", "herder-farmers", "farmer-herders", "farmers") which can be declined in different ways with respect to mobility ("nomads", "short-range nomads", "partially sedentary", "sedentary") (Gallais 1975, p. 189). In another renowned work of the French geographer on the inner Niger Delta (Gallais, 1984), he references further declinations of resource use related to fishing, rice farming and river trade. There are two aspects that we would like to emphasise. The first is the centrality of water and land in the Sahel: the different combinations of the two resources are the foundation of the many ways of life elaborated over time in the region. The second is that the 1975 volume closes with the dramatic outcomes of the year just ended, "the year of drought, distressing for the future" (Gallais 1975, p. 220). The word "condition" indicates in its etymology a degree of stability and a mode of being that continues over time. Thus, the "Sahelian condition" would seem to indicate constants in the relationship between community and environment, yet already in the years Gallais (1975, p. 222) observed, there were severe pressures that he described in terms of "population growth", "the extension of cultivated areas per person", "the increase in livestock" and "the increase in the area of pasture per animal raised". All of these factors were leading, even before the drought, to "the maximisation of resource use" (Gallais, 1975, p. 222). Drought exacerbates a previously fragile balance. Since the time of the publication of Gallais's work, 50 years have passed: the Sahel has been transformed and a whole universe of meanings and interpretations has changed, thereby opening the field to new declinations in the interactions between different constraints and, unfortunately, fragile possibilities. Indeed, everything has changed, as you will be able to read in these pages, which is why we have aimed to reframe the "Sahelian condition".

Our intent has moved in two different directions. The first has to do with mapping. Gallais's (1975) volume was supported by an important cartography, at different scales, presented both within the text and in a special pocket inside the cover. For Gallais, the tool of cartography was indispensable for understanding the unfolding of Sahelian territoriality and its evolving boundaries. We took up this suggestion accepting the challenge of developing a new cartography, thanks to the elaboration of information that comes from satellite remote sensing and the interpretation of the huge mass of data available on the web. The awareness achieved by geography in the critique of "cartographic reason" (Farinelli, 2009) and in the depth and diversity of the cultural meanings of maps (Rossetto & Lo Presti, 2024) over the 50 years since Gallais's work has led us to a metaphorical approach to mapping in an attempt to reproduce the world without perpetuating the patterns of domination over reality that tend to be implicit in many cartographic practices (see Chapter 9).

The second dimension of our purpose focuses on the meaning often applied to the concept of "condition". That is, we wanted to move beyond the idea that there is an essentialist nature of the Sahel—an Ur-Sahel or an original Sahel—with a stable and defined identity. Indeed, if there is a Sahelian "condition", it is one that lies precisely in the spatial movement of its components and in its mobility and change over time: the true "state" of the Sahel is immersion in the flow of change. This Sahelian condition includes a particular mixture of water and soil: irrigation. It is this combination that lies at the heart of the volume you hold.

Sahelian irrigated landscapes

The peculiarity of the research group of the University of Padua, which has been dealing with the Sahel for more than 30 years, is that it has meticulously examined the implantation and evolution of modern irrigation schemes almost everywhere in this vast region, reconstructing the events from the colonial era to the present. Chapter 1, which was written by Andrea Pase, Marina Bertoncin and Mariasole Pepa, is devoted to this theme and is thus placed in a primary introductory position, preceding the five divisions of this volume. In fact, the intent of the chapter is to identify as precisely as possible the central issue of the research. For this reason, the chapter unfolds extensively; the 100 years of large-scale irrigation schemes have a long and complex story to tell. These megaprojects are an intriguing example of the modern capacity to transform space and society. Triggering in the Sahel the terraforming process related to modernity and coloniality, following Ghosh's (2021) insights, they disrupted the previous ecological, territorial, social and political settings. In this way, they became "territorial traps" (Agnew, 1994; Bertoncin & Pase, 2017), capturing humans, water, lands and living beings, without, however, being effective in the economic growth goals they set out to achieve and indeed demonstrating an embarrassing "attitude of failure". Each of these failures represents an unfulfilled future; the disappointments of promised and unachieved development are no small part of the present-day Sahel's trouble. The materiality, everydayness and diverse practices of the Sahel's many irrigated landscapes are further explicated in Chapter 8, which attempts to map the different shapes caused by irrigation in the region, and in Chapters 10, 11 and 12, which delve into specific case studies.

Volume overview

This volume is divided into five sections, with the structure of each—including the number and length of the chapters—determined not by predefined criteria but by the need for the articulation and depth required to address the various topics.

Part I is entitled "Contextual Issues" and is divided into three chapters. The first chapter, written by Jeroen Warner, Andrea Pase and Angela Kronenburg García, presents environmental crisis narratives in the Sahel and critiques their intentions and interpretive assumptions. These narratives simplify reality by depoliticising how the region's environmental problems arose. Having a surprisingly long colonial pedigree, despite contradictory scientific evidence, the environmental crisis narratives have been ever persistent and continue to legitimise external interventions while tending to disregard local environmental knowledge. The case of Lake Chad illustrates well how these narratives work: by often clearly recommending megaprojects, such as transferring water from the hydrographic basin of the Congo River, as solutions. The chapter closes by asking how it might be possible to offer alternative narratives tuned with the complexity of reality.

Chapter 3, written by Martina Angela Caretta, Mariasole Pepa and Edmond Totin, focuses on the unevenly distributed impacts of climate change on water insecurity and the concentration of this burden on Africa. Climate change acts as a compounding factor that exacerbates existing vulnerabilities in the Sahelian region. Addressing water insecurity through effective and climate-smart pathways is critical to ensuring sustainable production and safeguarding livelihoods. This can be achieved by enhancing Indigenous knowledge in the water sector and increasing the adaptive capacity of the Sahelian population through various techniques ranging from behaviour-based (e.g. water conservation) to technical solutions (e.g. drip irrigation).

Chapter 4, written by Edoardo Baldaro, provides an overview of the current situation of insecurity in the Sahel. It traces the genealogy of the jihadist actors at the centre of the crisis and focuses on how they have used land and space management as key tools of conflict and contention. Different phases of the jihadist groups' presence in large portions of the territories of Mali, Niger and Burkina Faso are also highlighted.

Part II, "Cartographic Sketches of the Sahel", proposes in the space of five chapters a different and novel point of view for mapping the Sahel. Chapter 5 is the result of the contributions of many members of the research team: Andrea Pase, Federico Gianoli, Ludovica Crocitto, Luca De Felice, Michel Cherlet, Marina Bertoncin, Angela Kronenburg García, Mariasole Pepa and Carla Braga. The chapter addresses the very intensely disputed definition of the Sahel and its cartographic limits. Defining the borders of the study area is a necessary but decidedly political operation because it also involves an "ordering" of the Sahelian reality. The challenge involves constructing a cartography capable of rendering the areal definition of the Sahel that is visible and fluid at the same time.

Chapter 6 is written by Federico Gianoli, Mélanie Weynants and Andreas Brink. Using the convergence of evidence methodology as proposed by the third edition of the *World Atlas of Desertification* (Cherlet et al., 2018), this chapter addresses land degradation in the Sahel as a complex and multidimensional issue. The authors have identified patterns of human-environment interactions that are likely to lead to land degradation. The study underscores the need for sustainable land management practices to mitigate conflicts and ensure the long-term viability of the agricultural and pastoral livelihoods in the region.

The hydrological dynamics of the Sahel are discussed in Chapter 7, which was written by Luca De Felice, Federico Gianoli and Andrea Pase. High-resolution satellite data allowed for the analysis of the counter-season surface water phenomenon—outside the rainy season—where floods from major river systems like the Senegal, Niger, Chari and Nile bring water into the region from September to January. This distinctive water availability is marked by high intrinsic variability and significant human impact from irrigation, dam construction and infrastructure development, which disrupt natural hydrological processes and degrade vital ecosystems.

Chapter 8, by Ludovica Crocitto, Federico Gianoli and Andrea Pase, explores the two shapes that different irrigation systems "draw" in the Sahel: rectangles and circles. The two outcomes are caused by the different irrigation methods: rectangles derive from the orthogonal grid of irrigation channels in the systems that distribute water by gravity into the fields from reservoirs or pumping systems. Circles are formed by central pivot irrigation systems. By relying on remote-sensing techniques and the Google Earth Engine platform, the chapter identifies and monitors these irrigation systems and uses open-source GIS software to process the data.

Chapter 9, written by Andrea Pase and Federico Gianoli, presents a cartographic synthesis of the main aspects that are investigated in Part II. It highlights unusual ways of tackling the representation of the Sahel by following an approach that can be described as metaphorical. That is, the intention is to emphasise, on the one hand, the on-the-move and fluid dimension of these large spaces and, on the other, to underline how infrastructures and ways of organising the territory and of making it "productive" constrain the need for movement within rigid and coercive cages. An "interlude", by Chiara Gallanti and Chiara Forner, closes the reflection on cartography with an excursus on the Sahel as represented in the historical atlases preserved at the library and Museum of Geography at the University of Padua that span the period from 1493 to 1789. The interweaving of knowledge from ancient authors, legends, early European explorations and oral descriptions from native interlocutors shaped a cartographic idea of Africa and the Sahel region that significantly diverged from reality for a long time, for instance, in the identification of the path of major rivers such as the Niger or the Nile. Only from the mid-17th century forward did maps reflect accurate representations based on information verified by the cartographers.

Part III, "Complicated Muddles of Water and Land", presents three chapters on different irrigated landscapes in Sudan, the Lake Chad region and Burkina Faso. Chapter 10 brings together the efforts of the Padua and Sudanese geographers: Mariasole Pepa, Andrea Pase, Marina Bertoncin, Eltaib Saeed Ganawa, Abdelrahaman El Tahir A. Musa and Mohammed Hamed Hassan Abubakr. Through the lenses of water and land, their overview of the fieldwork carried out between February and March 2023—framed within the larger 30-year research path—provides an account of some of the social transformations, challenges and perspectives that emerged during that period, with a focus on the Gezira Scheme, the "Sudanese sugar belt" and major investments in pivot irrigation systems. In the chapter, the hopes and slippages associated with Sudanese irrigated agriculture emerge within the context of a region now radically destabilised by the ongoing war.

Chapter 11, written by Andrea Pase, Aboukar Mahamat, Dobah Marsala, Marina Bertoncin and Mariasole Pepa, describes recent developments in irrigated agriculture in the Far North Province of Cameroon and Chad. After a brief overview of the events related to irrigated agriculture in the Chadian region, the main results of the fieldwork conducted by Aboukar Mahamat and Dobah Marsala make it possible to identify the possible trajectories of the ongoing territorial transformations. These traces highlight the peasants' risks and opportunities as well as their disillusionment and hopes. Chapter 12, written by Moumouni Ouande, Lassane Yameogo and Francesco Staro, discusses the recent events of rural insecurity that challenge the existence of large-scale agricultural projects in the Sourou valley, which has been a major site in the economic development strategies of Burkina Faso since the 1960s. This rural instability is a multilayered crisis that is analysed in the chapter by focusing on land insecurity issues and the inefficiency of large-scale agricultural schemes.

Part IV, "Connections, Rights and Conflicts", proposes in three chapters specific in-depth lenses on water and land in the Sahel, even if not directly linked with irrigation issues. Chapter 13, written by Luca Pes and Ibrahima Poudiougou, is based on an ethnographic analysis related to the overlapping sets of rules, practices and authorities related to the management of land and water in the Inner Niger Delta. With an emphasis on the legal dimension of social relations, the study enquires about topics such as the exercise of legitimate authority, the negotiation of rights and duties, and the resolution of disputes.

In Chapter 14, Elvira Pietrobon focuses on the Malian city of Ségou on the banks of the River Niger. Over the past 30 years, the city has expanded rapidly, yet the river remains a constant presence for both city dwellers and peasant villages along its banks. Based on several surveys and interviews, the chapter's author offers an in-depth cartography of the practices and spaces along the Niger River in the three riverside villages that once comprised Ségou when it was the "multiple capital" of the Bambara kingdom. Chapter 15, written by Aboukar Mahamat, Dobah Marsala and Francesco Staro, explores the dynamics of the conflicts in the Waza-Logone floodplains in the Extreme North Province of Cameroon. Recent clashes between farmers and herders are the most visible phenomena in a broader process of growing inter- and intra-ethnic competition over natural resources. These conflicts are connected with the emergence of new practices of natural resource management, particularly the proliferation of canals and the individual appropriation of canals and fishing reserves.

Part V, "Towards a Fluid Research Agenda", closes the volume by offering two chapters of reflection and self-reflection. Chapter 16, written by Angela Kronenburg García, Mariasole Pepa, Andrea Pase, Dobah Marsala, Moumouni Ouande, Aboukar Mahamat, Ahmed Abdelwahid Abdelrahim and Francesco Staro, traces the research process since the sudden irruption of COVID-19. The resulting rethinking of the study framework involved a very relevant and innovative role for the resident researchers in the Sahel countries such as a redefinition of interpersonal relationships and forms of group management. The practices we adopted allowed us to reflect on the very meaning of a non-extractivist approach to academic research.

Chapter 17, written by Carla Braga, Evelien de Hoop, Angela Kronenburg García, Esther Miedema, Andrea Pase and Mariasole Pepa, offers a radical reinterpretation of the ways of thinking and acting relative to water and land. It starts from the insights that are emerging from below on the African continent and in the literature, particularly suggestions of decolonial thought and with a pluriversal worldview. The chapter advocates a dilation of ontologies to move beyond the pretended unique genealogy of Western thought and a broadening to more-than-human voices.

Concluding remarks

One of the outcomes of polycrisis is the difficulty on the part of Western decision-makers on how to re-engage with the Sahel at a time when Eurocentrism no longer works and we have entered a multipolar and "pluriverse" world. There is a disorientation as a result of what appears to be a loss of control over "pieces of the world": whole regions escape the understanding of the Western gaze and the Sahel returns to being just blank spaces on maps. This time, however, unlike in 19th-century atlases, these blank spaces are not fascinating and "tempting" (think of Conrad's Marlow in *Heart of Darkness*) but rather unfathomable and potentially dangerous (Pase, 2024). This perception no longer induces *doing* (such as the long-established narratives that traditionally supported large infrastructure projects) but rather *not knowing what to do.*

The minimal contribution we think we have made with this volume is aimed precisely at not abandoning the Sahel but rather persisting and maintaining the attention that it rightly deserves. We are, however, aware that the "doing" of any operational proposal that deals with polycrisis needs to overcome the colonial and high-modernist logic of "terraforming", and open up to dilated ontologies and vitalist approaches, to use Ghosh's words (2021). A necessary capacity is being able to identify and enhance, in the interstices created in the interconnection between the many levels of crisis, those spaces for innovation and change from below that the people of the Sahel are able to build and protect, even in the midst of a thousand difficulties. In this, the Sahel, in dialogue with other global Souths, can teach the West what it means to face a horizon of social and territorial metamorphosis in creative and original ways. The rivers will continue to flow.

This volume exists in close connection with its virtual "double": the AtlaSahel.it website. The website reports the results that are progressively obtained from the research project through a web GIS that encourages interaction with its readers. The website and the volume have two different communicative codes and two different time orders: the website is constantly evolving, but it is fragile because it is destined to die out as the driving force of the research team declines and as the available funds run out. This volume, on the other hand, is a stop-motion of the research project, but it will remain over time a legacy of what we have tried to understand and do.

Note

1 PRIN 2017 "Contested wetlands in Sahelian drylands: Which kind of development and for whom?" (2017B4JBWN).

References

Agnew, J. (1994). The territorial trap: The geographical assumptions of international relations theory. *Review of International Political Economy*, *1*, 53–80. https://doi.org/10.1080/09692299408434268

Amselle, J.-L. (2022). *L'invention du Sahel*. Croquant.

Beck, U. (2016). *The metamorphosis of the world*. Polity.

Bertoncin, M., & Pase, A. (2012). *Autour du Lac Tchad: Enjeux, tensions et conflits pour le contrôle de l'eau*. Harmattan.

Bertoncin, M., & Pase, A. (2017). Interpreting mega-development projects as territorial traps: The case of irrigation schemes on the shores of Lake Chad (State of Borno, Nigeria). *Geographica Helvetica*, *72*, 243–254. https://doi.org/10.5194/gh-72-243-2017

Cherlet, M., Hutchinson, C., Reynolds, J., Hill, J., Sommer, S., & von Maltitz, G. (Eds.). (2018). *World atlas of desertification*. Publication Office of the European Union.

Farinelli, F. (2009). *La crisi della ragione cartografica*. Einaudi.

Febvre, L. (1949). *La terre et l'évolution humaine: Introduction géographique à l'histoire*. Albin Michel. https://classiques.uqam.ca/classiques/febvre_lucien/terre_evol_hum_geo_histoire/febvre_la_terre.pdf

Gallais, J. (1975). *Pasteurs et paysans du Gourma: la condition Sahelienne*. CNRS.
Gallais, J. (1984). *Hommes du Sahel: Espaces–temps et pouvoirs: Le delta interieur du Niger, 1960–1980*. Flammarion.
Gerstenhaber, A. J. (2023). Salgado, the Sahel, and photography's biggest dilemma. https://www.thecrimson.com/column/f4–the–world–in–pictures/article/2023/3/2/f4–salgado–the–sahel/
Ghosh, A. (2021). *The nutmeg's curse. Parables for a planet in crisis*. Murray.
Henig, D., & Knight, D. M. (2023). Polycrisis: Prompts for an emerging worldview. *Anthropology Today*, *39*, 3–6. https://doi.org/10.1111/1467–8322.12793
Jullien, F. (2009). *Les transformations silencieuses*. Grasset.
Morin, E., & Kern, A. B. (1999). *Homeland Earth: A manifesto for the new millennium*. Hampton Press.
Morton, C. A. (2006). Review of the book *Sahel: The end of the road*. *Anthropological Quarterly*, *79*(1), 175–177. https://dx.doi.org/10.1353/anq.2006.0009
Olivier de Sardan, J. P. (2023). *L'enchevêtrement des crises au Sahel: Niger, Mali, Burkina Faso*. Karthala.
Pase, A. (2024). Emptying and filling. Maps of inland Africa. In T. Rossetto & L. Lo Presti (Eds.), *Routledge handbook of cartographic humanities* (pp. 235–245). Routledge. https://www.taylorfrancis.com/chapters/edit/10.4324/9781003327578-31/emptying-filling-maps-inland-africa-andrea-pase?context=ubx&refId=c53051a9-ebb8-49a2-b83c-5b5207eaf79b
Retaillé, D. (2018). Sahel, sahel. *Information Géographique*, *1*, 34–81.
Retaillé, D., & Walther, O. (2019). *Mapping the Sahelian space*. https://arxiv.org/abs/1906.02223
Rossetto, T., & Lo Presti, L. (Eds.). (2024). *The Routledge handbook of cartographic humanities*. Routledge. https://doi.org/10.4324/9781003327578
Salgado, S. (2004). *Sahel: The end of the road*. University of California Press.
Turri, E. (1990). La questione del Sahel. In R. H. Rainero (Ed.), *Storia dell'età presente* (Vol. III, pp. 513–567). Marzorati.

1 Setting the scene

Unfulfilled futures—One hundred years of large-scale irrigation projects in the Sahel

Andrea Pase, Marina Bertoncin and Mariasole Pepa

Introduction

There are images that have a lot to say. The first picture (Fig. 1.1) shows a cotton gin in action at the Sudanese Ginning Factory in Barakat, Sudan. A cotton gin (*gin* is an abbreviation of the *engine*) is a machine that automatically and speedily separates cotton fibres from their seeds, with much higher productivity than that of manual cotton separation. The photo was taken in late February 2023 during our fieldwork in Sudan. This factory is one of the three operated by the Sudan Gezira Board for processing cotton cultivated under the massive Gezira irrigation scheme. However, it is the second picture (Fig. 1.2) that helps us understand the exceptionality of what we are observing. The finger of Mustafa Ibrahim Mustafa, the general manager of

Figure 1.1 Cotton gin in Sudanese Ginning Factory, Barakat, Sudan.

Source: Mariasole Pepa, 2023.

DOI: 10.4324/9781003429111-2

Figure 1.2 Place and date of construction of the engine.

Source: Mariasole Pepa, 2023.

the ginning factory, points out the place and the date of the machine's construction: England 1925. Few machines from the 1920s in the world are still in working order. In the enormous, century-old industrial warehouse, a row of cotton gins operates, while another stands disassembled, its parts scattered on the floor: the machines are partly awaiting maintenance, while others are hopelessly blocked. Some parts are still the original ones; others have been remade, somehow imitating the original design. When a machine is permanently out of order, its working parts are used to keep other machines operational. These photographs reveal many remarkable aspects: the colonial footprint, the century-long duration, the cotton (i.e., the type of product cultivated and processed), the engine and the factory itself. Yet, a completely different reality exists outside the factory. A few months after these photographs were taken, further key events would unfold in the Gezira region.

The Gezira represents the heart of the agricultural modernisation that occurred during the colonial era in the Sahel. Immediately after the defeat of the Mahdists in Omdurman (1898), W. Garstin, a talented English hydraulic engineer with extensive experience in irrigation projects in Egypt, conducted a survey in Anglo-Egyptian Sudan between 1899 and 1903 to assess the "future irrigation possibilities of the whole Nile system" (Gaitskell, 1959, p. 37). The Blue Nile was identified as the most favourable area, providing suitable locations for the construction of dams: "on both sides of the river fertile plains extended for an immense distance, apparently only waiting irrigation to render them as productive as any land in the world. In particular, the Gezira plain seemed predestined by nature for the purpose" (Gaitskell, 1959, p. 39). The Gezira, a roughly triangular island situated between the White Nile and the Blue Nile, with Khartoum at its confluence, consists of large alluvial deposits formed by the Blue Nile flowing down from the Ethiopian highlands. These are excellent vertisols (the so-called *black cotton soils*). The slope is optimal for gravity irrigation. Preliminary land-surveying operations began in 1907, identifying the site for the dam just downstream from the city of Sennar and establishing the route for the main canal, which would extend towards the city of Wad Medani to supply the irrigation network. The Sennar dam would have allowed the establishment of a reservoir, capable of feeding the main canal, which in turn would have carried water to the major canals. Minor canals would then branch off from these major canals, which, in turn, would feed into channels called *abu ishreen*, then flow into smaller channels, called *abu sitta*, reaching the single irrigated plots (Barnett, 1977, p. 11). Initially, to avoid creating disputes with Egypt over Nile water shares, Garstin envisioned cultivating wheat for the nearby Arab market since it required less water in the lean months than cotton did, where smaller areas could be allocated (Gaitskell, 1959, p. 49). However, pressure from the British textile industry, seeking other sources of raw materials (as an alternative to Indian cotton), led to a change in this orientation. The British Cotton Growing Association, which represented the expectations of the textile industry, mainly concentrated in Lancashire, played an essential role in this regard (Barnett, 1977, pp. 3–6; p. 54). Experiments were conducted to assess the precise water consumption required for cotton cultivation. Parallel to the feasibility studies of the entire project, the first small experimental pumping schemes began at Zeidab in 1906, near the city of Atbara, and then also at Tayiba, in Gezira, in 1911 and in Barakat (near Wad Medani) in 1914 (Gaitskell, 1959, pp. 51–72). The First World War slowed down the progress of the project, but soon after, the huge construction work on the Sennar dam began, in which 20,000 workers were employed, and was completed in July 1925. The deployment phase of the extensive gravity irrigation project commenced in 1926, initially covering an area of 300,000 feddans.[1] This area was gradually expanded through subsequent extensions in 1929 and 1931 (Gaitskell, 1959, p. 98). The project was managed by the colonial government, which controlled the land and water resources in Sudan, and by large private corporations in the hands of American and British investors, which were in charge of overseeing the production): the Sudan Plantations Syndicate (founded in 1904 and so called since 1907; Gaitskell, 1959, pp. 51–52) and the Kassala Cotton Company (founded in 1924, initially with the aim of developing cotton cultivation in the Gash delta, near the city of Kassala; Gaitskell, 1959, p. 131). By 1950, the management of the major project had been entrusted to the Sudan Gezira Board, a governmental entity (Barnett, 1977, p. 6; Gaitskell, 1959, pp. 248–253). The construction of the Roseires Dam in 1966, on the Blue Nile upstream of Sennar, guaranteed an increase in the availability of water for agriculture. With the addition of a second main canal, the irrigated area of Gezira was thus greatly enlarged, concluding the work already begun for the Managil Extension (Bertoncin et al., 1995). Since its completion, the Gezira irrigation scheme has had the potential to cultivate 2,100,000 feddans, varying crops through rotation cycles.

Re-examining the two photographs (Figs. 1.1 and 1.2), we are able to compile a short list of implications and open questions related to the advent and spread of irrigated agriculture under large-scale schemes in the Sahel.

First of all, the colonial signature is evident, with the word *England* embossed on one side of the cotton gin. Likewise, the engineer Garstin, the skills and techniques and all of the initial directors of the Sudan Gezira Board were British, and the capital invested was mainly British. Then, there is the date of construction: 1925. More than a century (two actually, taking into account the very first experiments in Senegal; see below) has passed since the advent of modern irrigation in the Sahel. A two-century-long period leaves a deep mark on the past and the present of these territories; it is worth observing some signs of this *longue durée*, in Braudel's words. The machine in the pictures is used to gin cotton; a colonial crop par excellence, a true industrial crop. The initial crop selection, along with the subsequent crops chosen for cultivation under large-scale schemes, is crucial for understanding how the history of Sahelian irrigation is intricately linked to global markets and national and international agricultural policies. Rice, sorghum, wheat, maize, sugar, fodder, vegetables and fruit trees are other choices that involve different market horizons and establish diverse power frameworks. The modernisation of cotton processing, as witnessed by the cotton gin, is an outcome of a more general process of modernisation of agriculture, whose design is anything but trivial because its implications go far deeper than a simple technical innovation. The outdated industrial warehouse is a well-defined space within another enclosed space (the ginning factory), housed in the Sudan Gezira Board headquarters in Barakat, itself at the heart of a huge and well-cut irrigated space. All these containers are designed for the propagation of "modernity"; "backwardness", "tradition", "conservatism" and "unproductive practices" lie outside. If inside the ginning factory in Barakat the perception was that of stepping back a century, everywhere else significant changes were occurring in March of 2023. For decades, cotton production was drastically downsized because of the negative price dynamics and the invasion of the American bollworm (a major insect pest invading long-stemmed staple cotton). With the introduction of the genetically modified, short-stemmed cotton species (originally of Chinese origin and then from both Brazil and India), the cotton industry has been revitalised. Evidence of this development was the multiplication (in Barakat and, more or less, everywhere in the region) of the ginning factories managed by entrepreneurs, Chinese and Turkish above all. A second life had opened up for cotton. Past tense verbs are used because barely a month after our visits to Barakat and Gezira, on 15 April 2023 to be precise, war broke out between the Sudanese Armed Forces (SAF) and the Rapid Support Forces (RSF), a militia that was structurally part of the military forces of the deposed dictatorial al-Beshir regime. Initially, Wad Medani and the entire region of Gezira hosted refugees from clash-torn Greater Khartoum. However, on 18 December 2023, the fall of the state capital of Gezira into the hands of the RSF marked a dramatic turning point in the irrigated and productive heart of the country (see Chapter 10). Violence is rampant under the irrigation scheme; villages try to defend themselves, even by cutting canals to flood fields, in an attempt to prevent the movement of raiders. The RSF allegedly bombed Wad al-Noura, an agricultural community in the western region of Al-Jazira state, on 5 June 2024, as a direct punishment for the canal cuts and the consequent flooding. The village borders the White Nile state, 60 km from Managil, the only town in Gezira still controlled by the SAF. In a Sudan War Monitor report[2] dated 6 June 2024, a horrifying photo depicts at least 50 corpses of civilians about to be interred in a mass grave just excavated by a bulldozer.

The above-mentioned events are developed in this chapter, interpreted within a framework that we hope is capable of accounting for the underlying historical dynamics of modern irrigation in the Sahel, thus serving as lenses through which the current transformation can be viewed. More than 30 years of field research form the basis of this analytical effort; this chapter contains the legacy of generations of scholars who have studied vast areas, ranging from the Atlantic coasts of Senegal and Mauritania to the Red Sea coasts of Sudan (Bertoncin & Faggi, 2006; Bertoncin & Pase, 2008, 2017; Bertoncin et al., 1995, 2019). It provides an opportunity for synthesis and systematisation and, at the same time, the opening up of new research directions. The bibliography mobilised over these decades is truly huge and will necessarily be limited here to the essentials. The rest of this chapter begins by providing some references on the meaning and role of irrigation in the Sahel, starting with the Sahelian people enduring the uncertainty of water and land resources combined with irregular rainfall patterns. We will then focus on the colonial genesis of agricultural modernisation.

Living with randomness: A long history of confidence in combining water and land resources

In the Sahel, the irregular inter-annual rainfall is the major constraint. As shown in Chapter 5, the randomness of rainfall is a fact of life for the region's inhabitants and imposes constant mobility for water resources. The annual rainy season (between June and September, gradually narrowing down, moving northward) can be longer or shorter and more or less wet, with variability from year to year, with even prolonged sequences of dry or rainy years. The Sahelian peoples have devised practices and extended their knowledge over time to confront these irregularities. Mobility is the first solution: nomadic and transhumant pastoralism on one side and itinerant agriculture on the other. A second practice is the diversification of activities in order to increase the capacity to respond to environmental variability. Polyculture farming techniques are practised; agricultural, pastoral and fishing activities are integrated. Local and wide-ranging market networks are set up for the exchange of products. Many solutions have also been found to interact directly with water resources, including rainwater and floodwater management techniques to capitalise on water supply for the dry season, as well as regular irrigation practices. However, water in the Sahel does not only come from rain; some large rivers (mainly the Nile, the Senegal, the Niger and the Chari-Logone) run along or through the region, bringing water that is all the more precious to areas where rainfall is scarce (see Chapter 7). These rivers are fed by the heavy rains upstream in the Sudanese climate zone or on the highlands; when the floodwaters retreat, the humid lands are exploited for recessional cultivation, transhumant herd grazing and fishing activities. The result is a mosaic of rainwater or floodwater presence evolving over time and diversified in space, a scenario where the Sahelian populations have demonstrated their ability to manage water resources. The most immediate forms of recuperation are limited to taking advantage of the moisture from retreating floods along the edges of the rivers and lakes (e.g., Lake Guiers, Lake Chad or Lake Fitri) or in the accumulation of water in depressions and the consequent infiltration into soils, especially clay. The capacity for dialogue with climatic randomness is dependent on the flexibility and aptitude for identifying and acting on every minimal opportunity available of flowing waters or hydrographic bodies. This assortment of possibilities is expanded by the peasants' attempts at control, ranging from cleaning natural channels that distribute the flood, to building small dams (which follow the contours of the land and allow the water blade to be held back long enough to infiltrate the soil) or to constructing thresholds along watercourses (e.g., along the fossilised loops of the Komadougou Yobé). The purpose of these interventions is always to delay the runoff of rain or floodwater in order to prolong the duration of submergence and groundwater supply, thus increasing the availability of moisture in the soil during the dry season. Particularly elaborate is the case of the so-called traditional polders in the northeastern shores of Lake Chad, obtained by exploiting the long strips of lake waters that push into the dune sequence of the fossil *erg* (dune field) of Kanem. The waters of the lake were trapped by the construction of dams, consisting of double rows of acacia wood structures filled with sand. The dams, placed in the narrows of the lake's branches, were between 50 and 100 m long; the progressive drying of the trapped waters allowed the land to be cultivated. As the drying out of the polder proceeded, the crops extended until they reached from the edges to the centre of the area. The digging of wells and the use of *shadoof* (explained later in this paragraph) would eventually allow the higher lands to continue being irrigated. The lower, wetter soils were the most sought-after. The soils of the polders are naturally highly productive (a result of the lake deposits) and do not require any fertilisation, thus making it possible to have three harvests a year (one of wheat and two of maize) with good yields. The water from the lake that filtered through and underneath the dam, similar to the seasonal increase in the volume of water in the water table, made it possible to cultivate a scrap of land protected by the dams for a few years. The earliest evidence of this technique dates back to a period between the late 19th and the early 20th centuries (Bouquet, 1990, p. 19). If excessive flooding of the lake did not occur, the dams constructed in this way could last for up to eight to ten years, provided that hippopotamus raids or the passage of livestock did not weaken the structures. The agricultural system faced a crisis due to the salinisation of the soils and the gradual lowering of the water table. At this point, the banks were torn down to allow the lake water to enter the polder again. This intervention permitted the leaching of the soils and the recharging of the water table. In subsequent years, the dikes were closed again and the cultivation cycle resumed. We have dwelt on this form of hydro-agricultural infrastructure, albeit limited to a particular context because it bears witness to elaborate knowledge and practices that predate the colonial period and are capable of going beyond the capitalisation of flood and rainfall moisture. In fact, prior to agricultural modernisation, effective irrigation systems allowed for the expansion of cultivation possibilities by accessing water reserves in the dry season, whether shallow aquifers or surface water, which were permanent or, in any case, available for a sufficient time to serve the vegetative cycle of the planted crops. The water elevation systems ranged from the simplest, such as the direct withdrawal of water from wells with the use of pulleys, to the shadoof, where a mechanism allows the efficiency of human labour to be multiplied using a long pole counterbalanced on a pivot, with a bucket stuck at its end (de Planhol & Rognon, 1970, pp. 109–115). From the wells, a network of gullies brings water to small agricultural plots, measuring around a tenth of a hectare per shadoof. Labour for irrigation systems (e.g., to operate the shadoof) was often entrusted to the poorest and most marginal segments of the population, often slaves. The shadoof spread into the Sahel from the Nile valley and the Sahelian oases (Iyébi-Mandjek, 2000, p. 104; Roupsard, 1987, p. 246), testifying to the network of transversal relations that linked the territories of the different river basins of the Sahelian belt. Nomadic and pastoral groups played a key role in the transfer of hydraulic knowledge and techniques. They were particularly familiar with the management of water scarcity and were often also involved in the use of seasonal irrigation systems. Restricted to the Nilotic axis is the use of the *saqiya*, a water-lifting mechanism that uses a vertical wheel with scoops to lift water; a drive shaft connects the vertical wheel to a horizontal one, set in motion by animal power (donkeys, oxen or dromedaries). The banks of the main rivers were bordered by sequences of patiently irrigated lands, forming complex systems, as was the case in the lower reaches of the Komadougou Yobé, where local people, the Mobbers, had built irrigation networks fed by shadoofs, placed on several levels (Bouquet, 1990, p. 178), to extract every last trickle of water from the river, then also drawing from wells dug at the bottom, to raise it to reach the gardens on the banks. The entire lower course of the Komadougou was affected by the continuous work of shadoofs, as witnessed by the French geographer Bouquet, who admired the "maîtrise presque parfait" of the water along the river (1990, p. 263). These different forms of irrigated cultivation networks provided a means of livelihood and, at the same time, they were able to feed regional-scale trade networks.

The same need for water to irrigate the fields in the dry season was fulfilled very differently by agricultural modernisation, first brought by the colonisers and then pursued after independence. Before colonisation, the function of "deseasonalised" water use techniques (outside the rainy season) was complementary to that of other multiple forms of resource use. However, with the advent of modernisation, irrigation assumed a strategic role, becoming a singular element of social and territorial transformation.

Colonial irrigated agriculture: Reshaping the Sahel with water

The very first agrarian modernisation plans envisaged by European colonisers in the Sahel date back to the French governors of Senegal, J. Schmaltz (1816–1820) and especially J. F. Roger (1822–1826). Schmaltz's intention was to develop a plantation economy. Dagana, in Senegal River Valley, had been chosen as a pilot centre for two essential production prerequisites: good land and permanent water. However, the threats of the neighbouring population of the Maures on the one hand and the land claims of the traditional authorities on the other hand proved to be sufficient obstacles to scupper the plan (Ghedini, 2023, pp. 104–105; Maïga, 1995, p. 33). Governor Roger would be the one to put in place a series of cognitive operations, relating to climatic, hydrological, pedological and botanical aspects, and agricultural experiments. Particularly notable is the establishment in 1822 of Richard Toll, in the Wolof language "Richard's field", the expert to whom Roger had entrusted this important agrarian experimental station located in Senegal Valley (Maïga, 1995, pp. 34–38). The premature closure of this experiment in 1829 was due to the poor productive results and the passivity—if not hostility—of the people of the city of Saint-Louis. From the outset, scientific evaluation of a project's potential and experimentation were two cornerstones of the colonial approach to agricultural modernisation.

That being said, the actual starting point of modern agriculture in the Sahel takes us back to the engineer Garstin and Gezira, mentioned in the introduction. Beyond the extraordinary transformation of the lands between the White Nile and the Blue Nile, the largest and most impressive irrigation project in the Sahel also exerted a profound influence on the French colonies through E. Bélime, a hydraulic engineer employed in the colonial services. He would first play an important role in the resumption of irrigation development projects in Senegal Valley—100 years after the failure of Governor Roger's experiment—and, above all, would propose and support for 50 years the irrigation project that would later be named *Office du Niger*. In fact, the study of irrigation techniques applied by the British in Sudan had a prominent place in the formation of this engineer, and the Gezira project would remain a fundamental inspirational model for his proposals (Gallais, 1980, p. 56). Impressed by the reading of some studies on the area crossed by the Niger, even before the First World War, Gastrin obtained secondment to French Soudan with the firm intention of transferring to it the possibilities offered by the irrigation techniques that he had applied in his previous experiences. In the French Indies, from 1907 to 1911 and from 1912 to 1914, he also familiarised himself "with the traditional Indian irrigations taken up by the English in Asia as on the Nile" (Morabito, 1977, p. 54). However, Bélime would have to wait until the end of the conflict to begin his African itinerary; in 1919, he arrived in Senegal as an engineer in charge of hydraulic-agricultural inspection for the Afrique Occidentale Française (AOF). His vision of hydraulic transformation would be achieved towards the Senegal Valley and then mainly in Niger Valley, particularly in the Segou area.

As regards Senegal, Bélime filed two reports (Maïga, 1995, pp. 61–62): the first in 1922 and the second in 1934. In the 1922 plan, he advocated the choice of focusing on cotton cultivation in Senegal Valley, as opposed to the orientation towards groundnut in the rest of the region, convinced that American or Egyptian cotton could produce good yields if grown under irrigation. Evidently, this option entailed important hydraulic works, which would be presented in the 1934 report: first, the construction of retention basins between the cities of Dagana and Bakel, would store the floodwaters, with a further suggestion for identifying other possible sites for dams in the upper reaches of the river or along its tributaries. These works, as well as allowing the irrigation of the cotton perimeters, would also have ensured the navigability of the river and the production of electricity. Although at this point, Bélime left Senegal to devote himself entirely to Niger, the imprint of his ideas would continue to be seen in subsequent initiatives. Several studies and projects were put in place during the colonial period to develop navigation, irrigation and the exploitation of the hydroelectric potential of the Senegal River (Maïga, 1995, pp. 62–74). The plans remained on paper. During the colonial era, only two hydraulic projects were actually realised: Richard Toll's rice irrigation scheme and Guédé's controlled flood submergence perimeter (Maïga, 1995, pp. 74–80).

After Senegal, Bélime directed all his attention to the Niger, which he also called the "French Nile", with an evocative image. In fact, other engineers had already thought about and even implemented irrigation schemes for cotton along the Niger. For example, in the locality of Diré, towards Timbuktu, Marcel Hirsch had obtained a large concession (100,000 ha); however, only 200 ha were actually cultivated (Morabito, 1995, p. 114). Nonetheless, a leap forward in scale would be proposed by Bèlime, starting with the first project in 1920 when he hypothesised the irrigation of 1,600,000 ha, mainly for cotton, thanks to canals on both sides of the river and diversion and retention dams downstream from Bamako (Morabito, 1995, p. 114). An experimental station was already active in 1921, the Sotuba dam was completed in 1929 and an irrigation perimeter of 1,500 ha was started. However, the predeltaic part of the river course would soon be put on the back burner, in favour of the fossil delta of the Niger downstream of Segou, judged to be the most suitable area for the extension of irrigated crops. On the hydrographical left of the current course, a gently sloping sandy plain, crossed by the dead arms of the river, called *fala*, lent itself well to gravity irrigation, starting with a dam that would allow the water level to be raised during flooding so that the Molodo and Macina fala, suitably remodelled, could be reconnected. In this way, Bélime thought that it was possible to irrigate 960,000 ha over a 50-year period.

Meanwhile, the project's aims were being redefined. Governor Carde had pushed for a shift from cotton to food crops, especially rice, which was suitable for the inland delta's soils, also in view of the severe famines that had afflicted the region's population (Morabito, 1995, p. 116). The balance found comprised 510,000 ha for cotton and 450,000 ha for rice. In 1932, the Office du Niger officially started, Bélime would be its director-general until 1942. In 1937, work began on the Markala dam, the hydraulic cornerstone of the project, which was completed in 1947. In this way, the water level would be raised 5 m during floods and would be directed to two canals that fed the irrigated areas (Bertoncin et al., 2011). In 1935, the first areas in the Kokry-Kolongotomo and Niono began to be cultivated, with temporary dams and canals (Morabito, 1977, p. 58). There was a steady growth in irrigated hectares, although it was not until 1942 that cotton began to be produced in appreciable quantities. The allottees (people or families assigned one plot each in the irrigation project) came not only from the French Soudan but also from the neighbouring colonies of the AOF, characterising the project as a point of reference for the development policies of the entire vast inland region under French rule.

The last of the large basins to be involved in irrigation-driven agricultural modernisation was that of Lake Chad because it was certainly the most isolated and was so difficult to reach that the predominant theme in all projects was *désenclavement* (Cabot, 1955, 1965, pp. 163–168). The Chari and its main tributary Logone seasonally flooded the great plain, in what has been called the Chadian Mesopotamia. Here as well, the impulse for implementing modern irrigation arrived, albeit late and in undersized forms. After a few minor experiments, the planning of the Casier A perimeter, north of Bongor, Chad, began in 1951; here, the cultivation orientation was soon turned to rice. Through the construction of a side dam on the Logone River, the flood had to be controlled in order to cultivate 55,000 ha. The first work began in 1960, already at the time of Chad's independence, and led to the provision of only 2,000 ha managed through partial flood control. In 1954, across the river in Cameroon, the *Secteur expérimental de modernisation de la riziculture de Yagoua* (SEMRY) also started a rice-growing project, beginning with the construction of a 40-km embankment, between the cities of Yagoua and Djafga, to partially contain the flood (Roupsard, 1987, p. 266). The water was channelled into 1,500–2,000 ha, which increased to 6,500–7,000 ha in the second half of the 1960s (Brunet-Jailly, 1981, pp. 71–72).

It is important to emphasise that during colonisation, all possibilities of an "agrarian enhancement" through irrigation were explored, and with much experimentation, even the most peculiar ones, such as the Bol polders modernisation projects involving the construction of dams and pumping stations to make the polders permanent and more efficient (Bertoncin & Pase, 2008, pp. 342–353). Its most striking aspects are the vastness and total coverage of the process of territorial change, all implemented through large-scale irrigation projects whose main objective was to move the "backward" Sahel towards progress. The logic behind this process is the subject of the next section.

Ways to impose modernity and coloniality in the Sahel: The "Terraforming process"

Perhaps Albert Sarraut, for a long time the French minister of colonies, governor of Indochina and a profound connoisseur of the dynamics of colonisation, provides the best expression of the colonial logic. He presented a general programme of economic development of the colonies to the French Chamber in 1921, based on a considerable commitment of resources on the part of the metropolis. In 1923, he published *La mise en valeur des colonies françaises*; the volume is a true manifesto of this line of colonial intervention (Sarraut, 1923). The Sarraut plan, as it would be called, aimed at the valorisation of both human resources and natural resources, with the former indispensable to the exploitation of the latter. The scarcity of the Indigenous population is in fact one of the major weaknesses, especially for African colonies, and so is the lack of education in the local labour force; in this way, the natives would first of all not be "subjects but auxiliaries" (Guy, 1923, p. 267), this being their only possible role in the colonial logic. Secondly, there would be a "methodological division of labour" among the different French colonies, requiring them to contribute by concentrating on productions for which it was "suited". The plan essentially consisted of a detailed programme of indispensable works (first and foremost, railways and roads) for the development of the colonies: in reality, these works would prove to be too costly to sustain. Note how implicit in the concept of *mise en valeur* was a devaluation of what preceded colonial intervention and of the forms of resource use activated by African populations. Without external intervention, clearly, those territories would have no value, and their resources would not be adequately exploited. Leroy-Beaulieu's (1908, p. 689) definition of colonisation is explicit in this respect: "the methodical action of an organised people upon another people whose organisation is defective". Defective—that was the point—African societies and territories lacked organisation, and colonial and methodical action had the task of addressing this deficiency.

The colonisers' action had a premise and an unfolding. First, the premise was a survey, entailing the exploration of the "new" spaces; the enumeration of the bodies, objects and qualities they contained; and the construction of the cartographic representation. In short, it involved the cognitive appropriation of the reality that the colonisers intended to dominate. The colonial survey actually had two components: the first and more obvious was the exploration of the unknown, about which Europeans previously had no information. The second aspect, somewhat overlooked, was nonetheless the more interesting in terms of revealing the very meaning of the operation. In neo-Latin language, in the military context, the word for survey is "recognition", that is, gathering, by means of direct observation, the information needed to set up a tactical, logistical or technical action. The etymology precisely attests to finding something already known ("re-cognition") as useful for military action. This re-finding of what is already known due to previous cognitive experience, albeit generated in a very different social and environmental context, implies a retracing of the usual cognitive frameworks; there is no readiness to welcome the unusual and the unexpected. If something was different and unrecognisable, for the colonisers, it was only in the sense that it was less strange and backward, in any case, useless. The contribution of "scientific technicality" and thus of "colonial research" is indispensable for recognition (Bonneuil, 1991, pp. 41–42). The cognitive appropriation of new spaces and their contents is expressed in the production of representations, that are maps and enumerations: on one hand, a topographical survey and mapping, on the other hand, the description of social and natural contents or, rather, the reduction of qualitative contents (evidently simplified) to number or to quantity. Through cartography, the political geometry of modernity is imposed; the map makes it possible to reason about the space with the definite rules and certainties of geometry, exploiting all its capacity for simplification and abstraction. In contrast, enumeration measures what space contains, the stock of resources available for colonial planning—natural and human resources, according to Sarraut. Enumeration is indeed an "abstract and summarised representation … but already satisfactory enough to allow an intervention that want[ed] to be effective" (Raffestin, 1981, p. 79). Regarding the numbers (how much population, how much water, how much land, etc.), one can operate with the supposed effectiveness of arithmetic: thanks to the numbers, one can calculate the effect of each action. The colonisers' action is therefore, before becoming an actual intervention, geometric design and arithmetic calculation, the action planned on the representations that *savants pour l'empire* could provide (Bonneuil, 1991). The calculation of costs and benefits, the economic and demographic estimation, made it possible to foresee the outcomes of the interventions. The numbers offered a sure foothold of mathematical certainty. It enabled projection into the future; indeed, in a certain sense, it committed the future to correspond to the wishes of those who had constructed the project. The map made it possible to mobilise operations, visualising the resources, the resistances and the essential points for the colonisers' strategy. The maps drawn and the enumerations made listed all the elements of the future territory. The territory that would be (that the coloniser wanted and believed it could become) was already represented and thus made present; the dream of colonial power was anticipated on maps and calculations. The dynamic of prevision was essential in the colonial project; thus, the imagination of the future and the planning followed the recognition.

If recognition and imagination constituted the premise, the unfolding was the project in action, the chain of acts operated in order to give form to the colonial reality. This could not be achieved immediately and everywhere, but colonisers must necessarily concentrate their forces on a few points considered strategic and exemplary at the same time. With respect to vast areas and often remote Sahelian rural regions, colonial governments identified large-scale irrigation schemes as "a small, relatively self-contained, utopian space where high-modernist aspiration might more nearly be realized" (Scott, 1998, p. 257). These well-localised contexts served as exemplary models of what the whole modern and colonial reality could become: "the miniaturization of high-modernist development offers a visually complete example of what the future looks like" (Scott, 1998, p. 258). The juxtaposition of modernity and colonialism is not accidental; decolonial studies are those that warn about how there is a temporal and material coincidence between the advent of modernity and coloniality, that is, the establishment of a cognitive and power system capable of trapping the entire globe in the power plots of European powers (Mignolo, 1992, 2009). From this point of view, irrigation projects take on a "quasi-experimental" value (Sautter, 1987, p. 5); they are laboratories where it is possible to observe the unfolding of modernity and coloniality, starting from what we might call "degree zero". The starting point is the emptying of all prior social and environmental contents; on the tabula rasa thus created, the machine of territorial transformation, the *mise en valeur* (and then development) apparatus is activated. The large-scale irrigation projects fully express the willingness to affect a social and territorial body in order to modify it according to a design and through a precise sequence of interventions. The future is foreseen; energies are channelled (a verb quite appropriate to the context) towards a precisely defined goal.

The first action is to isolate the area of the development project from its surrounding region; that means separating it, building a caesura, and marking a distance. Inside is the future, the expected, the progress at hand. Outside there is only the past, what is to be overcome. It is significant that in 1929, Carde, the then-governor of French Soudan, referred to the Office du Niger precisely as an "island of prosperity" (Morabito, 1995, p. 117). It is also worth noting how almost 100 years later, the same metaphor is used; for example, the 2011 World Bank report on the Bagrépôle project in Burkina Faso describes it as an "island of success and promising cluster of initiatives and enclaves that can (…) pull along the rest of the economy" (World Bank, 2011, p. 33; see Chapter 12). This continuity is truly astonishing. The large-scale irrigation projects, as islands of modernity, involve a hollowing out of every previous social content instituted: land rights, local knowledge and uses of resources. At that point, modern technologies for territorial and social transformation were applied to the contoured reality: the land was levelled for irrigation; the hydrographic network was modified with the planning of channels and building of river banks; new agronomic techniques requiring centralised organisation of work were introduced; and strong social, political and economic controls were established. According to a relevant debate in the geographical literature, the megaprojects can be interpreted as "enclave spaces" (Jeffrey et al., 2012; Sidaway, 2007; Vasudevan et al., 2008). The historical reference points to the enclosures or the fencing off of common lands by private parties, with the consequent loss of the local people's collective rights over the use of the soil, water and other resources present (wood, fish etc.) (Blomley, 2007; Grossi, 1992; Polanyi, 1944). In all respects, the implementation of the megaprojects is a process of dispossession of the lands and rights of the local people in favour of the colonial government and as a function of the market. Territorial isolation is achieved by acting on the distance (Bertoncin & Pase, 2023); it is about pushing away and bringing closer. The elements that are expelled (and thus abraded from the closed project space) are customs and the commons, social structures linked to lineage, polyculture and the integration of different uses of resources, in short, what appears to Europeans as backwardness or even primitiveness. All these elements remain nearby the project yet are clearly separated from it. On one hand, the project area thus distances itself from what used to be close by, or rather, was one (Bertoncin & Pase, 2022). On the other hand, the large-scale irrigation scheme succeeds (with greater or lesser effectiveness, depending on the situation) in bringing closer what was far away. Large-scale irrigation schemes are connected by networks of circulation and communication with economic and political centres; from the outside, engineers and technicians, representatives of modern knowledge, arrive and depart, living in their well-equipped compounds. These islands are part of an archipelago of modernisation united by well-travelled routes: from these routes come knowledge, machinery and external people. In turn, the islands feed these routes with the cash crops they produce. Large-scale irrigation projects can work because the energy and resources indispensable to the modernisation effort are concentrated on them. What Scott calls "high-modernist agriculture" is based on trust in "the superior technical efficiency of large-scale farms, the importance of mechanization to save labour and break technical bottlenecks, the superiority of monocropping and hybrids over polycropping and landraces, and the advantages of high-input agriculture, including commercial fertilizers and pesticides" (Scott, 1998, p. 271). The means used to impose agricultural modernisation is with "large, integrated, planned, projects rather than piecemeal improvements"; "the greater the industrial content of a scheme and the more its environment could be made uniform (through controlled irrigation and nutrients, the use of tractors and combines, the development of flat fields), the less was left to chance. Local soils, local landscape, local labour, local implements, and local weather appeared to be almost irrelevant to the pre-packaged projects" (Scott, 1998, p. 271). The diffusion of this model of agricultural modernisation in the Sahel coincides with the multiplication of these "territories in miniature"; "by concentrating the material and personnel resources of the [colonial] state at a single point, miniaturization can approximate the architecture, layout, mechanization, social services, and cropping patterns that its vision calls for" (Scott, 1998, p. 258). The idea is that their visible order and productive effectiveness allow these islands of modernity to become the centres for the spreading of cultural and social innovations in the entire Sahelian rural world. In this way, two processes are set in motion by the construction of the islands of modernity: simplification and naturalised transformation. Developers carry out a drastic simplification based on the assumptions of science and modern agronomy techniques, along with social engineering applied to the lands and the Sahelian people. This "one-commodity machine" (Scott, 1998, pp. 10–19; 2010) that is the large-scale irrigation project is built precisely on a "heroic simplification" (Scott, 1998, p. 8) of the reality on which it imposes itself. The oversimplified "thin spaces" of modern irrigated agriculture contrast with the surrounding "thick spaces"; "they are thin in the sense that they cannot reasonably plan for anything more than a few schematic aspects of the inexhaustibly complex activities that characterize 'thick' [spaces]" (Scott, 1998, p. 261). The only option for the colonisers, if they intend to maintain control, would be to act on a few well-defined characteristics regarding the quantity and distribution of water, the quality of the soils, their agronomic preparation, the labour force and its organisation, and the choice of the most productive and profitable crops. The "visual order" of modern agriculture contrasts with the "working order" (Scott, 1998, p. 275) of Sahelian societies, or rather, that order hidden in the spatial and temporal multiplicity and differentiation of local practices. The thickness of Sahelian spaces precisely lies in the multiplicity, variability and differentiation of practices and knowledge and in being mobile, changeable, blurry, constantly redefining and indistinguishable from the "colonial other" and from the outside except as "chaos". On the contrary, agricultural modernisation projects imply a clear visual order in the geometric regularity of the channels and the fields and in the planning of all internal processes. Isolation and simplification form the premise for the transformative attempt, which extends over the entire material, living and human reality under colonial rule. The aim is to introduce order where randomness, fatality and lack of organisation would otherwise have dominated, in the eyes of Europeans. A colony is an example, perhaps in the purest form, of the tendency of the modern state, "through its territorial form, to naturalize (at once to mask and to normalize) its own transformative, intensely patterning effects upon sociospatial relations" (Brenner & Elden, 2009, p. 354). Naturalised transformation comes through masking and normalisation. About masking, Sack (1986) has already shown how modern territoriality precisely allows the rapport between controllers and controlled to be masked, making the territory itself appear as the control agent. The bounded area becomes the justification of the control exerted through it. In this way, the attention is shifted from the social, political and economic relationships to the presumed objectivity of the bounded territory. The management's control over what occurs within the project area, its interference with the social organisation and even the cultural aptitudes of its inhabitants are automatically justified. The territory of large-scale irrigation projects functions like a mask to conceal the social transformation that colonisers seek to put in place. Regarding normalisation, in large-scale irrigation projects, the farmers lose their personal and social status; the project, at least at the beginning, takes no interest in ethnic differences, family membership and cultural references. Moreover, in large irrigation projects, "where authorities decide when to release the water, how to distribute it, and what water fees to charge, or of agricultural plantations, where the workforce is supervised as if it were in a factory setting", the farmers are subjected to "a radical de-skilling" (Scott, 1998, p. 286). The nullification of farmer know-how is a tremendously powerful instrument in the hands of the management.

If naturalisation certainly means making obvious or self-evident something that is, on the contrary, the result of a precise power strategy, it also and exactly has in it the reference to nature, that is, the colonial state's ability to change social identities along with natural conditions. Referring to the concept of terraforming, as Ghosh applies it to European colonialism (Ghosh, 2021), allows a better understanding of this process. The term *terraforming* is borrowed from the science fiction writer Jack Williamson, who, in his 1942 novel, forged this neologism, joining *terra* (land) with *forming* in the sense of making or moulding, "hence it could be glossed as 'land-making' or 'land-molding'" (Ghosh, 2021, p. 53). Initially applied (in science fiction) to projects designed to reshape other planets in order to make them habitable by humankind, for Ghosh, the term is useful in explaining how Europeans acted to conquer the new lands by radically altering the environmental context. In this way, the "New Worlds" became suitable for colonialist settlers yet at the same time hostile to those who previously inhabited it. The process modified the environment in its physical form: moving rivers, building cities and roads and establishing plantations, as well as modifying the flora and fauna by introducing new plants and animal species and, first and foremost, for the purposes of conquest, viruses and bacteria. The aim entailed "the undermining and elimination of the ways of life of those who had inhabited those lands for many thousands of years.

The project of terraforming was therefore fundamentally conflictual; it was in itself a mode of warfare, of a distinctive kind" (Ghosh, 2021, p. 55). Settlement colonialism is certainly not at play in the Sahel, as is the case in the Americas or other areas of Africa, but the underlying mechanism is the same. From this perspective, undertaking a large-scale irrigation project is an act of terraforming, whose intent is precisely to mould a different, new, modern and performing reality. The conflict, the warfare masked through it, has several components: biopolitical (in the sense given to this term by Ghosh), ontological and epistemic. It is biopolitical because it calls into question human and non-human forms of life (first, animal and plant) and even agents such as soil and water[3] because it "included massive biological and ecological disruptions" (Ghosh, 2021, p. 55). In other words, it takes the form of a total conflict, running through human and "other-than-human beings and entities: pathogens, rivers, forests, plants, and animals" (Ghosh, 2021, pp. 57–58; see also Tsing, 2015). It is ontological in the sense that what is at stake is the very legitimacy of the existence of different ways of living and being in the world, different from modern and colonial ones. Its clearest sign is the systematic denigration of everything considered "backward" and "primitive" in the Sahel. Finally, it is epistemic because at the bottom of it all is a cultural clash between different ways of viewing the world. In Ghosh's thought, the contrast is between a technicist approach (called "high-modernist" in the words of Scott) that establishes the colonial, extractivist economy that perceives the planet "as [an] inert entity that exists primarily to be exploited and profited from" (Ghosh, 2021, p. 257) and a vitalist, "anti-mechanistic" one. The latter recognises the kinship between humanity and "the entire spectrum of nonhuman kin, including rivers, mountains, animals, and the spirit of the land" (Ghosh, 2021, pp. 235–237). The technicist approach, which underpins modern and colonial logic, has, moreover, served its apprenticeship precisely in those swamps where "the Mediterranean found its New World, its own Americas in the plains" (Braudel, 1995, p. 67). From the 16th century onwards, the wetlands of the Mediterranean plains have been hit by a powerful process of terraforming, the reclamation actions whose primary intent was to separate the wet from the dry (the very logic of modern territorial organisation aspires to solidify what is fluid and amphibious; Farinelli, 2009, pp. 66–67), which included imposing order in the hydrographic disorder by constructing large land estates, meanwhile cancelling the inhabitants' rights (hunting, fishing, grazing, cultivation and trade) (Bertoncin, 2004). Clearly, this was a biopolitical, ontological and epistemic conflict, with winners and losers. The losers had to abandon their lands and emigrate, but where? Some of them to the New World. It is all bound together.

The technicist approach and its main proponent, the modern and colonial state, have at their disposal a powerful tool to silence vitalist visions that incorporate the voices of "other-than-human beings and entities" (Ghosh, 2021, pp. 57–58). In the interpretation of the Italian geographer Franco Farinelli, the geometric and geodetic cartography first allows representation of space and then intervenes by abstracting its contents, such as its qualities for human collectives, and beings that occupy the space. The zenithal representation is capable of rendering the territory smooth and thin, flattening out every roughness, and every autonomous specification in order to reduce everything to geometric forms, in the conviction that "nothing essential can escape the geometric sign" (Farinelli, 1992, p. 27). Through the map, "everything that is alive is transformed into something motionless and lifeless" because "the act of cartographic representation" corresponds precisely "to the sacrifice of the living, to the deposition on a flat surface of a piece of corpse" (Farinelli, 2004, p. 77). Farinelli places this "opposition between life and geometry" at the very beginning of "science and western consciousness" (Farinelli, 2004, p. 78). The view from above that legitimises human dominance over nature, together with the emergence of the modern perspective (which detaches the subject from the object) and the imposition of Cartesian rationalism (whose *ego cogito* connects with the *ego conquiro* of European penetration into the new world; Andrade, 2018; Dussel, 1977; Zúñiga, 2023), paves the way for a cheap and sure environment for capitalism (Moore, 2016; Patel & Moore, 2017), which is deeply hostile to vitalist sensibilities and practices.

In the colonial terraforming process in the Sahel, large-scale irrigation schemes assume an exemplary role as traps that allow the colonisers to capture in their nets the lives and essences of their dominated spaces. The next section is devoted to an analysis of the components and functions of these traps. It should be noted that the logic that informs the colonial terraforming process has remained active after the Sahelian countries' independence, which Quijano (2000, 2007) refers to as the "coloniality of power", in other words, a colonial domination that continues to be exercised through modernity to maintain the subalternity of Africa and the Global South. This is a long-lasting legacy, although the orientation of the large-scale irrigation schemes will no longer be aimed at the colonial metropolises' interests but rather at the objectives of food self-sufficiency and consensus-building of the new elites and actors. Within this frame, control over water and land through mega-irrigation projects has remained after independence in the name of development, purporting this modernity apparatus, which, in reality, is just another means for continuing the process of colonial terraforming and economic exploitation (Ndlovu-Gatsheni, 2012; Ndlovu-Gatsheni, 2023).

Making modernity: Development apparatuses as traps

The Sahelian large-scale irrigation projects can be regarded as exemplary cases of development apparatuses (Ferguson, 1994; Sidaway, 2007). As Foucault (1980, pp. 194–196) states, an apparatus is a network of connections that is present in various heterogeneous elements (discourses, institutions, physical structures, rules and scientific proposals); this network is oriented to improving the relations of specific forces towards an expected change. It is therefore an assemblage of different elements deliberately oriented by the decision-makers to affect reality in order to steer it in a certain direction; in large-scale irrigation schemes, as described in the previous section, the orientation is towards modernisation and indeed development. The supposed technical neutrality of the apparatus is in reality a way of concealing changes in the relations of power on the field and thus takes the form, in Ferguson's words, of an "anti-politics machine" (1994). Development discourses have reduced issues such as poverty and food insecurity to technical problems that can be fixed by technical solutions, hiding social hierarchies, class dynamics and the gender spectrum and de-politicising and legitimising development interventions (Ziai, 2015). To understand how this specific form of development apparatus works, we have tried to identify four dimensions in order to group its various heterogeneous elements: physical, relational, cognitive and operational (see also Table 1.1 at the end of this chapter). The physical dimension concerns the materialisation in space of a large-scale irrigation project and thus contains all the infrastructures that radically alter the situation prior to the deployment of the apparatus, acting "rapidly, intentionally, and profoundly in very visible ways" and requiring "coordinated applications of capital and state power" (Gellert & Lynch, 2003, pp. 15–16; Haller et al., 2022). The relational dimension identifies how social and political relations (in a broad sense, as a decision-making process) change because of the implantation of the project, understood as an institution, both formal (a set of written regulations and explicit roles) and informal (a set of suggested or rather, imposed behaviours and attitudes). The cognitive dimension concerns the knowledge systems that first allow the planning of the project and then support its management; it deals with scientific knowledge (from hydrology to pedology, from botany to agronomy) and technical skills (from topographical surveys to construction engineering, from production techniques to those of product transformation and conservation). Finally, the operational dimension involves the organisation, clearly identifying the leadership, management structures and labour and defining the operational protocols that enable the production machine to function. The board is entrusted with overall management, that is, all the preceding, accompanying and final stages of production: land levelling, ploughing, water distribution, sowing, supplying fertilisers and pesticides harvesting, processing, and storing and selling the products. Ideally, mechanisation affects all these stages. In parallel, the board is in charge of maintaining water networks, pumping stations, machinery and rural tracks. Finally, the board must also check if all production

Table 1.1 Dimensions of large-scale irrigation projects as development apparatuses.

Physical dimension	*Relational dimension*	*Cognitive dimension*	*Operative dimension*
Catchment of water (dams, pumping stations), network of canals, administrative and operational centres/ buildings, new settlements, road and communication infrastructures	Projects' institutions (formal and informal), social roles and status derived from them	Before: survey, feasibility studies, topographic and thematic maps, planning During: technical and scientific knowledge mobilised	Organisational structure (who commands, who monitors, who obeys) Organisational procedures (who does what, how and when)

steps and methods are respected. To perform these tasks, the board has at its disposal a large number of technicians and employees at various levels: engineers, agronomists and administrators; operators of agricultural machinery and pumping systems; maintenance workers; and agronomic advisors and supervisors. The plots of irrigated land within the specific sectors of the schemes are instead entrusted to allottees, that is, farmers whose task is to follow the growth of the crops and fulfil the water shifts. Providing defence against granivorous birds is their duty as well. They must obey the agronomic advisors and overseers. Each agricultural season, they also have to pay for land allotments, water supply, fertilisers and pesticides and all mechanised work provided to them by the board. Land allotments are not permanent and must be periodically renewed; if fees are not paid or if the plot is poorly maintained, they may lose their allotment. This general outline of work organisation has many variations from case to case and evolves over time.

Consistent with the modernisation objectives described in the previous section, each of the four dimensions isolates and selects what is internal and what is external and extraneous (if not the enemy) to the development apparatus (see Table 1.2). The isolation of the development apparatus has the most visible aspect in its physical dimension and is played out with both the fences and the network of canals themselves, which become physical barriers preventing access and free movement. No less important are the immaterial limits that define what is admitted, what

Table 1.2 Modes of isolation; identification of included and excluded elements.

	Physical dimension	*Relational dimension*	*Cognitive dimension*	*Operative dimension*
Isolation framework	Fences and "water walls" (canals)	What is legitimate and what is unlawful	What is useful/ modern and what is useless/ backward	What works and what does not work
Who/what is included	Convenient water, crops selected, machinery, formal settlements, allottees/ technicians and officials	Relationships established by the project and functional to it (vertical or horizontal), written rules	Technical and scientific expertise	Production, processing and marketing of products, clear job descriptions
Who/what is excluded	Floods and wetlands, vegetation and natural fauna/ local crop varieties, informal villages, herds	Local/contextual social and political relations, habits, customs	Contextual/tacit knowledge	Polyculture, transhumance, other productive and commercial practices

is useful and modern and what works. These defining criteria are the measures that decide what is permissible and what is excluded in the apparatus. Various elements are listed in the table, but what counts in the end is that inside, there is the laboratory of modernity; outside is the domain of backwardness, whose destiny is gradually shrinking in the face of the advancing modernising wave. In the next sections, we shall illustrate how this claim of clear-cut separation is an illusion and how the boundaries that are thought to be so clear-cut are actually porous, flexible and continually crossed.

In development discourses, the emphasis is on the modernisation and productivity gains that irrigation will allow; the machine serves to provide a path from uncertainty and backwardness. It is the way or, at least, the main instrument of progress; its supposed benefits make it indisputable. A critical approach immediately identifies the hidden dimension of this operation: the radical modification of power structures in favour of the modernisers (and, in the first long phase of the history of modern irrigation in the Sahel, of the colonisers). With the modernisation of agriculture, development institutions did not just bring new agricultural machinery but also primarily introduced new class relations in rural areas through the "enclosure of land" (Federici, 2018, p. 21). To try to explore this concealed side, it is useful to refer to Agamben's (2009, p. 13) observation regarding apparatuses and their main functions: capturing living beings, forcing them to do something they would not otherwise do and directing their will and even their desires. From this perspective, an apparatus is "literally anything that has in some way the capacity to *capture*, orient, determine, *intercept*, model, control, or secure the gestures, behaviours, opinions, or discourses of living beings" (Agamben, 2009, p. 14; italics for emphasis). In other words, an apparatus can be conceived as a trap. In effect, a good definition of a trap is an apparatus that allows capture. A development apparatus, in our case, a large-scale Sahelian irrigation project, performs the functions of capturing, orienting, determining, intercepting, modelling, controlling and securing spaces, humans and "other-than-human beings and entities" (Ghosh, 2021, pp. 57–58). Reality (natural and human) prior to the project is enclosed and isolated from the outside, only to be totally reshaped by it. Natural elements, societies and cultures are "combed" by the project to make them congruent with the idea of modernisation; a mighty process of normalisation is put in place to make every aspect adherent and functional to development goals (Table 1.3).

We can divide this normalisation act into the four dimensions of the development apparatus: physical, relational, cognitive and operational. In the physical dimension, in the well-defined spaces of large-scale irrigation schemes, the transformative power of modern technology significantly alters the land's profile and level, reshapes the hydrographic network and modifies vegetation. This process enables a geometric partitioning of space and total control over what enters or leaves it (water, people, plants, animals and crops). Relationally, the development apparatus introduces specific socio-political relationships. Large-scale irrigation projects become arenas where modern institutions (e.g., the state itself, ministries,

Table 1.3 The trap in action.

	Physical dimension	*Relational dimension*	*Cognitive dimension*	*Operative dimension*
Who/what is standardised and how it is standardised	Spaces, waters, human and other-than-human beings and entities Geometric partition of the space (canals, fields and roads), water regulation, levelling of land, planning of production spaces and settlements, selection of crops and cattle	Governing of relations: legitimised attitudes/ behaviours	Forced/imposed knowledge, vulgarisation (diffused through simplification) of modern expertise	Command and control Water shifts, techniques and timing of agricultural operations, market management

engineers, technicians, international construction companies and aid organisations) operate, thus marginalising and delegitimising local institutions, which are often perceived as irrelevant or obstructive to modernisation. The former relies on written rules, legal statutes and formal procedures, while the latter is grounded in customs that lack legal recognition or are considered merely auxiliary. The project hierarchy invalidates local authorities, establishing institutional roles to mediate and manage power relations within the irrigation project. However, one of the paths pursued by project management was sometimes the co-optation of the pre-existing authorities, which were also obviously subject to change over time, as happened in the colonial era in the framework of indirect rule. Cognitively, the development apparatus is rooted in scientific expertise from the Global North, which marginalises local knowledge about soil, water and resource use approaches (e.g., fishing or transhumance). Technical knowledge must be disseminated to farmers through training, implying a necessary simplification and vulgarisation. Operationally, the procedures are predetermined and precise, involving water schedules, enforced monocultures or specific crop rotations, agricultural calendar decisions and control over production stages. All activities within the project perimeters, including farming practices and product processing, are dictated and imposed from above through a command- and-control system.

Conceiving a developmental apparatus as a trap highlights two consequences. First of all, because of its constricting, squeezing and oppressive characteristics, it is often linked with violence. Large-scale irrigation schemes result in a closed, suffocating, homogenised and uniform horizon. There is no place for originality, creativity or even slight deviations from the prescribed path. In some ways, they embody a "total institution", to quote Goffman's (1961) term with respect to asylums, which are then more broadly extended to include other situations of constrictive closure such as in relationships in a contest of high level of control from above. In this regard, the large-scale irrigation schemes are spaces cut off from their surroundings, where people are forced to comply with a predefined set of rules and behaviours, strictly controlled by staff members, with no room for free determination, in a daily, continuous and extended interaction over long periods, spanning a lifetime. The second consequence is alienation: a feeling of being a stranger in one's own home, residing in a land that is no longer familiar, that has no sense of belonging, devoid of memory, and, consequently with no desire to care for it. Paradoxically, these sentiments translate into the management complaining that the farmers "do not participate", "are lazy" and that they even sabotage the infrastructure and production processes.

Not only are human lives trapped; so are plants, animals, water and the Earth itself. These mega-projects are one of the most all-encompassing experiments in the terraforming process; nothing is exempt. Natural forces and social energies are all forced in one direction: the most appropriate environment for the triumph of modernisation. Everything is arranged. Modernity must work, but something is not operational. In fact, many things get jammed, falter or stagger; the gap between the expected growth (of production, jobs and earnings) and the actual figures widen. The frequent stalling and sometimes failure of large-scale irrigation projects is the subject of the next section.

Taking a turn for the worse: The fate of large-scale irrigation projects

The footprint of a large-scale irrigation project is deep, but its significance and true legitimacy lie in its capacity to fulfil future expectations, guaranteeing a promised production and income and thus allowing the benefits of modernity to trickle down to the everyday life of the population. However, what often occurs is that large-scale irrigation projects do not work as expected. History has repeated itself so many times in the Sahel. The projects quickly grind to a halt, spiralling into disruptive dynamics linked to financial, technical and organisational shortcomings: canals without maintenance decay, machinery stops operating and pumping stations are at a standstill. Modernity appears overwhelmed. The circulation networks, roads to markets and large cities and rural tracks are left without maintenance and become hardly practicable. External crisis factors, such as the states' indebtedness and the consequent structural adjustments, the exposure to the dynamics of the global market and political insecurity, make the contexts in which large-scale irrigation schemes operate unstable. However, to attribute all responsibilities to these external factors is to fail to ponder the internal fragilities of the model. Therefore, it is important to focus on the internal causes of failure. Too often, the plans hypothesised by the developers were not realised, to the point that it is possible to speak of the "systemic failures" of the projects (Scott, 1998, pp. 263–264; see also Ika, 2005). Three internal causes of these failures can be identified. The first is the myopia of modern irrigation planning (in space and in time). The second concerns the underestimation of the intrinsic fragility of complex systems with elevated interrelations among the constituent elements. The third involves the rigidity of action segments and the consequent inability to deal with climatic variations and social complexity (Hirschman, 1967; Scott, 1998). The first element of the modernisation process is its "weak peripheral vision", that is, short-sightedness in space as in time. The management is incapable of recognising what happens outside a limited spatial and temporal horizon, underestimating "long-term outcomes (soil structure, water quality, land-tenure relations) and third-party effects" (Scott, 1998, pp. 263–264). Moreover, the negative effects are inadequately taken into account, such as the by-products of the project outside its perimeter (such as water pollution, removal of water for other uses and impact on local power structures) or over time (for instance degradation of soils, maintenance costs). The second aspect concerns the selection of a few variables considered sufficient to set the machinery of agricultural development in motion; it appears enough, at least in the first instance, to have land, water and technical capacities. This means first setting aside many other elements that come into play and then underestimating the inherent interdependence between all the factors involved. The "various components are difficult to fit into place at the same moment: besides the construction of the irrigation works, land has to be distributed and settled, new crops have to be grown and new markets found" (Hirschman, 1967, p. 43). There are uncertainties regarding both the production process and the product and its commercialisation. To these are added the difficulties tied to the dependence on uncertain supplies of resources (water distribution, agricultural inputs, spare parts and fuel) and the human factor. A short list could include the risk of internal or nearby conflicts (whether they take the form of strikes or even sabotage); the changes that the project brings to the political, social and administrative status quo (conflicts between the authorities overseeing the projects and pre-existing authorities); and political interference at the local or the national level. The final element helps us understand the internal drivers of failures. As mentioned above, the mobility and flexibility that the Sahel population have shaped over time while responding to climatic randomness include: polyculture, shifting cultivation, pastoralism and integration of different activities. Modern irrigation proposes a tightening up of physical structures (dams, embankments for containing floods and field perimeters) equal to those of agronomic practices (monoculture, centralised organisation, subdivision by roles and phases, and precise timing). Confronting climatic randomness is thought to be possible through radical territorial changes and the use of modern technologies and scientific knowledge; the rallying cry is to adapt the environment to the project through the presumed thorough control of the water, land and agronomic input. However, this rigidity is costly, has little efficacy and is impracticable in the long run.

Despite the fact that reports pile up on the desks of ministries and donors warning of the risk of failure or, on the contrary, masking a de facto collapse or, at best, attesting to the limping pace of modernisation, after 30 years of fieldwork, we have observed the strenuous attempts to revive these major development apparatuses. In the end, they have been able to demonstrate a strong, long-standing obduracy even in increasingly deteriorating conditions. Against the backdrop of the now evident repetition of stumbles and falls of large-scale agro-irrigation modernisation and in the context of the progressive withdrawal of states (caught in the debt trap and forced into structural adjustment policies) and donors (disillusioned, at least in appearance, by mega-intervention policies), since the 1980s, a new cloud of discourses has enveloped existing large-scale irrigation schemes, whose keywords can be, respectively, rehabilitation of the schemes and empowerment of the beneficiaries.

Rehabilitation is the most institutional and conservative reaction to project failures. Basically, the idea is that something has gone wrong but that it is possible to give large-scale schemes a second chance. Lessons can be learned from the mistakes of the past. Various proposals such as a massive maintenance overhaul or a redefinition of the infrastructure from the ground up, with new technologies such as electrification of pumping stations or laser levelling of soils, and a massive injection of new economic and technical resources would have, in the opinion of the megaproject boards, undoubtedly made it possible to resolve the situation. Words like these have been repeated many times by the management of the mega-schemes, and many rehabilitation proposals have been submitted to ministries and major development agencies. The possibility of being heard depends on the conditionality of emergencies—above all, the risk of famine. However, the fate of the rehabilitation proposals has been either to remain only on paper or with a piecemeal implementation such as first investing in new tractors, then in new pumping stations, and the next time in redefining the levelling of soils in one section or another and later by introducing new crops for different markets. To borrow the words of Tomasi di Lampedusa (1958), "Change everything so as not to change anything": neither those in command, nor the tight meshes of control, nor those who really gain from these large apparatuses.

The second strategy proposed was the empowerment and participation of the farmers. Accused for a long time of being lazy and not up to the modernisation challenge proposed to them, the farmers are rediscovered as an opportunity: it will be through the devolution of responsibilities and tasks to them that the great project is revitalised and open to a new, more sustainable possibility of functioning. Cooperatives and Water Users Associations (WUAs) multiply; the tertiary network (close to the fields) of irrigation canals is entrusted to them for both maintenance and agricultural organisation. This is also the case for all mechanised operations such as land levelling, ploughing, sowing and harvesting. In fact, the strategy appears to be that of using the WUAs and the new redistribution of tasks as a gateway to a silent privatisation of large projects, entrusting them to those with the most resources at their disposal. Small and medium-sized agricultural entrepreneurs have slowly, though not formally, accumulated plots and enlarged their scope of action, often acquiring the means of production, such as tractors and combined harvesters. They are allowed greater freedom in cultivation and organisational choices, redefining a new supporting role for the board, the state and international donors. This phenomenon occurred in Sudan with the Gezira Act of 2005. The paradox is that power comes to the farmers at a time when little or nothing seems to work any longer and there is no more power to be exercised; the extreme solution seems very much like abandonment. Those who benefit most are the banks and large commercial companies that hold agricultural credit, often granted in forms close to usury (interviews from the fieldwork in February–March 2023).

A conclusion that can be drawn from these policies to revitalise large-scale irrigation schemes is that they are either too big or too tied to the very idea of modernisation to fail completely. It happens, of course, as is the case with the South Chad Irrigation Project (SCIP) (Bertoncin & Pase, 2017), but more often than not, large-scale schemes manifest great obduracy, a stubborn attachment to endure. What remains are first the hydraulic infrastructures; somehow, they require minimal attention to prevent dams from collapsing (a risk, for example, reported for the Maga dam in Cameroon) or canals from completely filling with sediment (as in Gezira). The state cannot forget the large-scale irrigation schemes altogether, at least because their complete blockage would have potentially catastrophic effects. A second dimension continues, that is, the organisational structures of management, the boards. Even when subjected to drastic staff reductions (this has happened everywhere, for example, in the Sudan Gezira Board, the Office du Niger, SEMRY), they always show themselves ready to start again with extraordinary maintenance and rehabilitation projects. The Sudan Gezira Board is exemplary from this perspective. It was forced in the early 2010s to drastically cut its staff, reducing it to a minimum, yet as soon as new windows of opportunity appeared of a possible return of donors during the transitional government, the board members were able to rethink and be ready to move forward with the new funding being planned. These management structures function in an accordion-like manner; if forced, they shrink to a minimum, ready to expand again when the context allows. This was the case of the rehabilitation projects envisaged for Gezira in early 2023 when Ukraine and Russia were at war. Transporting Ukrainian wheat to the global market became difficult for the African markets in particular. As described in Chapter 10, at that juncture, the major international donors recognised Gezira as the breadbasket for the entire region. Then, the war erupted.

The overt or latent failures of large-scale irrigation schemes have another outcome worth emphasising: the focus on what does not work risks obscuring the trends underway at a local level, such as those power dynamics where failure presents an opportunity for consolidating their grip on the territory. This is one of the aspects highlighted in Chapter 15. These and other outcomes of the major project crisis are discussed in the next section.

Lives escaping broken traps

The traps capture living and inanimate beings. Unsuccessful results exacerbate contradictions. The landslide of delegitimisation begins. The traps give way, crack and are no longer able to separate the modern and advanced from its surroundings—the backward and retrograde. Local societies attempt to overcome these failures. Breaches open up in the walls of what were once "total institutions", according to Goffman's interpretation (1961). Previously locked-up lives look outwards, while others come in from outside; timidly and gradually, everything begins to change. Different processes are activated when the traps built by the development apparatuses are unravelled (see Table 1.4).

The first elements to give way are the compactness of the large-scale irrigation projects and the gradual collapse of the difference between inside and outside. A sign of the first phenomenon is the progressive reduction of areas of production due to the lack of maintenance and efficient machinery. The irrigation scheme becomes fragmented into parts that are still functioning and others that are progressively being abandoned. Additionally, the clear boundary that separated the inside from the outside deteriorates in all respects: the proliferation of cultivated areas outside the project but with the water pilfered from it; the illegal sale of portions of the harvest; or of the agronomic input distributed by technicians; or the illegal entry of herds into the irrigated areas following the harvest.

If the foot of the development apparatus, drained of energy by its failures, is lifted, this leaves room for new possibilities for action and reinvention of modes of production, allowing unprecedented margins of freedom for the farmers and small entrepreneurs who are beginning to emerge. The context is obviously extremely difficult; freedom comes from the de facto abandonment of central organisational structures, not from the willingness to broaden the base of shared choices. Having overcome the initial effect of disorientation inevitable upon leaving despotic total institutions and thus the difficulty of perceiving themselves without the project that previously determined everything in their lives, farmers begin to react, with various forms of adaptation and innovation. The formal and informal institutions of the

Table 1.4 The trap, when it breaks.

	Physical dimension	*Relational dimension*	*Cognitive dimension*	*Operative dimension*
What is broken/ who and what undermines isolation	Abandoning parts of the irrigation scheme and its infrastructure Entry of transhumant herds and wandering animals, illegal exit of water, fertilisers, agricultural products	Loss of meaning of formal and informal institutions, penetration (from the outside) of new relational networks	Contamination between tacit and expert knowledge, mixing as a possibility of creativity	Unexpected uses of project infrastructure, innovations of crops and agronomic practices through imitation and contact

large-scale schemes begin to lose their meaning; their strength lies in representing the model of a successful future. When this future is lost in a maze of inefficiency, even the formal rules and informal behaviours linked to agricultural modernisation appear increasingly incoherent and inadequate. In contrast, other institutions and approaches start to gain access from the outside and begin to build new relational networks. Over time, skills related to irrigation techniques and modern agriculture are learned by the villagers. With the gradual grip loss of development apparatuses, a process of cross-pollination between tacit and expert knowledge takes place. On one hand, irrigation skills proliferate and expand beyond the original schemes, either in the immediate vicinity—perhaps by drawing water illegally or, with unofficial permits, from the canals themselves—or further afield by creating small irrigation schemes from below, using small motor pumps. On the other hand, the former resource-use practices are also beginning to be practised within the large schemes, such as fishing in the canals, closing them with fishing nets and hindering irrigation (as in SEMRY II), or herds coming in to graze, no longer being driven out, or cultivating crops other than those that were planned. This mixing of techniques and knowledge becomes a laboratory of creativity, both outside and inside the project area, in fact, awakening, at least partly, from the hiatus that had been brought about by the implementation of the development apparatus. Thus, different arrangements are configured, but not all have the same capacity to be successful. The relational networks that will somehow be able to interact positively with the crisis context are those of the strongest actors, linking state officials with local authorities and big city traders with rural elites. The fissures damaging large-scale projects open up margins of action for those with greater economic means or greater social and political clout.

In reality, there is a wide range of situations in which large-scale irrigation schemes in the Sahel find themselves. If the overall horizon is that of a progressive slide from the certainties of modernisation, in some situations, projects, or at least parts of them, continue to function and are effectively rehabilitated, and cultivation and organisational innovations are tried. In other situations, projects never advance beyond their stunted pilot phase (as is the case of the Baga Polder in Borno, Nigeria). To better understand how a large project is reshaped by these dynamics dealing with failures, it may be useful to examine what is happening in the context of the Office du Niger, which can be described as a patchwork of modernised, partially rehabilitated and degraded areas. As Jamin and Coulibaly (1995, p. 134) testify, "irrigation and irrigated agriculture are today completely appropriated by farmers and therefore transformed, reinterpreted and integrated into their production system". Where a complete rehabilitation has been carried out, intensifying crop cycles and introducing organisational changes, the results are good and the board's grip on the project remains firm. Elsewhere, in the Office du Niger itself, side by side with the performing sectors, in areas that are only partially or have not been fully rehabilitated and where disintegration processes are therefore active, farmers implement numerous deviations or transgressions. In such situations, it is impossible to respect the intensive rice-growing model proposed by the staff, so farmers attempt ways to make the most of the situation and respond more effectively to the constraints, first and foremost, the loss of efficiency of the development apparatus. The first example is the widespread use of the so-called *hors casier* (out of the irrigation scheme) cultivation (in the area surrounding the project perimeters, generally located along the adduction or drainage canals) to compensate for the lack of plots inside the Office du Niger and to supplement the farmers' rice-growing income. Officials tolerate this practice by further subjecting it to payment of the water consumption tax. These lands constitute important earning possibilities for farmers; a *hors casier* plot is more than half the size of the one owned in the official perimeters and represents a space of freedom from the agricultural and land constraints imposed by the Office du Niger (Jamin & Coulibaly, 2005). In other cases, it is not uncommon to see farmers pirating water from canals, through underground pipes, to avoid paying the tax. A second example is the practice of horticulture in the lands surrounding the villages (again taking water from the project canals), an activity long hindered by the Office du Niger because it is regarded as competing with rice. Horticulture is undertaken especially in the Niono area, where the asphalting of the Niono-Markala Road axis has facilitated the trade of vegetables in the markets of Ségou and Bamako. A third example involves all those secondary activities that flank irrigated agriculture, such as livestock farming, planting rain-fed crops, seasonal fishing (practised also in the irrigation and drainage canals) and collecting firewood, forming a complex and articulated production system. As the French anthropologist J.-P. Olivier de Sardan (1995) says, two principles are at play in the relationship between development projects and beneficiaries: the principle of selection (whereby of the project's package of measures, only some are adopted) and the principle of diversion (*détournement*), where farmers avail the opportunities offered by a development apparatus to put them at the service of their own objectives. Selection and diversion become increasingly evident, precisely with the downward spiral of projects. In the eyes of planners, transgressions undermine the coherence of development interventions but actually open the way to new negotiated paths, such as the hybridisation (*mestizo*) of land practices mentioned earlier.

In the time of the cracks of large-scale irrigation schemes, farmers have proven themselves capable of finding innovative solutions. Even on the ruins of completely failed projects, such as the Baga Polder and the SCIP, which were supposed to take water from Lake Chad through two intake channels, farmers were able to find escape routes. In the lower zones of the areas affected by projects, now a bare stretch of dry clay, where water accumulates from rainwater or even inside the irrigation canals, no longer in use, some farmers have returned to cultivating *musakwa*, a local offseason grain. Along the intake channels, with water flowing again after the great drought of the early 1980s had dried them out completely, farmers have set up small pumps to irrigate the nearby fields. Thus, for kilometres and kilometres along the two intake channels, a continuous expanse of cereal and vegetable crops is irrigated directly by farmers during the dry season. The harvests are sold in the large markets in the federal state capital of Borno, Maiduguri (Bertoncin & Pase,

2017). Remnants of the failed projects find a new use through highly autonomous initiatives: canals still remain as sources of water, but outside of what was intended to be an irrigated area, embankments serve as rural roads to reach the fields and transport production.

Here, it may be useful to recall Agamben's (2018) reasoning about how it is possible to deal with times of catastrophe, that is, those phases in which a given system (political, social, economic or cultural) appears to be on the verge of breaking down. The Italian philosopher borrows a typical expression from the Neapolitan mask of the Commedia dell'arte called Pulcinella: "*ubi fracassorium, ibi fuggitorium*" ("where there is a catastrophe, there is also an escape route"). In the face of a disaster, one must look for a way to flee. "The [proverbial] secret of Pulcinella is that, in the comedy of life, there is no secret but only, at every moment, a way out" (Agamben, 2018, p. 125). Among his skills, Pulcinella has the capability of always being simultaneously inside and outside the situation in which he finds himself; in every drama, he discovers "a shortcut or a byway" (Agamben, 2018, p. 43). He deviates and surprises; he is never where others would like him to be: he cannot be trapped. Similar to Pulcinella, in times of crisis, farmers are in and out—of a project, of modern agricultural practices or of development policies. "What we are dealing with here is the liberation of that which remains captured and separated by means of apparatuses, in order to bring it back to a possible common use" (Agamben, 2009, p. 17). The profanation of apparatuses, using their component parts in a desecrating and detached way compared to the model (Agamben, 2009, pp. 17–19, 24), is the means of liberating themselves from the grip of the (broken) trap. In this way, the farmers re-establish a "possible common use" of what had been imprisoned by the development apparatus and thus subtracted from the local population.

However, more issues have emerged. The malfunctioning of the traps and the silting up of agricultural modernisation are symptoms that the entire terraforming process that began in the Sahel during the colonial era has not achieved the expected results. Everything contributes to blocking the gears. Animals contribute: granivorous birds, such as cuela-cuela, pounce on the seeds before harvesting; cows and goats leave their tracks on the banks and erode the banks of canals; and fish proliferate in the dams' reservoirs and even in canals, providing opportunities for favourable alternative activities. So do sprouting weeds that clog the canals and drains, as well as plants such as Prosopis, perhaps initially imported to contrast desertification, that become invasive and advance on all soils where there is a minimum of moisture; only if eradicated completely can they be stopped (Bertoncin & Pase, 2008, pp. 278-280). The smallest organisms play their part as well: mosquitoes carrying plasmodium and malaria, and blackflies with filarial worm and onchocerciasis. Other-than-human entities—rains, floods and the soil—also make their mark. During the rainy season, mud makes the roads within the perimeters of the schemes impassable. Villages are isolated. Exceptional floods break the banks, inundating the perimeters. Pronounced lower levels leave the pumps dry. Other-than-human beings and entities oppose the standardising design of modernisation, declaring their autonomy and decrying boundaries through their resistance. Even machines—whether agricultural, earth-moving or transport—and hydraulic pumps make their contributions, with their increasingly frequent breakdowns, the shortage of spare parts and the wear and tear due to extreme working conditions.

However, the lesson is far from being learned. The technicist and high-modernist approach, with its simplification and reduction of reality to a few elements to be lined up to constitute productive apparatuses, reappears as soon as the conditions seem right. In the next section, we discuss these new attempts.

Land and water for Sale: New traps

Stronger than any denial in the field, the terraforming paradigm resurfaces even in this new millennium, taking on different modalities but retaining the presumptuous assumption of being able to completely redesign the destinies of large spaces to make them better—effective and with improved performance. New and in many ways different traps have been designed, constructed (sometimes only partially) and activated to once again select and isolate variables useful to the high-modernist production effort. Proposals for large-scale irrigation projects once again reappear in the form of huge domestic or foreign investments, yet in areas that are different from those already affected by the former schemes. Sometimes alongside them, sometimes far from them, the sites identified as suitable for these new investments are, in any case, outside the locations where large-scale irrigation schemes had been established earlier. In fact, the longevity of the major irrigation schemes has produced a hodgepodge of rights (formal and informal), interests and practices that discourages major investors from taking them on. They are too complex, messy and full of intricate legal and customary relationships that it would not make sense to invest in them. These new investments resulting from the land rush started in the first years of the 21st century and accelerated with the soaring prices of agricultural commodities in 2007–2008. This phenomenon has different definitions: "global land grab" is used more frequently, but others, such as "international agricultural investments" or "large-scale land acquisitions" are preferred by global institutions such as the United Nations and the World Bank (Allan et al., 2013; Deininger, 2011; Hall et al., 2015; Li, 2011; Mehta et al., 2012; Pase et al., 2022; Schoneveld, 2017; Wolford et al., 2013). In that context, private companies (national or international) acquired lands to produce cash crops. Additionally, countries that have scarce water resources and lack suitable soil for agriculture have committed funds for these large-scale agricultural investments to cultivate crops abroad and ensure domestic food security or simply for business purposes. Recently, alternative approaches to direct land control have emerged, offering strategies different from traditional land grabs, which can often provoke local community protests (Oliveira et al., 2020). For instance, contract farming has become a diffuse practice in African agriculture (e.g., under the Gezira scheme), intensifying the exploitation and marginalisation of smallholder farmers (Martiniello, 2020). The most common way of acquiring land is not the purchase of property but rather with a long-term loan. In exchange, investors offer technology, access to the international commodities market and job creation, as well as compensation—more or less effective—to the affected local communities. With a view to expanding their economies, the governments of the host countries facilitated this process, often with the hope of gaining immediate benefits for the ruling elite. The phenomenon has global dimensions and particularly affects Africa, which, once again, is presumed as an empty space, available for foreign projects (Pase, 2024). In the Sahel, of course, it is not enough to acquire land, the key resource is water; spaces will therefore be selected where water is available and at low or minimal cost. We have selected three examples describing this new wave of irrigation projects: two in Sudan, the White Nile Sugar Company (WNSC) and the spread of foreign investments in fodder cultivation, and the Malibya Project in the area of Office du Niger in Mali.

In the early 2000s, before the eruption of armed clashes and the spread of insecurity in the Segou region, the Malian state had granted large land loans to an international aid organisation, for interventions that targeted the dissemination of small and medium-sized agricultural entrepreneurship, or directly to foreign investors, near the Office du Niger irrigation schemes. The main one was the 100,000-ha area granted to the Societé Malibya Agriculture, strictly linked to the Libyan government under Gaddafi (Bertoncin et al., 2010). To put this in perspective, consider that the actual productive irrigated area within the Office du Niger is less than 90,000 ha. The objectives stated in the convention signed by the two governments in 2008 are self-sufficiency and food security (of both countries), the development of the agro-industry and livestock farming for meat. Between 2008 and 2010, a Chinese company (China Geo Engineering Corporation, CGC) carried out preliminary works, particularly the 40-km main canal, which draws water from the Fala of Boky Wéré (*fala* is a palaeochannel of an ancient branch of the Niger River), and the asphalt road alongside it. These works have caused considerable hardship to the populations settled in the area: displacement of villages and destruction of field, markets and burial grounds, cutting off connections in the lands of the same village with the lengthening of routes, particularly transhumance corridors. Once again, the project took control over an assumed empty space, while in reality, the territory involved was very much alive and full of agrosilvipastoral activities. Villagers held many protests, and even organised in claim committees (Larder, 2015). The first phase of the project would have foreseen the cultivation of 25,000 ha. However, the project came to a halt, first due to the fall of Gaddafi (Adamczewski & Hugon, 2013) and then due to the political instability in Mali. In this area, one can observe something similar to the incidents along the Baga Polder and the SCIP intake channels. In the lands bordering the 40-km-long canal and in the terminal area, satellite photos show extensive cultivation, organised in small, irregular plots that pump water from the canal. The grassroots initiative has once again proven capable of coping with the failures of large projects and finding viable alternatives.

Around the same time, a large sugarcane cultivation perimeter was being developed in Sudan, in the federal state of White Nile, south of Khartoum, to produce the raw material for the sugar factory that was simultaneously being built on-site. The WNSC began operations in 2006, but production launched in 2012. As its name suggests, it draws its water resources from the White Nile. Commissioned by the government of al-Beshir as a demonstration of its desire to modernise and open up to the markets, being a mixed company of public and private capital, it had a high symbolic and propagandistic value: from the day of its inauguration, a gigantic portrait of the president greeted visitors in the management hall. The Kenana Sugar Company (KSC), the most successful of its kind in Sudan (see Chapter 10), oversaw both the planning and later the construction phase. The KSC holds a direct share of 30%. The Arab Authority for Agricultural Investment and Development holds 13.9%, while the remaining 56.1% is divided among the national and the White Nile State governments, the Abu Dhabi Development Fund, the Saudi Development Fund and some Sudanese banks. The total area of the WNSC is 165,000 feddans; only part of the planned 125,000 feddans is under cultivation. Its production capacity is supposed to reach that of the KSC, but the WNSC is encountering several difficulties and is far from achieving its goals. The people who previously settled in the area occupied by the irrigation scheme have been forced to move to villages set up by the companies or to other neighbouring villages. The affected families have lost their houses (built according to deeply rooted techniques), land rights and the usual sources of livelihood: pastures for transhumant herding and fields for rain-fed agriculture. In return for these losses, the affected households receive various forms of compensation (see Cernea, 2003; Feldman et al., 2003; Kanbur, 2003). The first form of compensation for the inhabitants expelled from the villages consists of the allocation of building plots in the new settlements (a plot is 30 × 20 m). Then, they receive building materials (two trucks of soil to make bricks and, in some cases, bundles of reeds and wooden poles), along with a small stipend available to them (1,500 SDGs per house). The displaced inhabitants consider registered ownership of the land and the house an advantage, but the amount given to them is far from what would be needed to rebuild (from our interviews in Elfirai and Elmokhtar villages). The expropriated land, for those who are able to testify to actual ownership (a fact far from being taken for granted), is compensated in a ratio of 10 to 1—that is, for every 100 feddans expropriated, 10 are given back in infrastructure land. This land is recorded in the land register; currently, the owners have a 20-year lease with the company. The company then manages the production of sugarcane in these areas. At the end of the 20 years, if the lease is not renewed, the land will revert to the owners. The last direct compensation is employment, whether permanent or seasonal; a share of the staff positions is reserved for the local people. In particular, for every 30 feddans expropriated, one family member can apply for a job. The inhabitants of the relocated villages complain that these compensations are insufficient: besides the material aspect, there is also tearing apart and forced removal from places of historical and ancestral sentiment. The cemetery areas of the ancient villages, unlike in the excavation of the main canal of the Malibya project, are respected under the schemes; no cultivation takes place in them. Nonetheless, when families are relocated 18–20 km away, how will their contact with the land of their ancestors be maintained? There has been no shortage of incidents of the local people protesting and interventions by the forces of law and order. Already in crisis due to many causes, among them saline soils, drainage difficulties, poor soil levelling, an abundance of weeds in land only recently irrigated and the government's lack of funds to invest in equipment and personnel, the WNSC is now in the midst of a war zone.

In the early 2000s as well, the third case of new large-scale irrigation schemes began, linked to investments by the Arab Gulf countries in Sudan. Against the background of the rising prices of agricultural commodities, several drivers, both regional and internal, pushed this arrival of Arab investments: from within Sudan, first the regime's need to re-establish political relations after a prolonged phase of isolation and then, following the separation of South Sudan, the drastic reduction of oil revenues. On the side of the Gulf countries, there is an increasingly strong drive to find agricultural production opportunities abroad, given the excessive overuse of domestic water resources. This led to 2013, when al-Bashir passed new investment legislation (Sudan Republic, 2013)) to vigorously attract foreign investment to Sudan by offering significant tax exemptions and strong bureaucratic support to investors (Hassan, 2015; United Nations Conference on Trade and Development—UNCTAD), 2015). Investments are concentrated in the River Nile State, north of Khartoum, and are essentially aimed at the agro-industrial production of alfalfa (*Medicago sativa*) (Bertoncin et al., 2022). This crop is considered the queen of forages due to its high productivity and nutritional value superior to other forages; it is highly sought after by the dairy industry in Arab countries. Irrigation is carried out with central pivot systems. The water is distributed inside pipes placed high above the ground, connected by mobile towers and with sprinklers positioned along their entire length. The outermost tower dictates the speed of the entire system, while a device checks that each tower is aligned with the adjacent one. One pivot can irrigate extensions from 34 to 63 ha. This agrarian system can produce up to ten harvests per year: the harvest is repeated every 30–35 days, and the average production of each cultivated land amounts to 2–3 t/ha, considered a very high quantity. The fodder is entirely destined for livestock (dairy and slaughter) in Saudi Arabia and the United Arab Emirates. The logic followed by these investments leads to a further step into high-modernist agriculture, which we have called the ultra-thin spaces (Bertoncin et al., 2019). Infrastructural interventions are very limited. Investors prefer areas close to communication routes and pre-existing electricity and telecommunication networks, eliminating the need to invest in expensive infrastructure. With the use of central pivot irrigation, expensive land levelling and soil preparation work are avoided. No settlements are planned in the projects, except for the few compounds needed for technical staff. The areas are often fenced off, and the local population is completely excluded. No anchorage to the local area and its people is desired. The investor demands unencumbered land from the state. This does not mean that conflicts do not arise; in fact, bargaining processes are generated by the people who believe they have rights to the land where they have always grazed their cattle. In this situation, the state appears more like a broker, selling (or selling off) its resources in exchange for access to markets and rents, particularly for the elites involved. These ultra-thin spaces are apparently among the extreme outcomes of both the terraforming process and the dynamics of nomadic and uprooted global capitalism (Bauman, 2002; Delueze & Guattari, 1980). In the ever-changing and hyper-accelerated flows of late neo-liberalism, the manipulation of the Earth loses its drive towards constructing a different world. It is no longer important to shape social and territorial reality in order to give it a new form and to propose a different future; instead, what counts is the immediate present, expelling, along with long-term horizons, all engagement and mingling with people and nature. Somehow, it is as if modernity's desire to affect society and space by determining new, stable and grounded geometries has lost faith in itself. The ultra-thin irrigated spaces are telling examples of the tendency of global capital to be constantly on the move, identifying and exploiting the transient and immediate combination of the best productive variables to obtain the greatest possible profit in the shortest possible time.

Between Atbara and Berber, just before the war, the Zadna Company, directly linked to the Sudanese Armed Forces (SAF), was building a gigantic project that aspired to activate 1,500 pivots; 32 large pumps, already constructed and truly impressive, were intended to feed water into the main canal to reach, after 8 km, the area to be irrigated. This case—the appropriation of resources by the elites who held so much of the country's economy in their hands, in addition to the army—is explicit. All these investments are also now at risk because of the war that has been ravaging Sudan since April 2023. This is the fate, for example, of the Al Waha irrigation scheme, 50 km south of Khartoum, on the left bank of the Blue Nile. Built-in 2010 by one of Sudan's major economic actors, DAL Group, to supply fodder to one of its large dairy companies and to export the fodder to Saudi Arabia, it had over 100 pivots of 60 ha each, characterised by high productivity. It was looted and is now completely destroyed.

Violence, albeit in different forms, has affected all three of these new traps. In the next section, we attempt to understand the possible relations—and the logic behind them—of the large-scale irrigation schemes and the atmosphere of unrestrained violence in the Sahel today.

Holes in the water: How the denial of the future paves the way for violence

As we are finishing writing this chapter, the RSF are advancing inside Gezira and towards Sennar; the productive heart of Sudanese agriculture is under siege. Hydro-agricultural structures are directly affected. In the introduction, we mentioned the episode of Wad al-Noura where, in early June 2024, the inhabitants' attempt to keep the RSF away by cutting the canals and flooding the land was punished with a massacre of civilians. Again, from the valuable Sudan War Monitor updates, we know that on 12 July 2024, as the RSF approached the crucial town of Sennar, the army attested its defensive line "along an irrigation canal, but the RSF succeeded in seizing crossings at the Maiuorno water plant and bridge".[4] On Sunday, 14 July

2024, Sudanese activist groups, such as the Wad Medani Resistance Committees, denounced the RSF as "deliberately and systematically" destroying and looting the agricultural infrastructure of Gezira State: the Gezira, the Rahad scheme, the Guneed and the Sennar sugar factories. Thus, for the first time in a hundred years, the Gezira Agricultural Scheme will be forced to fully stop its operations (https://sudantribune.net/article288228/). In the opinion of many activists, these actions form the premise of a human-made famine to blackmail the population with the threat of hunger. Certainly, the war in Sudan is also a conflict for the grabbing of economic resources by the RSF and the army leadership since the power-sharing after the fall of al-Beshir did not work. Land and water, along with gold, are among the main prey.

In this section, we attempt to provide reasons for the possible link between the violence in the Sahel and the failure of efforts to modernise agriculture. One connection may be traced from the spatial overlap between one of the regions where the failure of large-scale irrigation schemes is most evident and the place of insurgency of one of the most violent organisations of Islamic fundamentalists, Boko Haram, and its branches that have been formed over time. The spatial-temporal overlap is striking. The megaprojects of the SCIP and Baga Polder proposed in the 1970s were intended to bring a modernising turn to the state of Borno. The dream was to build a true agro-town, a powerful agro-industrial system capable of offering a future of development to the region, as well as to counter the flight from the countryside to the state capital of Maiduguri. The projects in the 1980s failed completely; nothing remains of that dream. A decade or so after these events, in Maiduguri and, since then, in the entire surrounding region, with particular virulence on the shores of Lake Chad, the fundamentalist movement has spread and violence has exploded, from the terrorists, the police, the army and, finally, the self-defence forces. Unemployment and the absence of other opportunities certainly push young people to join the insurgent forces. We can legitimately ask whether the deconstruction of social and economic structures caused by the implantation of these irrigation projects and their subsequent complete failure are not among the many factors that triggered this spiral of violence. More generally, the holes in the water or water-holed projects that pierce the Sahel are among the most obvious proof of the failure of the modernisation desired after independence. Joblessness and the lack of an (albeit minimal) income plant the seeds for more extreme solutions where joining terrorist groups and the minimum wage they guarantee, even with looting, become the last resort. However, less immediate and deeper levels merit analysis. The first concerns the question of the future and the memory of the past, the second deals with modern traps on different scales, and the third refers to the extreme results of the terraforming process. These analytical findings, although addressed separately, are inevitably interconnected.

The first aspect involves the two opposite temporal directions of events—oriented to the future and coming from the past—and the related system of expectations, on the one hand, and entrenchments, on the other hand, that sustain and give meaning to the lives of human groups. The idea of the "regimes of historicity" proposed by Hartog (2003) is useful in this regard. Quoting Lorenz, Hartog's thesis is that "Western thinking about history is basically characterized by a succession of three 'regimes of historicity'—from a dominant past orientation until the French revolution to a dominant future orientation until the 1980s and then a dominant present orientation in the years since" (Lorenz, 2017, p. 112). Applied to our object of study, this proposal is useful in understanding the transformative and modern future-building orientation of large-scale irrigation projects in the colonial era and after independence, just as it helps to frame the narrowing down to the present time—indeed the here and now—of the ultra-thin spaces we mentioned in the previous section. The problem lies in the fact that the future pointed out by the developers and waited for by the farmers has turned out to be in vain; expectation has plummeted to disillusionment. A theft of the future, promised but unfulfilled, and a theft of the past have occurred simultaneously. In their desire to impose modernity, first, the colonisers and then the developers have totally devalued the knowledge and history of the people inhabiting the regions involved in their planning, defined as backward, if not primitive. Nothing has been worth considering. Thus, a rift from the past has also been created. The Sahelian peoples have found themselves at once robbed of a modern future that never arrived and of a past that counted for nothing. Forced into an "inhospitable present" (Bodei, 1995, pp. 14–15), the ways of a connection to the past and an opening to the future appear to be foreclosed. This is one of the premises of that "angry wind" that the American traveller and writer J. Tayler (2005) felt blowing in the Sahel during the early 2000s. Trapped in the *cul de sac* of a present closed to all except for illegal and dangerous migration routes across the desert and the Mediterranean, the drive towards violence can appear somewhat seductive to the people.

Besides the trap of the inhospitable present, other traps are at work in the Sahel, as a legacy of the conjunction of coloniality and modernity. We have mentioned one—the trap of the development apparatuses. However, this trap lies within another trap, which Agnew (1994) called the "territorial trap" in a seminal article that has left its mark on debates in contemporary political geography (Agnew, 1994, 2010, 2015; Brenner & Elden, 2009; Elden, 2013; Newman, 2010; Reid-Henry, 2010; Shah, 2012). According to Agnew (1994), the emergence of the modern territorial state is based on three geographical assumptions. First, sovereignty is defined in territorial terms, meaning that the state is regarded as a unified and stable sovereign space. Second, political space is divided between the internal politics of the state and the external realm of international politics. Third, the state is perceived as a uniform container of society, shaping the latter according to its own objectives. Although these assumptions might seem fundamental and self-evident, they have created an intellectual trap by overlooking the need to explore other forms of political organisation, such as analysing political space through networks, flows and place-making (Agnew, 2015). The weight of the territorial trap is even heavier in the Sahel, where state genesis is inextricably linked to colonial appropriation (Pase, 2011). The fragility and weak legitimacy of state structures are evident in the troubled political paths after independence (see Chapter 5 for current outcomes).

On different scales, the Sahel has thus been trapped in apparatuses (developmental and state-led) that have been based on a drastic simplification of precolonial paths. A violent tone accumulates in the construction of these traps, where coloniality and modernity overlap in erasing all specificities and every autonomous determination. These traps are in the hands of the few who approach them to obtain rents for the ruling classes, who drive development and state apparatuses (Magrin, 2013). The excluded masses observe these appropriative and opportunistic dynamics, progressively increasing their distrust. Much of the resources (including land and water) and connections guaranteed on a global scale are controlled by circumscribed elites, who often make investments abroad using the rents they obtain locally. Every apparatus has its profiteers: this is the ultimate and bitter feeling of many of our interlocutors during our fieldwork. Among the escape routes from the traps imposed on the Sahel, violence can, at some point, be viewed as a quick shortcut.

Yet another level of violence is implicit in the terraforming project, once again, within the large-scale irrigation schemes. The presumptuous will to create a new world, a new Africa, has adverse impacts on human and other-than-human beings and entities, trusting that they can be formatted and wedged into a predefined, fixed and ordered design. In this sense, too, violence has deep roots in agricultural modernisation, which ignores the different ways of behaving (the cultural differences), interacting between humans and nonhumans and using resources in common forms, not according to the proprietary, enclosed and exclusionary logic of modernity.

The compression of all these different levels of violence intensifies anger and nurtures armed clashes without form or limit, shocking countries around the world with images of millions of inhabitants dragged into an abyss of utter destruction, as is currently occurring in Sudan. The poor people torn from their daily lives become "bare lives" (Agamben, 1995, pp. 121–123), bodies exposed without any defence against extreme violence. To avoid being caught up in the deepest despondency, there is a need to explore different ways of thinking in order to turn things around.

Pluri-Diverse irrigated landscape: Going towards different futures

Based on our analysis of the relations between water and land and between humans and nonhumans in the Sahel, we feel that new directions need to be investigated, using different epistemologies and a pluralistic approach in order to imagine different outcomes from the raging violence today. Quoting Ndlovu-Gatsheni (2023), a counter-hegemonic project of "reworlding" from the Global South is needed to combat the modernist and colonialist model for weaving (epistemic) freedom and liberation. We identify four key layers. The first layer is more practical and immediately relevant to the agrarian question in its different dimensions (land issues, organisational and political). The second layer concerns different perspectives on the issue of Sahelian futures. The third layer entails dismantling the coloniality of development and territorial traps. Finally, the fourth layer involves the shift from the technicist to the vitalist approach, à la Ghosh.

The first layer, the agrarian dimension, takes us across the fields, right to the middle of the irrigated landscape, immersed in its great differentiation. The components of the irrigated landscape are always the same: a permanent water reservoir to draw from, a withdrawal system (by gravity or pumping), a distribution network and a drainage grid. From one side of the Sahel to the other, one is always at home in this landscape: water arrives on the fields, flows, is sprayed from above, drips and crops grow. The mosaic of irrigated agriculture in the Sahel is now reassembled in different directions, which no longer presuppose the absolute centrality of the large-scale project. We propose identifying it as pluri-diverse, defined as extremely differentiated in its shapes, sizes and spatial declinations. Interests and actors are changing. Compared to the imposition of the large spaces completely organised by the regular grids of the great irrigation, today the multiplicity and the differentiation from situation to situation and from place to place prevail. This phenomenon is occurring under the large-scale schemes, which we have observed as hosts to hybridisation and mixing today. It is present in the neighbouring areas transformed by contact; it is in all places where there is even the slightest possibility of bringing water to the lands. The advent of a low-cost Chinese motor pump (together with that of motorbikes and cheap mobile phones, almost always from China) is leading to what we could call a self-organised dusting in space, scattered wherever irrigation is possible and feasible, capable of directly linking production sites with markets. The legacy of the large-scale irrigation schemes (as well as that of the small village irrigation projects that so many NGOs have undertaken in recent decades) has brought expertise that today percolates along every possible route, with water flowing from the thousands of points where it can be accessed: groundwater, rivers, temporary streams and rainwater storage points, natural or excavated by government or cooperative initiatives. Here, margins seem to open up for an innovative and creative capacity, based on simple technologies, in contexts where irrigation does not become exclusive but is complementary to other productive activities. The places most accessible to markets are favoured because they are either close or well connected to large cities by good roads. There is obviously no lack of inequalities (class, age and gender) and power dynamics; those who can invest in irrigated horticulture and fruit-growing activities are above all those who have financial resources and are part of effective relational networks. The major traders, government officials and the key interconnecting figures between state and local political power are the ones who play a huge part. Labourers, women, the unemployed and uneducated youth remain on the margins. Thus, there is plenty of room for political, educational and cooperative initiatives that can counterbalance processes that otherwise concentrate the wealth that can come from agriculture in the hands of the few. It is also true that these emerging agricultural perspectives can reactivate pre-existing or otherwise autonomous approaches to production and exchange, without necessarily responding to purely market-driven logics (van der Ploeg, 2008).

The second layer, Sahelian futures, involves imagining the future. As we have observed, "futures past" (Koselleck, 2004) have disappointed expectations. For Nigerian historian Toyin Falola (2022), African futurism embodies the final stage of decoloniality, premised on the integration of indigenous knowledge and techno-scientific innovation with Afro-spirituality, youth creativity and folklore to construct alternative futures. Hence, there is a need to think about the future. If the present is desperate, if there is no direction to be given, the boisterous, ever-increasing energy of the younger generations becomes destructive, constrained to narrow horizons, with no opening to the future—at least not an encouraging one, possibly only an eschatological and otherworldly one (very often of a religious-fundamentalist kind)—capable of justifying the violent horror of looting and plunder. From this point of view, it is important to try to make sense again of the past, the precolonial history and the cultural differences that have always smouldered beneath the Western mask. It may be the way to reknit, the present, and the past with the future, in an original way that is not dependent on external imprints. It is an emerging possibility, most often through unconventional artistic forms and expressions. It is clearly not a matter of seeking authentic, essentialist cultural identities, albeit opposed to those of colonial inheritance; cross-pollination between the many pasts and many imaginable futures is inevitable and desirable in a present that is thus open to the unexpected.

The third layer, dismantling, deals with the tension to escape traps at different levels: the traps of large-scale development apparatuses and the territorial traps of states. As mentioned, the Sahel has witnessed the project of modernisation materialise in "spatial miniatures" where the transformative capacity has been concentrated, thus rendering the "visual order" of high modernism evident and instructive (Scott, 1998, pp. 133–134). At another level, the overall texture of political space is caught in the territorial traps of the states inherited from colonisation—traps within traps. In these traps, everything—living beings (human and nonhuman) and material entities (water and land)—should have found its place. The traps have actually obstructed the future, blocking different outcomes for a long time. The flow of animals, plants, people and waters has been prevented, thus hindering that dimension of movement, of network construction, which Mbembe describes, precisely for the Sahelian spaces, as "the *logic of moving sands*" (2021, p. 183, italics by the author). Many ways on the ground have been attempted to escape these traps; we have observed the transgressions with respect to the norms and constraints of large-scale irrigation schemes, but others concern states. In this sense, think of what transpires along the borders of states (see Chapter 5), where new regional polarisations are reconfigured beyond state boundaries, as nodes in a dense weaving of economic, social and cultural relations that transcend the divisions between states. Informal free zones are formed; routes are established, circumventing financial and police checkpoints. At the local level, especially in the most sidelined areas, borders often remain only on paper and are not obstacles to the daily confidences of life. Even porosity is exploited to vote on both sides of the border, creating de facto situations of multiple citizenship. These situations have been interpreted in different ways. There are those who see signs of the construction of new territories, of other centralities alternative to the postcolonial states. Precisely on these paths of cross-border economic and social thickening, Igue (1995) identifies one of the outcomes of the unfinished territorial legitimisation of Sahelian states. These dynamic border regions that are in the process of being established or strengthened could perhaps represent a form of soft response to the traumas and the ruptures generated by colonial partition (Igue, 1995, p. 57 ff.). In the Sahel, national peripheries would thus be configured as autonomous areas straddling several states, potentially able to break up the state spaces themselves and overcome, albeit informally, the territorial design inherited from colonisation. Thus, we would witness the re-emergence of territorial arrangements predating colonisation or at least the restoration of the long-standing ties that unite areas separated by colonial partition. Contrary to those who perceive these cross-border movements as ways of contesting the territorial trap of the state, for Bennafla (1999), they do not undermine the state's territorial order. In fact, the resulting "territorial landmarks" (markets, urban aggregations and axes of circulation) are "too fleeting to threaten territorial integrity" since they always hang on the specific conditions of markets and power relations between states (Bennafla, 1999, p. 33). Moreover, in many situations, the state proposes itself as the primary actor in these border spaces, with supporting policies (infrastructures in particular) for the trades that cross them. Populations would then be able to "play, manipulate and take advantage of the socio-economic differentials created by state borders" (Bennafla, 1999, p. 36). This is not rejection but rather an opportunistic attitude; local populations redraw state borders, managing to build significant economic activities around them. Whether they are signs of a spatial reconfiguration capable of overcoming the rigidity of territorial traps or are open arenas where states are capable of redefining themselves, the unfortunate fact remains that these cross-border spaces are often the first to be affected by the spread of violence. It is true in the region bordered by Mali, Niger and Burkina Faso, in the Lake Chad region and in the area between Wadai and Darfur. The upheavals in these remote regions are then able to reach the heart of the states, bringing war to the capitals, first in Chad and today in Sudan. If we look beyond the current acute crisis in many Sahelian states, different ways out of the territorial traps can be devised at the state level, such as strengthening interstate cooperation and forcefully opposing the presence of foreign interests and the looming competition of global markets. Outside the state and development machinery, at the local level, an important signal is the consolidation of civil society, for example, the Sudanese struggle to topple the al-Beshir regime and the subsequent troubled period.

The last layer, shifting, describes the overcoming of oppressive situations of violence in the long-term dimension, namely the very forms of the terraforming process described earlier. The constriction of living and inanimate entities has occurred at various levels: from large-scale irrigation projects (the "miniature" of high modernism, quoting Scott, 1998) to the remodelling of entire river systems and their hydrological functioning through the construction of dams to store water and manage flooding (see Chapter 7). As seen in Bélime's plans for the Senegal River or in the construction of the Jebel Aulia and Sennar dams on the Blue Nile, this scale of action is also characteristic of the colonial logic of *mise en valeur*. By acting in the drylands and manipulating the most precious resource there (water), these modifications have profoundly altered environmental systems, such as the river and lake banks and wetlands associated with annual flooding, interfering both with plant and animal life and with resource use practices respected over time (Kronenburg García et al., 2022). However, this gigantic terraforming process has demonstrated all its limits; the biopolitical, ontological and epistemic battle that seemed to have been won once and for all by Western modernity, has thus been reopened. Today, other actors (human and other-than-human beings and entities) have the right to be heard. Other cultures (inevitably *mestizo* but profoundly different from the one linked to coloniality and modernity) are demanding to be recognised. Finally, different ways of understanding the world and perceiving life and its relationship with nature are gaining courage and voice (see Chapter 17). Other "destabilising geographies" (Ulloa, 2024)—black, feminist and Indigenous—are emerging from the pencils of many different topographers.

What has been one of the main laboratories of terraforming—the organisation and re-ordering of water and land—can perhaps become a place for experimenting with new mixtures of water, land, plants, animals and human societies, in a gradual move towards the vitalistic approach suggested by Ghosh (2021).

In several indigenous movements, almost everywhere in the world from Ecuador to New Zealand, it is possible to grasp "the influence of that subterranean river of vitalism, which, after having been driven underground for centuries, is now once again rising powerfully to the surface around the world" (Ghosh, 2021, p. 238). These movements strive to retrieve their lost voices, reinstate their autonomy and even legal subjectivity to rivers and forests, ground and water, plants and animals.

Even in the midst of this tremendous age of deconstruction and violence that the Sahel is experiencing, elements of innovation are ever-present. It is a matter of being able to recognise them when they appear, making them visible and allowing them to exist.

Returning to the finger pointing to a place and a date in the opening image (Fig. 1.2), we have tried to follow the direction indicated by the finger, one that goes beyond the colonial legacy towards the rising moon, to perhaps (we hope with all our hearts) illuminate the end of such a long "dark night" (Mbembe, 2021).

Notes

1. 1 Feddan (the unit of measure used in Sudan) = 0.42 ha.
2. Sudan War Monitor is a collaborative network of journalists and open-source researchers tracking the events of Sudan's war (https://sudanwarmonitor.com/).
3. Feral Atlas provides multiple examples to reflect on the relationship between nonhuman entities and human infrastructure projects (https://feralatlas.org/).
4. https://sudanwarmonitor.com/p/battle-begins-for-eastern-sudanese?utm_source=post-email-title&publication_id=1721782&post_id=146525700&utm_campaign=email-post-title&isFreemail=true&r=2ub7bo&triedRedirect=true (last access September 2024).

References

Adamczewski, A., & Hugon, P. (2013). The Libyan crisis and its impact on property transactions in Africa. The Malibya Project in Mali. *Afrique Contemporaine*, *245*(1), 105–107. https://shs.cairn.info/journal-afrique-contemporaine1-2013-1-page-105?lang=en

Agamben, G. (1995). *Homo sacer. Il potere sovrano e la nuda vita*. Einaudi.

Agamben, G. (2009). *What is an apparatus?* Stanford University Press.

Agamben, G. (2018). *Pulcinella*. Seagull.

Agnew, J. (1994). The territorial trap: The geographical assumptions of international relations theory. *Review of International Political Economy*, *1*, 53–80. https://doi.org/10.1080/09692299408434268

Agnew, J. (2010). Still trapped in territory? *Geopolitics*, *15*(4), 779–784. https://doi.org/10.1080/14650041003717558

Agnew, J. (2015). Revisiting the territorial trap. *Nordia Geographical Publications*, *44*(4), 43–48.

Allan, J. A., Keulertz, M., Sojamo, S., & Warner, J. (Eds.). (2013). *Handbook of land and water grabs in Africa. Foreign direct investment and food and water security*. Routledge. https://nordia.journal.fi/article/view/64824

Andrade, L. M. (2018). L'ego conquiro comme fondement de la subjectivité moderne. *La Revue Nouvelle*, *1*(1), 30–35.

Barnett, T. (1977). *The Gezira scheme. An illusion of development*. Frank Cass.

Bauman, Z. (2002). *Society under siege*. Polity Press.

Bennafla, K. (1999). La fin des territoires nationaux? État et commerce frontalier en Afrique centrale. *Politique Africaine*, *73*(1), 24–49.

Bertoncin, M. (2004). *Logiche di terre e acque. Le geografie incerte del Delta del Po*. Cierre.

Bertoncin, M., Bicciato, F., Bonollo, L., Croce, D., Faggi, P., Mariani, L., Minoia, P., & Pase, A. (1995). Irrigazione, stato e territorio in Sudan: Il gioco della posta in gioco. *Terra d'Africa, 1995*, 8–40.

Bertoncin, M., & Faggi, P. (Eds.). (2006). *Cosa resta nel piatto? Fallimenti e promesse dell'agricoltura irrigua nella valle del Senegal*. L'Harmattan Italia.

Bertoncin, M., & Pase, A. (2008). *Attorno al lago Ciad. Sguardi diversi sullo sviluppo*. L'Harmattan Italia.

Bertoncin, M., & Pase, A. (2017). Interpreting mega-development projects as territorial traps: The case of irrigation schemes on the shores of Lake Chad (State of Borno, Nigeria). *Geographica Helvetica*, *72*, 243–254.

Bertoncin, M., & Pase, A. (2022). Geographical proximity questioned. In A. Torre & D. Gallaud (Eds.), *Edward Elgar handbook on proximity relations* (pp. 204–219). Elgar.

Bertoncin, M., & Pase, A. (2023). Handling distances as a key factor in social power dynamics. In L. Biasiori, F. Mazzini, & C. Rabbiosi (Eds.), *Reimagining mobilities across the humanities (vol. 1): Theories, methods and ideas* (pp. 73–87). Routledge. https://doi.org/10.4324/9781003278665

Bertoncin, M., Pase, A., & Quatrida, D. (2010). Al margine del campo. *Geotema*, *41*, 50–59.

Bertoncin, M., Pase, A., Quatrida, D., & Turrini, F. (2019). At the junction between state, nature and capital: Irrigation mega-projects in Sudan. *Geoforum*, *106*, 24–37. https://doi.org/10.1016/j.geoforum.2019.07.002

Bertoncin, M., Pase, A., & Turrini, S. (2022). Water, land and Arab investments in irrigation projects: Continuity and innovation in Sudan. In J.-N. Bach (Ed.), *Routledge handbook on Horn of Africa* (pp. 657–669). Routledge. https://doi.org/10.4324/9780429426957

Blomley, N. (2007). Making private property: Enclosure, common right and the work of hedges. *Rural History*, *18*(1), 1–21.

Bodei, R. (1995). *Libro della memoria e della speranza*. il Mulino.

Bonneuil, C. (1991). *Des savants pour l'empire. La structuration des recherches scientifiques coloniales au temps de la "mise en valeur des colonies française", 1917–1945*. ORSTOM.

Bouquet, C. (1990). *Insulaires et riverains du lac Tchad: Une étude géographique* (T. 1–2). L'Harmattan.
Braudel, F. (1995). *The Mediterranean and the Mediterranean world in the age of Philip II* (Vol. I). University of California Press.
Brenner, N., & Elden, S. (2009). Henry Lefebvre on state, space, territory. *International Political Sociology*, *3*, 353–377.
Brunet-Jailly, J. (1981). L'introduction de la riziculture irriguée en pays Mousgoum *Revue de Géographie du Cameroun*, *2*(2), 71–95.
Cabot, J. (1955). La mise en valeur des régions du Moyen-Logone. *Annales de Géographie*, *LXIV*(341), 35–46.
Cabot, J. (1965). *Le bassin du moyen Logone*. ORSTOM.
Cernea, M. M. (2003). For a new economics of resettlement: A sociological critique of the compensation principle. *International Social Science Journal*, *55*(175), 37–45. https://doi.org/10.1111/1468-2451.5501004
de Planhol, X., & Rognon, P. (1970). *Les zones tropicales arides et subtropicales*. Colin.
Deininger, K. (2011). Challenges posed by the new wave of farmland investment. *Journal of Peasant Studies*, *38*(2), 217–247. https://doi.org/10.1080/03066150.2011.559007
Delueze, G., & Guattari, F. (1980). *Mille plateaux. Capitalisme et schizophrénie*. Les Éditions des Minuit.
Dussel, E. (1977). *Filosofía de la liberación*. Edicol.
Elden, S. (2013). *The birth of territory*. The University of Chicago Press.
Falola, T. (2022). *Decolonizing African studies: Knowledge production, agency, and voice*. University of Rochester Press.
Farinelli, F. (1992). *I segni del mondo. Immagine cartografica e discorso geografico in età moderna*. Nuova Italia.
Farinelli, F. (2004). Sui tipi non cartografabili. In Istituto Geografico Militare, *Italia – atlante dei tipi geografici* (pp. 77–79). IGM.
Farinelli, F. (2009). *La crisi della ragione cartografica*. Einaudi.
Federici, S. (2018). *Re-enchanting the world: Feminism and the politics of the commons*. PM Press.
Feldman, S., Geisler, C., & Silberlingg, L. (2003). Moving targets: Displacement, impoverishment and development. *International Social Science Journal*, *55*(175), 7–13.
Ferguson, J. (1994). *The anti-politics machine: 'Development', depoliticization and bureaucratic power in Lesotho*. University of Minnesota Press.
Foucault, M. (1980). *Power/knowledge: Selected interviews and other writings, 1972–1977*. Pantheon Books.
Gaitskell, A. (1959). *Gezira. A story of development in the Sudan*. Faber and Faber.
Gallais, J. (1980). L'office du Niger. In Traoré, M. (Ed.), *Atlas du Mali* (pp. 56–57). Jeune Afrique.
Gellert, P. K., & Lynch, B. D. (2003). Mega-projects as displacements. *International Social Science Journal*, *175*, 15–25.
Ghedini, G. (2023). *Da "selvaggi" a "moretti". Schiavitù, riscatti e missioni tra Africa ed Europa (1824–1896)*. PUF.
Ghosh, A. (2021). *The nutmeg's curse. Parables for a planet in crisis*. Murray.
Goffman, E. (1961). *Asylums: Essays on the social situation of mental patients and other inmates*. Anchor Books.
Grossi, P. (1992). *Il dominio e le cose. Percezioni medievali e moderne dei diritti reali*. Giuffrè.
Guy, C. (1923). La mise en valeur des colonies françaises. *Annales de Géographie*, *32*(177), 265–271.
Hall, R., Edelman, M., Borras, S. M. Jr, Scoones, I., White, B., & Wolford, W. (2015). Resistance, acquiescence or incorporation? An introduction to land grabbing and political reactions 'from below'. *The Journal of Peasant Studies*, *42*(3–4), 467–488.

Haller, T., Pase, A., Warner, J., Hashimshony-Yaffe, N., Kronenburg García, A., & Bertoncin, M. (2022). Mega-infrastructure projects in drylands. From enchantments to disenchantments. In A. Kronenburg García, T. Haller, H. van Dijk, C. Samimi, & J. Warner (Eds.), *Drylands facing change: Interventions, investments and identities* (pp. 112–131). Routledge. https://doi.org/10.4324/9781003174486
Hartog, F. (2003). *Régimes d'historicité. Présentisme et experiences du temps*. Seuil.
Hassan, A. (2015). *Overview of the salient elements of Sudan's Investment Encouragement Act 2013. Paper presented at the Fourth Annual Conference on Structural Reform, Transformation, and Sustainable Development in Sudan: Economic, Political, and Social Perspectives*, University of Khartoum, Khartoum.
Hirschman, A. O. (1967). *Development projects observed*. The Brookings Institution.
Igue, J. O. (1995). *Le territoire et l'État en Afrique. Les dimensions spatiales du développement*. Karthala.
Ika, L. A. (2005). La gestion des projets d'aide au développement: historique, bilan et perspective. *Perspective Afrique*, *1*(2), 128–153. chrome-extension://efaidnbmnnnibpcajpcglclefindmkaj/https://perspaf.org/fileadmin/Articles/Volume1/Numero2/PA_Vol1_No2_pp.128-153.pdf
Iyébi-Mandjek, O. (2000). Cultures maraîchères. In C. Seignobos & O. Iyébi-Mandjek (Eds.), *Atlas de la province Extrême-Nord Cameroun* (pp. 102–106). IRD.
Jamin, J.-Y., & Coulibaly, Y. (1995). Déviances ou innovations? Les riziculture paysannes de l'Office du Niger (Mali). In A. Cheneau Loquay & A. Leplaideur (Eds.), *Les ricicultures de l'Afrique de l'Ouest – Partie II: Les modèles irrigués exogènes* (pp. 121–135). CIRAD.
Jeffrey, A., McFarlane, C., & Vasudevan, A. (2012) Rethinking enclosure: Space, subjectivity and the commons. *Antipode*, *44*(4), 1247–1267. https://doi.org/10.1111/j.1467-8330.2011.00954.x
Kanbur, R. (2004). Development economics and the compensation principle. *International Social Science Journal*, *55*(175), 27–35. https://doi.org/10.1111/1468-2451.5501003
Koselleck, R. (2004). *Futures past. On the semantics of historical time*. Columbia University Press.
Kronenburg García, A., Pase, A., Haller, T., Artur, L., Nogueira Lisboa, S., Metrak, M., Bertoncin, M., & Suska-Malawska, M. (2022). Wetlands in drylands: Large-scale appropriations for agriculture, conservation and mining in Africa. In A. Kronenburg García, T. Haller, H. van Dijk, C. Samimi, & J. Warner (Eds.). *Drylands facing change: Interventions, investments and identities* (pp. 53–72). Routledge. https://doi.org/10.4324/9781003174486
Larder, N. (2015). Space for pluralism? Examining the Malibya land grab. *The Journal of Peasant Studies*, *42*(3–4), 839–858. https://doi.org/10.1080/03066150.2015.1029461
Leroy-Beaulieu, P. (1908). *De la colonisation chez les peuples modernes* (T. II). Alcan.
Li, T. M. (2011). Centering labor in the land grab debate. *The Journal of Peasant Studies*, *38*(2), 281–298. https://doi.org/10.1080/03066150.2011.559009
Lorenz, C. (2017). 'The times they are a-changin'. On time, space and periodization in history. In S. Berger, M. Carretero, & M. Grever (Eds.), *Palgrave handbook of research in historical culture and education* (pp. 109–131). Palgrave Macmillan.
Magrin, G. (2013). *Voyage en Afrique rentière. Une lecture géographique des trajectoires du développement*. Publications de la Sorbonne.
Maïga, M. (1995). *Le bassin du fleuve Sénégal. De la traite négrière au développement sous-régional auto-centré*. L'Harmattan.
Martiniello, G. (2020). Bitter sugarification: Sugar frontier and contract farming in Uganda. *Globalizations*, *18*(3), 355–371. https://doi.org/10.1080/14747731.2020.1794564
Mbembe, A. (2021). *Out of the dark night*. Columbia University Press.
Mehta, L., Veldwisch, G. J., & Franco, J. (2012). Introduction to the special issue: Water grabbing? Focus on the (re)appropriation of finite water resources. *Water Alternatives*, *5*(2), 193–207.

Mignolo, W. (1992). The darker side of the Renaissance: Colonization and the discontinuity of the classical tradition. *Renaissance Quarterly*, *45*(4), 808–828.
Mignolo, W. (2009). Dispensable and bare lives: Coloniality and the hidden political/economic agenda of modernity. *Human Architecture: Journal of the Sociology of Self-Knowledge*, *7*(2), 69–88. http://scholarworks.umb.edu/humanarchitecture/vol7/iss2/7
Moore, J. W. (2016). The rise of cheap nature. In J. W. Moore (Ed.), *Anthropocene or Capitalocene? Nature, history, and the crisis of capitalism* (pp. 78–115). PM Press.
Morabito, V. (1977). L'Office du Niger au Mali, d'hier à aujourd'hui. *Journal des Africanistes*, *47*(1), 53–82.
Morabito, V. (1995). Cinquante années de riziculture irrigue à l'Office du Niger (Mali). In A. Cheneau Loquay & A. Leplaideur (Eds.), *Les riziculture de l'Afrique de l'Ouest. Partie II: Les modèles irrigués exogènes* (pp. 113–119). CIRAD.
Ndlovu-Gatsheni, S. J. (2012). Coloniality of power in development studies and the impact of global imperial designs on Africa. *The Australasian Review of African Studies*, *33*(2), 48–73.
Ndlovu-Gatsheni, S. J. (2023). Beyond the coloniser's model of the world: Towards reworlding from the Global South. *Third World Quarterly*, *44*(10), 2246–2262. https://doi.org/10.1080/01436597.2023.2171389
Newman, D. (2010). Territory, compartments and borders: Avoiding the trap of the territorial trap. *Geopolitics*, 15(4), 773–778. https://doi.org/10.1080/14650041003717541
Oliveira, G., McKay, B. M., & Liu, J. (2020). Beyond land grabs: New insights on land struggles and global agrarian change. *Globalizations*, *18*(3), 321–338. https://doi.org/10.1080/14747731.2020.1843842
Olivier de Sardan, J.-P. (1995). *Anthropologie et développement. Essai en socio-anthropologie du changement social*. Khartala.
Polanyi, K. (1944). *The great transformation. The political and economic origins of our time*. Beacon Press.
Pase, A. (2011). *Linee sulla terra. Confini politici e limiti fondiari in Africa sub sahariana*. Carocci.
Pase, A. (2024). Emptying and filling. Maps of inland Africa. In T. Rossetto & L. Lo Presti (Eds.), *Routledge handbook of cartographic humanities* (pp. 235–245). Routledge. https://doi.org/10.4324/9781003327578
Pase, A., Hofman, I., Kronenburg García, A., Haller, T., Cirillo, D., Giger, M., Abebe, M., Hurni, K., & Bertoncin, M. (2022). Large-scale agricultural investments in drylands: Facing some blind spots in the grabbing debate. In A. Kronenburg García, T. Haller, H. van Dijk, C. Samimi, & J. Warner (Eds), *Drylands facing change: Interventions, investments and identities* (pp. 73–92). Routledge. https://doi.org/10.4324/9781003174486
Patel, R., & Moore, J. (2017). *A history of the world in seven cheap things: A guide to capitalism, nature and the future of the planet*. University of California Press.
Quijano, A. (2000). Coloniality of power, Eurocentrism, and Latin America. *Nepantla: Views from the South*, *1*(3), 533–580.
Quijano, A. (2007). Coloniality and modernity/rationality. *Cultural Studies*, *21*(2–3), 168–178. https://doi.org/10.1080/09502380601164353
Raffestin, C. (1981). *Per una geografia del potere*. Unicopli.
Reid-Henry, S. (2010). The territorial trap fifteen years on. *Geopolitics*, *15*(4), 752–756. https://doi.org/10.1080/14650041003717509
Roupsard, M. (1987). *Nord-Cameroun: Ouverture et développement*. C. Bellée.
Sack, R.D. (1986). *Human territoriality: Its theory and history*. Cambridge University Press.
Sarraut, A. (1923). *La mise en valeur des colonies françaises, avec onze cartes en noir et en couleurs*. Payot.
Sautter, G. (1987). Libres réflexions sur les aménagements ayant pour objet la maitrise de l'eau par ou pour les agricultures. *Les Cahiers de la Recherche Développement*, *14-15*, 5–14. https://revues.cirad.fr/index.php/crd/article/view/36508

Schoneveld, G. C. (2017). Host country governance and the African land rush: 7 reasons why large-scale farmland investments fail to contribute to sustainable development. *Geoforum*, *83*, 119–132. https://doi.org/10.1016/j.geoforum.2016.12.007

Scott, J. C. (1998). *Seeing like a state. How certain schemes to improve the human condition have failed*. Yale University Press.

Scott, J. C. (2010). The trouble with the view from above. *Website Cato Unbound*, http://www.cato-unbound.org/2010/09/08/james-c-scott/trouble-view-above (accessed 3 July 2015).

Shah, N. (2012). The territorial trap of the territorial trap: Global transformation and the problem of the state's two territories. *International Political Sociology*, *6*(1), 57–76. https://doi.org/10.1111/j.1749-5687.2011.00144.x

Sidaway, J. D. (2007). Spaces of postdevelopment. *Progress in Human Geography*, *31*(3), 345–361. https://doi.org/10.1177/0309132507077405

Sudan Republic (2013). *National Investment Encouragement Act*. chrome-extension://efaidnbmnnnibpcajpcglclefindmkaj/https://www.a-mla.org/en/country/pdf/717 (last access Mars 2025).

Tayler, J. (2005). *Angry wind: Through Muslim Black Africa by truck, bus, boat, and camel*. Houghton Mifflin.

Tomasi di Lampedusa, G. (1958). *Il Gattopardo*. Feltrinelli.

Tsing, A. (2015). *The mushroom at the end of the world: On the possibility of life in capitalist ruins*. Princeton University Press.

Ulloa, A. (2024). Destabilising geographies in Colombia: Trajectories and perspectives. *Transactions of the Institute of British Geographers*, *49*, Article e12588. https://doi.org/10.1111/tran.12588

United Nations Conference on Trade and Development (UNCTAD) (2015). *Investment policy review. The Sudan*. Geneva: United Nations. www.unctad.org/en/PublicationsLibrary/diaepcb2014d5_en.pdf (last access September 2024).

van der Ploeg, J. D. (2008). *The new peasantries, struggles for autonomy and sustainability in an era of empire and globalization*. Earthscan. https://doi.org/10.4324/9781849773164

Vasudevan, A., McFarlane, C., & Jeffrey, A. (2008). Spaces of enclosure. *Geoforum*, *39*(5), 1641–1646. https://doi.org/10.1016/j.geoforum.2008.03.001

Wolford, W., Borras, S. M. Jr, Hall, R., Scoones, I., & White, B. (2013). Governing global land deals: The role of the state in the rush of land. *Development and Change*, *44*(2), 189–210. https://doi.org/10.1111/dech.12017

World Bank. (2011). Project appraisal document on a proposed grant in the amount of SDR 71million (US$115 million equivalent) to Burkina Faso for a Bagré Growth Pole Project. Retrieved May 31, 2024, from http://documents.worldbank.org/curated/en/581451468020361287/Burkina-Faso-Bagre-Growth-Pole-Project

Ziai, A. (2015). *Development discourse and global history: From colonialism to the sustainable development goals* (1st ed.). Routledge. https://doi.org/10.4324/9781315753782

Zúñiga, M. J. (2023). Ego conquiro. Esbozo de la crítica al colonialismo en Enrique Dussel. Res pública. *Revista de Historia de las Ideas Políticas*, *26*(1), 107–115. https://doi.org/10.5209/rpub.81870

Part I

Contextual Issues

2 Environmental crisis narratives in the Sahel and the solutions they call for

The case of Lake Chad

Jeroen Warner, Andrea Pase and Angela Kronenburg García

Introduction

The Sahel tickles the imagination. Primed by a flood of documentaries and NGO campaign videos, the mind's eye presents endless, desolate dried-out landscapes, traversed by hungry tribes and animals. This fallacious, exotic and popular image has its counterpart in policy images reflected in countless think-tank analyses and policy plans for the Sahel. The dominant policy narrative is one of an environmental crisis. "Over the last half century", the UN notes,

> the combined effects of population growth, land degradation (deforestation, continuous cropping and overgrazing), reduced and erratic rainfall, lack of coherent environmental policies and misplaced development priorities, have contributed to transform a large proportion of the Sahel into barren land, resulting in the deterioration of the soil and water resources.[1]

Moreover, this environmental crisis is seen as being at the basis of, or as exacerbating, many of the other crises and challenges in the Sahel (food insecurity, poverty, migration/displacement and violent conflict) subsuming them under an all-encompassing yet digestible "environmental fragility" paradigm.[2]

Overlapping with the Sahel belt is the site of another powerful image, the supposedly disappearing Lake Chad, once the world's sixth largest lake. Its 90% shrinkage since 1963, from 25,000 ha to less than 1000 ha, has been repeated many times, along with imaginaries of violent conflict over water and terrorism, compounding problems for its riparian countries and beneficiaries—Cameroon, Central African Republic, Libya, Nigeria, Niger, Sudan and, indeed, Chad. Magrin (2016) rates the lake in a class of its own as a global environmental icon, a geosymbol.

Why are these images of environmental and developmental doom so persistent, even in the face of clearly contradicting and overwhelming scientific evidence? Strong candidates for explanations have come from critical postdevelopment literature published since the late 1990s. Authors such as Jonathan Crush, Arturo

DOI: 10.4324/9781003429111-4

Escobar, James Ferguson and Timothy Mitchell influentially started asking existential questions about the rationale for development imaginaries, and other authors (e.g., Leach et al., 1999) showed non-equilibrium rangeland trends for the Sahel that were very different from the dominant equilibrium ecology narrative. Their data countered the conventional wisdom on desertification.

Drought and desertification discourse has long captured the common imagination of the Sahel region. The desertification narrative emerged from or at least was reinforced by the great Sahelian droughts of the 1970s and the 1980s. It was shaped in the international arena by "organizations working in the fields of international cooperation, human rights, regional development, economic regulation and environmental questions" (Gangneron et al., 2022, p. 117) and has been repeatedly presented as inexorable, a modern-day Dust Bowl, a self-inflicted mess due to overexploitation that could only be reverted by radical policies. The widespread acceptance of this narrative in development circles led to draconian measures, including sedentarising nomads (Scoones & Nori, 2023) and mega-infrastructural projects (Haller & Weissman, 2024), such as the Great Green Wall (Haller et al., 2023) and the Trans Africa Pipeline (TAP). Critical authors showed the resilience of the Sahel, which started to regreen in the 1980s (e.g., Brandt et al., 2015; Giannini et al., 2008; Hickler et al., 2005), and they surmised that chronic poverty in the Sahelian belt did not occur despite but *because of* such interventions. These authors observed that much of the science underlying environmental crisis narratives about the Sahel was dominated by like-minded expatriates. Thus, focusing on how ideas shape policies and how policies shape projects is in order. After all, new ideas cannot gain much political force without facilitation by institutions and power constellations already in place (Hall, 2020), and the Sahel crisis discourse has been successfully institutionalised in international policy practice. Development narratives "structure options, define relevant data and rule out alternatives" (Cooper & Packard, 1997, p. 24) for decision-making, establishing the authority to legitimise interventions while silencing local environmental knowledge (Glenzer, 2002). When applied to a particular region, regional discourse formations develop (Peet & Watts, 1993) as a relation between dominating and dominated forces. The Sahel literature shows the persistence of such policy narratives, which have a surprisingly long pedigree, rooted in Eurocentric beliefs about the tropics as a dangerous, chaotic place—whether ridden by disease, poverty or disaster (Bankoff, 2001).

However, narratives are not static; they develop over time, mutate, merge and sometimes disappear and reappear. Counternarratives develop, challenging the hegemonic account, as does the current decolonial narrative. The "social life" of concepts (Molle, 2008, p. 131) can be illustrated by the current evolution of the Sahel crisis narrative. The critical view is persistent but currently challenged in turn by a decolonising narrative, exposing it as perpetuating Northern ideas and practices.

As we shall explain, due to their reductionism of complex realities, environmental crisis narratives are powerful fairy tales with predictable but powerful quests. Such fairy tales or stories provide a "deep structure", filling uncertainty and complexity gaps in understanding the world we live in, but, crucially, they point to solutions and the way forward.

Crisis narratives occupy a particularly potent place in the discursive field as they bring urgency to the story. Bilateral and multilateral agencies in the North are interested in representing the Global South as areas that need development and intervention (Mitchell, 2005). However, NGOs and national ministries also peddle crisis narratives, responding to the perverse incentives for promoting accounts of environmental degradation—disasters bring fundraising opportunities and employment (Swift, 1996).

We review this literature and illustrate it by presenting the case of Lake Chad, which, in the environmental development discourse, is in (climate and security) crisis and needs to be saved. As noted about desertification, the incredibly shrinking lake turns out to be oscillating between growing and shrinking, and inexorable "water wars" narratives (Warner et al., 2023) have not been borne out, yet a megaproject has been put forward as a solution to these problems.

We first introduce (crisis) policy narratives, followed by a brief overview of their evolution during and after colonial times and underpinning by colonial and postcolonial science. Next, we turn to a number of Sahelian narratives and the (often infrastructural) megaprojects they engendered, before zooming in on the case of Lake Chad. In closing, we briefly discuss the emerging decolonial literature and its potential to offer an alternative account closer to the complexity of reality.

Nature of policy narratives

Language, categories and labels are not neutral. How they are understood and interpreted depends on the context in which they are used. Discourse privileges some concepts and relations between concepts at the expense of others (Peet & Watts, 1993). Environmental narratives, crisis narratives included, fit particular policy agendas.

As alternative agendas vie for (hegemonic) control of the discourse, legitimising myths and beliefs are created to influence perceptions of the problem (Warner & Thomas, 2015). Discourse coalitions bring to the fore certain issues and relegate others to the background to promote their particular solutions. Specific narratives and ideas underpin knowledge constructions, imposing their view of realities on others (Hajer, 1995).

This issue points to the role of power relations. Power relations are largely about the capacity to convince other actors to mobilise and coordinate their efforts in a certain way because it is in their perceived interest to do so.

We cannot live without narratives, myths and fairytales (van Eeten, 1997) to make sense of the world. Narratives fill in the blanks to simplify and rationalise decision-making. They are sense-making constructs that seemingly explain how situations and contexts have come to be, meant to persuade others to take certain kinds of action (Czarniawska-Joerges & Joerges, 1988).

Policy narratives are "scenarios (stories and arguments) that stabilize the assumptions for decision-making in situations of high turbulence and dynamics" (Roe & van Eeten, 2004, p. 36). Narrative framing is a strategy used to fixate on certain meanings at the expense of others by developing accounts with a view to making a certain discourse on future actions hegemonic and legitimate (Haller & Galvin, 2011). Some narratives are actively promoted internationally, to the disadvantage of others. Once a narrative has become hegemonic, everyone will have to draw on the hegemonic storyline for their contribution to be taken seriously (Hajer, 1995).

Narratives are populated with characters (heroes, villains and victims) and their relationships, entangled in plotlines, with complications, reactions and resolutions. These plotlines unfold in particular settings and have resonances. van Eeten (1997) has shown how narratives and counternarratives often mirror each other—one person's hero is another person's villain and vice versa.

Crisis narratives

The crisis narrative is a "rhetorical strategy by means of which the shadows of past catastrophes, or an impending one, are invoked to authorize particular forms of political power, or the use of collective power and resources, while depoliticizing the catastrophe in question" (Vázquez-Arroyo, 2013, p. 740). As Rahman et al. (2023, p. 24) note:

> A crisis narrative implies an urgent demand for action, serving to justify immediate decision making in the face of a looming danger. In contexts of an uncertain future, such narratives serve to stabilise policy-making processes, renew the authority of policymakers, and foster decisive solutions to unpredictable events. However, a key problem with crisis narratives is that they tend to come at the expense of complexity (Roe, 1995). Labeling a situation as a "crisis" helps focus attention and mobilises resources, but it also risks sidelining alternative ideas and responses, as crisis narratives are often constructed by powerful policy actors to advance particular interests.

Crisis is both socially produced and discursively constituted "in its prevalence, disruptiveness and (appearance of) inevitability" (de Rycker & Don, 2013, p. 499).

Crisis is socially constructed—not only in the sense of intersubjectively giving meaning to an overwhelming situation but also materially propelled by the responses to an earlier crisis. The *crisis* identification and labelling, often by outsiders, can trigger and legitimise measures that lead to a new crisis, Rocheleau et al. (1995) claim. For this contribution, we follow these authors in that crisis construction has two meanings:

- discursive framing, the mobilisation of representational discourses that attempt to exert social power by telling one particular story of crisis, and
- mobilising factors that create an actual crisis—the discursive construction of one crisis over another leads to policies or acts of resistance that contribute materially to the next crisis.

In this context, Lewis (2010) indicates that crisis narratives not only simplify and depoliticise reality but also evidence amnesia, a forgetfulness about what actually led to the crisis in the first place. Crisis narratives are used to centralise or consolidate existing forms of dominance and social hierarchy.

Daoust and Selby (2023) present the key elements that make the (in their case, Lake Chad) crisis narrative so compelling to spread:

- visualisation: striking, media-genic images;
- climatisation/securitisation of issues, making them extraordinary challenges with one way out;
- validation by think tanks (SIPRI, Clingendael, Adelphi); and
- evasion and organised ignorance (undergirded by only scant primary research).

Catastrophisation, that is, invoking an impending catastrophe, raises an issue from the normal democratic debate to the level of life and death for immediate intervention with extraordinary measures (Warner, 2013; Warner et al., 2023). Framing a situation as a crisis gives the story an edge over alternative narratives and fast-tracks measures and resources, at the expense of a considerable loss of accountability. Crisis scenarios "generate extra-ordinary consensus, open up new avenues of legitimacy and stretch thresholds for accepting sacrifice" (Krätli, 2013, p. 1); when collective survival is at stake, people will accept more than they do normally, even if it hurts them.

Other elements of dominant narratives are intuitiveness and an appeal to simple, causal and explanatory beliefs (Molle, 2008), clothed in neutral scientific language (Dreyfus & Rabinow, 1982; Shore & Wright, 2003). Although the basis may not be all that scientific (Sullivan, 2000; Verhoeven, 2014), scientists' involvement provides an aura of "scientistical" objectivity and truth and a buffer against counterevidence (Ribot, 2004, p. 74).[3]

In the next section, we examine the production of this science.

Colonial and postcolonial Sahel science

Critical post-development analysts claim that environmental and development discourse reinforces a separation of the Northern *us* from the Southern *them*. In this discourse, homogenised *others*, obviously lower in status than *us*, are not only alienating; they also come from a dangerous place that requires control or a vulnerable place in need of saving.

Bankoff (2001), for example, shows that the early discoveries of the non-European world instilled delight and awe at the exotic landscapes and abundant resources encountered. However, as large groups of the settlers that roamed the world fell ill and perished, the focus shifted to unfavourable images of danger—weather extremes, wild animals and rough terrains.

The misery found in the tropical world came to be perceived as self-inflicted by the godless Indigenous peoples. The marriage of the theological and the secular is epitomised in the Malthusian narrative, named after the reverend scientist who in the late 18th century predicted that population growth would outpace natural resource availability and that this would lead to conflict over scarce resources. The currently hegemonic thinking about environmental conflict still reflects this thought.

> The belief that deserts and drought were results of divine punishment for indigenous peoples' sinful ways was widely embraced … colonial officials, military officers, missionaries and scientists accused the indigenous colonial subjects of creating their own misery by undermining their natural resource base.
>
> *(Benjaminsen & Hiernaux, 2019, p. 208)*

With the rise and more systematic application of ecology as a science, this morality tale moved from a religious to a scientistic narrative after the severe African droughts and famines of 1910–1915. A self-regulating ecological equilibrium was assumed, and humans were identified as the culprits of the disturbance of that equilibrium, responsible for the desiccation of certain regions (Glenzer, 2002). Among colonial scientists, the belief that Africans were carelessly destroying their environment through practices of slash-and-burn agriculture, bush fires and overgrazing started to become normalised. Some researchers extrapolated "very localized and short-term studies to the larger region in order to explain broad-scale environmental changes", while neglecting to incorporate local knowledge into their models (Moseley & Laris, 2008, p. 60). As the latter authors indicate, these colonial interpretations tended not to be based on systematic investigation but on the direct transposition of ecological models derived from temperate climates to the tropics.

The beliefs expressed in colonial discourses reflected some self-interest: "… discrediting of native land uses was used to justify land rights dispossession of these peoples and to prepare for other land uses such as settler agriculture, state-led irrigation schemes, large-scale forest exploitation and forest and fauna conservation areas" (Benjaminsen & Hiernaux, 2019, p. 209).

In the post-1945 Cold War context, development came to be merged with security. Immiseration would risk the spread of communism, so abject poverty needed to be rooted out. Many non-European societies came to perceive themselves as backward and in need of development intervention.

With the end of the Cold War as a grand scheme and the declining global sway of development as a driving force, the development/security nexus did not let up; indeed, as noted by Duffield (2007), "their" development was supposed to benefit "our" security. In the early 1990s, the US Pentagon started to prepare for new wars as the "green wars" (including water wars) discourse took off. A new domain, environmental security, saw the light, strongly dominated by a scarcity-conflict assumption rooted in Malthusianism and arising in the 1960s and the 1970s, when the World Bank CEO Robert McNamara put population growth central to the crisis discourse (Glenzer, 2002). In this line of thought, farmer–herder conflicts in the Sahel were represented as "competing for scarce resources", a dispute between two social groups contending for access to resources (Homer-Dixon, 1994, p. 10).

Successive UN Secretary-Generals, presidents and politicians, such as, of late, European Commissioner Frans Timmermans, have claimed that the next war would be about water. Countless semi-academic books titled *Water Wars* (starting with Starr & Stoll, 1988) were published, inciting the Clinton administration to create an environmental security undersecretariat under US Defense Secretary Warren Christopher (Thomas, 1997, p. 397).

After US Vice President Al Gore released his *Inconvenient Truth* film in 2006, a seemingly inexorable climate crisis narrative has come up, merging with or even overtaking developmentalism—curiously reflecting the worldview of the US military establishment (Daoust & Selby, 2023). While academics hardly agree on this nexus, a link between climate change and conflict is now commonly inferred in international policy circles, especially around the Sahel, while Lake Chad remains one of the key hotspots of the literature on water wars, next to the Middle East.

In those days, the two popular representations of the Third World, especially in the US, were the failed state (incapable of providing basic services) and the rogue state (supporting terrorism threatening the West); after 9/11, these two were collapsed into weak states harbouring terrorist elements (Bachmann, 2008).

The (post)colonial narratives have now mostly been superseded by climate narratives, again largely originating in the Global North. The climatisation of security has led to a narrow, reductionist focus on climate as the explanation for every problem. Assumed links among climate disaster, migration and war have become commonplace, with nothing much in the way of evidence. As Krätli and Toulmin (2020, p. 47) note, "When closely examined, the correlation between the intensity of conflict and the intensity of climatic events appears weak and inconsistent. Most importantly, the majority of conflict incidents never escalate into violence". As Buhaug (2010) found, climate variability is a poor predictor of armed conflict. Rather, exclusion and frustration drive conflict (Buhaug, 2010), a finding supported by a growing body of research.

Sahelian green and blue megaprojects

Sahelian policy narratives tend to call for megaprojects as solutions to environmental crises. Many of these have proven to be castles in the air but continue to attract donors and backers (Mitchell, 2005; Warner et al., 2023). The Great Green Wall project is a case in point. It was launched in 2005 by three African heads of state (O. Obasanjo of Nigeria, A. Wade of Senegal and M. Gaddafi of Libya) and approved by the African Union in 2007. The idea was to reforest a 7000-km-long and 15-km-wide strip of land from Senegal to the Horn of Africa, located between the isohyets of 100 and 400 mm of rain per year (Haller et al., 2023, pp. 116–117). The "wall" was meant to stop the advance of the desert, which threatened to overwhelm the region, according to the desertification narrative (Cherlet et al., 2018; Morel, 2006). The Sahelian countries promoting the Great Green Wall project intended to redesign development (or perhaps better, re-engage it) in line with the new dictates of environmental sustainability (Mugelé, 2018). The image of a "green wall" (not new actually, having already been applied to other African contexts; Goffner et al., 2019; Mugelé, 2018; Reenberg, 2012) wields great evocative power, effectively mobilising and rallying donor countries and major international institutions for support. Often described as "pharaonic", it was intended to bring improvements over a huge area, with impressive targets ("100 million hectares of land restored, 250 million tons of carbon sequestered and 10 million jobs created") on the horizon of the year 2030 (United Nations Convention to Combat Desertification [UNCCD], 2020, p. 6). The impressive rhetoric has been matched by scant results. The 2020 UNCCD report states that after 15 years, only 4,000,000 ha of land has been restored in the region. The only country really committed to the project has been Senegal, so much so that some researchers speak of the "*Grande muraille sénégalaise*" (Magrin & Mugelé, 2020, p. 10).

Another idea for a megaproject is the TAP, proposed by "a not-for-profit organization working in collaboration with the 11-country Pan African Great Green Wall (PAGGW) agency", according to the agency's website.[4] The motivations for the project lie in three keywords that align seamlessly with the climate security crisis narrative: *crisis* ("Safe, clean water is vital to human existence but for tens of millions of people across 11 countries in the Sahel region of northern Africa a reliable fresh water supply has always been a distant dream"), *desertification* ("Chronic Food & Water Shortages. Drought causes water-related disease and death, and the loss of farmland to increasing desertification results in chronic food and water shortages. Lack of safe water has historically proved to be the single greatest obstacle to human survival, development and prosperity across this vast and vital region") and *severe drought* ("Persistent in the Sahel Region. The African continent is warming faster than the global average due to climate change. Some 60% of Africa is dry and the Sahel region is the driest, suffering severe drought conditions year after year. Regional conflicts over fresh water supplies are predicted to rapidly increase, also resulting in large-scale migration to Europe. The problem is immediate and worsening, the challenge immense"). The project proposal intends to respond to this "immense challenge" by building an 8800-km pipeline across the entire Sahel ("the largest humanitarian civil engineering project of the 21st century") to bring fresh water, desalinated in four solar-powered plants, two located on the Atlantic Ocean and two on the Red Sea, to Sahelian populations and enterprises. The designers, all Westerners, firmly believe that "upon completion, and in perpetuity, TAP will distribute an unlimited supply of safe, free, clean water for communities and people across the Sahel, and cost-efficient water for use in agriculture, forestry and commercial entities in all 11 countries in the region". Presented since 2015 at various conferences, notably organised by the PAGGW agency, it received "enthusiastic endorsement" (https://transafricapipeline.org/, accessed July 22, 2024).

The Great Green Wall and the TAP are two projects that promise a heroic, salvific solution to the crises of the Sahel, which aligns with a long-standing history of megaprojects believed to be capable of changing reality, a reality full of problems awaiting rational solutions, such as the large-scale irrigation systems discussed in Chapter 1. We now turn to another massive project that, different from the former two that (aspire to) traverse and expand across the Sahel, is located in the geographical heart of the Sahel and seeks to address the problem of a "disappearing lake".

Saving Lake Chad

Lake Chad is … *strange*. It is an endorheic lake, depending essentially on the flooding of only one river, Chari. The bottom is shallow, around 3 m; therefore, the variation in annual flooding rapidly affects the lake's water level. It has a long track record of great changes in shape and size. It is assumed that around 6000 BP, the lake, with its water level 310–320 m above the mean sea level, occupied a large area (over 300,000 km^2), the so-called Mega-Chad, a large inland sea of fresh water. In historical times, the lake has receded a great deal and decreased in surface area, while maintaining its pattern of continuous movement. The classification proposed by Tilho (1928), already in colonial times, provides three levels for the lake. *Normal Chad* is determined by reaching an elevation of water 281–282 m above sea level; the lake is composed of two large basins (south and north), with a water surface area of around 20,000 km^2. The open waters are extensive, particularly in front of the Chari and Komadugu Yobè rivers, and navigation is also possible between the islands of the archipelagos. The height of the water table above 284 m determines the situation known as *Great Chad*; this was observed in the 19th century by European travellers in the region but never in the 20th century. *Small Chad*, from an altitude of 280 m or less, finally occurred several times in the 20th century, alternating with *Normal Chad*: in the period between 1904 and 1917, around the 1940s and from 1973 until the second half of the 1990s (Lemoalle & Houdret, 1996, p. 23). In the *Small Chad* situation, the two cuvettes are separated, and there is no longer a single water surface.

Although well known to hydrologists, this continuous and strong fluctuation of the lake (which can change from a large area of open water to a collection of ponds) disappeared from the environmental narratives about Lake Chad. It is now common to mention that Lake Chad has been shrinking since 1963. This presumed shrinkage is represented as a crisis of climate security that needs saving by both (Northern) funders and consultants, as well as the national leaders of the region, and one of the ways to do so is through a megaproject that transfers water from the Congo basin to the lake. The UNEP calls Lake Chad one of the ten most "water-impoverished" regions in the world. Daoust and Selby (2023) show that there are plenty of arguments against the three key elements of the Lake Chad crisis narrative: (1) the lake is shrinking (2) due to climate-induced drought, (3) which leads to radicalisation and human insecurity.

Dismantling the crisis narrative

"*The lake is shrinking* ...". Daoust and Selby (2023, pp. 1292–1293) link the spread of the disappearing lake idea to the effectiveness of aerial and satellite photography. It is precisely visualisation that proves to be a highly effective persuasive tool. In fact, the visual effect of the great droughts has "lingered in memory" in the narratives about the dying lake (the years 1983–1984 and especially 1984–1985 were the most dramatic for the Sahel; only 6.7 km^3 of water reached Lake Chad, which is one-sixth of the average supply; Olivry et al., 1996, p. 221). However, there has been a positive trend since the mid-1990s, leading to large areas being back underwater, even in the most exposed part, the northern cuvette. Taking the maximum flow rate of Chari River at N'Djamena as a rough but indicative reference, compared with an average of 1870 m^3/s in the 1980s and 1930 m^3/s in the early 1990s (Olivry et al., 1996, p. 232), the Agrhymet data show that from 1998 onwards, years with a maximum flow rate of over 2500 m^3/s up to over 3000 m^3/s alternated with years with a flow rate around or even below 2000 m^3/s. Pearce (2022) claims that the massive irrigation schemes in place contribute to the shrinkage of the lake. In reality, the anthropogenic contribution is minimal. given the struggles and failures of large-scale irrigation schemes (e.g., the South Chad Irrigation Project and Baga Polder in Nigeria; see Chapter 1) and the weak impact of dams on Chari River and its main tributary, the Logone (see Chapter 7). The river system with the most anthropogenic modifications is Komadugu Yobé, which, however, has a minor contribution to the lake.

"... *due to climate-induced drought* ...". Many observers appear to prefer a linear "drying out" explanation for Lake Chad's shrinkage, taking a nicely stable lake as a reference point (Gangneron et al., 2022). As noted, cycles of shrinkage after prolonged drought and expansion after a return to regular flood patterns have always characterised Lake Chad. Fluctuations in the level of a relatively shallow lake can be observed not only within a year (flood/recession) but also from year to year, with wet and dry phases over a few decades and beyond. Developments around Lake Chad are thus dynamic and far from straightforward. However, the World Bank has clearly adopted the linear climate-conflict narrative.[5]

"... *which leads to radicalisation and human insecurity*". As Daoust and Selby (2023, pp. 1293–1294) note, this assertion is linked to the spread of violence in the region, especially in the 2010s, due to the Islamist insurgents belonging to Boko Haram and the movements that grew out of it, as well as the violent repression by the Nigerian army. Radicalisation and violence are discursively linked to climate change, asserting that scarcity of resources (water) exacerbates conflict and insecurity. However, such an assertion ignores a longer and more complicated history of conflict and violence that involves border disputes (Abdouraman, 2008) and clashes among the armed forces of the countries adjoining Lake Chad (Nigeria, Niger, Cameroon and Chad), as well as the failure in reforming a joint force in the face of increased activities by armed gangs in the 1990s. Given this background, Treszakai (2018) presents the Lake Chad crisis as a two-level diplomatic process, a game of chess in which the moves at both the domestic and transboundary levels strongly influence each other. These events are not *water wars* per se but successfully illustrate the potential of an environmental violence narrative to a willing policy audience. The climate argument only reinforces this insecurity and instability narrative. For national leaders, climate is a convenient scapegoat, diverting attention from violent counterinsurgency in countries such as Nigeria, and presenting themselves as victims. Saliently, the Nigerian government has accused Boko Haram of systematically poisoning water streams, which, despite the lack of verification, has ended up being a reality in the literature (Nett & Rüttinger, 2016).

The solution

The first hypothesis put forward to save the lake through a transfer of water from the Congo basin to the Lake Chad basin emerged in the second half of the 1980s. The Italian company Bonifica's Transaqua project first presented in 1988 (Mimbale 2021), was imposing; it envisaged a radical transformation of the Central African region, with the opening of a large waterway between the Congo, Chad and Niger basins, as well as the production of electricity and the extension of irrigation with a view to developing an agro-industry. Sayan et al. (2020) note that Bonifica actively promoted an alarmist discourse—calling their proposal "The Sahel Belt, an impending tragedy"—and, in 1988, produced a short video aired on national Italian TV, juxtaposing the drought-stricken Sahel with the water-abundant Congo River spilling unused water into the Atlantic Ocean. As noted, the plentiful (water) resources that European discoverers found were sources of early fascination; in this context, we may point to the fascination with the origin of the Nile. Africa's water abundance is still a modern obsession, generating a plethora of gigantic infrastructure projects (Warner et al., 2019). In its most visionary version, this was the case for the *Atlantropia* formulated in the 1930s by German architect H. Sörgel, who envisaged linking an enormous Lake Congo, to be created, with what, thanks to the new gigantic water inflows, would become Chad Sea (wider than the 6000 BP Mega-Chad), perhaps even to be joined with the Mediterranean (Magrin & Lemoalle, 2015, p. 156). A proposal of a lesser impact compared to that of Transaqua was later presented by the Nigerian National Electric Power Authority, limited to recommending the transfer from Ubangi River, one of the largest tributaries of Congo River, to a tributary of Chari River, the Fafa (Lake Chad Basin Commission [LCBC], 2000, p. 3). The formulation of the project has these main components: a dam on the Ubangi River at the Palambo site, a 150- to 170-km long tunnel/canal to connect the basins, the widening of Chari River's bed and a connecting canal to the Benue River. The 49th session of the LCBC's Council of Ministers in Yaoundé, 2002, therefore decided to proceed with a feasibility study—entrusted to the Canadian research bureau CIMA International—which was concluded in 2011 (CIMA International, 2012; Magrin & Lemoalle, 2015). The CIMA project proposal contained two options, related to the construction of two dams: one on the Ubangi River at Palambo and the other on the Kotto tributary at Bria (see Fig. 2.1; CIMA International, 2012, pp. 24–47). From the first by pumping and from the second by gravity, water could reach the Chari River, suitably reprofiled. The solution that appeared most reasonable and realistic at the time was to divert water from the Bria dam and use the Palambo dam to regulate the flood, allow navigation on the Ubangi River and enable hydroelectric production (Magrin & Lemoalle, 2015, pp. 156–157).

In keeping with the environment–security nexus, Transaqua is not only expected to promote development for poverty reduction but also to reduce the risk of terrorism. During the 2015 United Nations Climate Change Conference (COP21) in Paris, Mahamadou Issoufou, President of Niger, "underlined the close link between the drying up of Lake Chad and terrorism in an area threatened by Boko Haram since 2009" (quoted in Magrin, 2016, p. 205). The arguments for proposing this salvaging of Lake Chad pertain to the climate security crisis narrative (Bøås 2019) and are effective to the extent that the drastic shrinking of the lake can be visualised, which (as argued earlier) has not been demonstrated since the mid-1990s. Following this narrative evidently makes sense to some parties: Western engineering companies, the governments of the countries concerned with the lake and the authorities involved in overseeing water usage, such as the LCBC (Daoust & Selby, 2023). It is in the latter's interest to engage Western narratives in order to attract funds and to direct international attention to the conditions in the region. This is not just an adaptive or opportunistic attitude. Even if the narrative is out of focus with respect to (hydrological, first of all) reality and the solution of transferring water between river basins is questionable, certainly the request for attention is understandable. By speaking the language that can be heard and understood by donors, perhaps a request for consideration might align with the interest of international decision-makers.

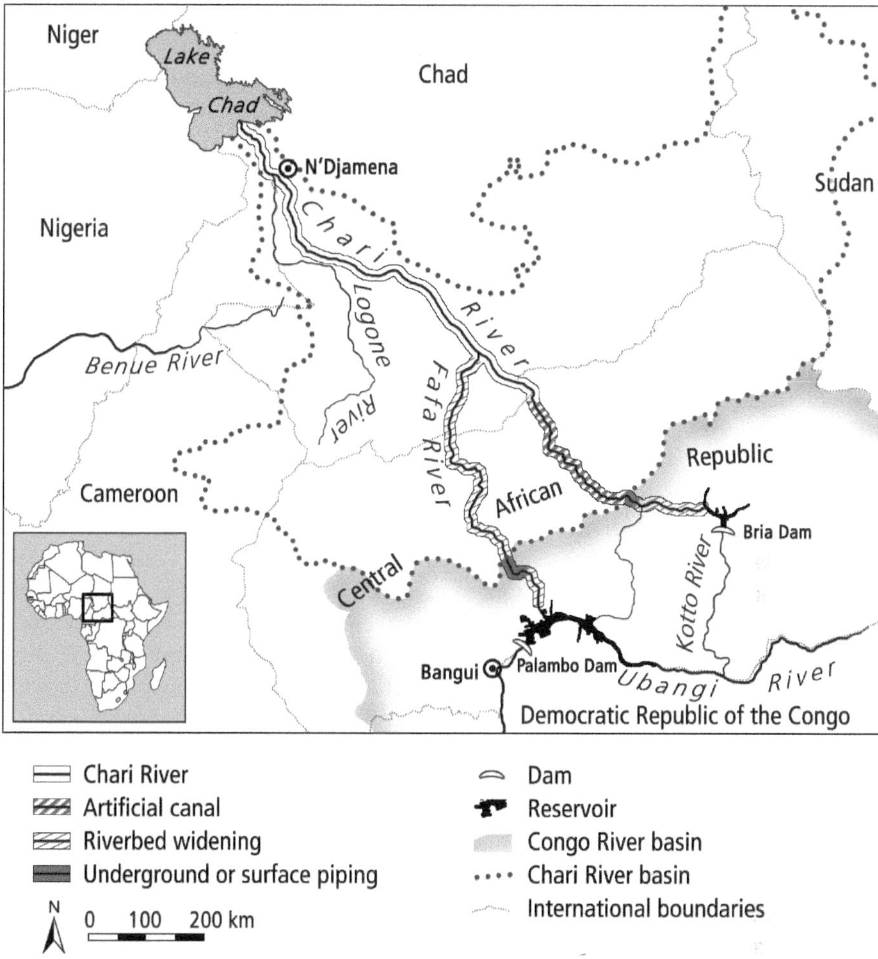

Figure 2.1 The proposed water transfer from Ubangi River to Lake Chad, as drafted by CIMA International (CIMA International, 2012, p. 23; kindly redrawn by F. Ferrarese).

Decolonising African crisis narratives?

It is perhaps past time that a counternarrative be written, critiquing not only how development is conceived but also what happens in the kitchen—the science. The current wave of decolonial analysis puts itself squarely opposed to the continuing assumption that the problems are created in the Global South and the solutions are devised in the Global North, and stands in antithesis to all contrasting elements celebrated by the hegemonic development discourse: North versus South, capitalism versus anticapitalism, patriarchy versus feminism, technocracy versus Indigenous practices, rationality versus identity and even old versus young.

This new counternarrative can also be heard in discussions on water and climate security. For example, Khandekar et al. (2023) decry the current "hegemonic knowledge systems, ontologies and epistemologies, underpinned by technocratic, capitalist, extractivist, and patriarchal systems" and argue that "framings, interpretations, and methods of doing water science need to emerge from, be shaped by, and ultimately serve the needs of marginalised communities from the Global South". Relevant to the Lake Chad case discussed here, Nagheeby and Amezaga (2023) lambast the hydrodiplomatic initiatives in conflict areas, such as Lake Chad, all originating in the North, for taking an approach based on material interests and rationality (e.g., the Harvard school's mutual-gains approach) rather than an identity- and equity-based strategy on negotiations. They advocate taking emotions, harms and resentments seriously. While this matter needs more thought and practical elaboration, such contributions are likely to point a way forward to different modes of conflict resolution.

Nonetheless, tensions remain. In rebutting dominant discourses, counternarratives risk reproducing its lack of nuance in some aspects, though this is understandable—after all, for a counternarrative to gain traction and have an impact, it needs to speak in the same (simple) language in order to resonate with its intended audience. There is an apparent tension between the desire to detach from liberal universalist notions and the wish to appeal to other global totalising notions of justice and sustainability. The pairing of the Global North with colonial science and the Global South with decolonial thought is another example. Not only the modernist mainstream but also the political ecologists and critical analysts exposing Eurocentric assumptions (present company included) are from the Global North themselves or, where authors are from the Global South, their ideas are often shaped in Northern institutions. It is also notable that not only are many influential decolonial authors from the Global North and at times repeat its tropes (e.g., Deborah Rose invoking resource wars as the wars of the future), but the elements of the discourse often echo critical and feminist theories, as taught at European and American universities.

To date, there is no clear consensus on what exactly constitutes "African solutions to African problems"—as unpacked by an African scholar (Kaweesi, 2023) in (you guessed it) a European journal. Questions such as "where do you speak from?" and "where do you write?" make good sense on a decolonial horizon. Even though we (the three authors of this chapter) are Westerners (by training if not by residence), it still seems useful to us to try to point out a possible direction towards the construction of alternative narratives.

Small steps towards a decolonial narrative that embraces complexity

Its elusiveness and constantly changing environmental conditions (Bertoncin & Pase, 2012) may make Lake Chad precisely the right place to imagine a different narrative. We have mentioned how dominant narratives are based on a simplification of factual complexity in order to be persuasive and thus be able to shape policies. Hegemonic narratives are all the more dangerous the simpler they are. The extreme effectiveness of the narrative ("the shrinking lake due to climate change, leading to competition over resources and insecurity") is evidenced by the choice of a megaproject; the high-impact option of large-scale water transfers makes it possible to conceal the strategic interests of the proponents.

The first requirement of a decolonial counternarrative would be to avoid crushing complexity. Decolonial counternarratives on the Lake Chad situation would do well to proceed right in the opposite direction: the layered articulation of times, spaces, living beings, practices and actors.

We can observe many times (temporalities) at play on different scales. The *short time*, annually, of the high and low waters of the lake is a function of the flooding of the Chari River. There is the *inter-annual variation*, with the cycles of drought and rainy years and their effect on the height of the lake's waters. In the *long term*, the alternating advance of water and desert leaves its traces in the composition of the soils (clay and sandy) and in the geomorphology (e.g., with the dikes of the Kanem erg). These time scales are also impacted by climate change, currently expressed as an increase in rainfall, but often concentrated in extreme episodes (flash floods) (see Chapters 3 and 7).

Many spaces fit one inside another, thus identifying multiscalar dimensions and transscalar interactions. One cannot understand Lake Chad without reading it in relation to the desert, on the one hand, and the Sahelian belt, on the other hand. At the regional level, the lake is part of a whole set of wetlands linked to the expansion of permanent and temporary river floods throughout the vast and flat Chadian basin. On yet another scale, it is necessary to distinguish the northern from the southern basin and their shores characterised by very different forms.

Finally, the lake is an extremely complex mosaic of differentiated spaces: islands, marshes, open waters, sandy extensions of the erg from the north and constantly changing shorelines, depending on the height of the lake's waters. All these spaces, amphibious and multiple, together open on vast horizons and close in particular niches, are teeming with living beings: insects, fish, mammals and birds; grasses, reeds and plants; and microorganisms and algae. Human communities, also differentiated and constantly reshuffling, interact with a plurality of practices in the changing temporal, spatial and living scenarios: recessional agriculture and multiple forms of irrigation, collective and individual fishing and smoking of the catch, transhumant farming and pasture search, and mobility by water (pirogues) and by land (on foot or with Chinese motorbikes and the most varied vehicles). So many actors are interested and involved in the framework of the lake: states (but which ones and with what kinds of force and presence?), so-called traditional authorities (e.g., the sultanates, but even here, we may ask: with how much ancient heritage, colonial invention and continuous adaptation?), illegal and insurgent forces, the LCBC and international organisations, various organisations of cooperation (local and foreign NGOs), farmers' and fishers' associations and so on.

This, then, is the lake that is meant to be saved: both the lake itself and saving it are anything but simple. The narratives we construct and choose to embrace on saving Lake Chad should really try harder to somehow reflect this remarkable depth and the many imbrications embedded in it.

Notes

1 https://www.un.org/africarenewal/magazine/december-2013/sahel-one-region-many-crises (accessed September 26, 2023).
2 https://www.oecd.org/dac/Environmental_fragility_in_the_Sahel_perspective.pdf (accessed September 26, 2023).
3 Ferguson (1990) has shown the importance of policy discourses in depoliticising development problems into technical problems that can be solved by the development industry. In the process, territories are simultaneously turned into objects in need of intervention.
4 https://transafricapipeline.org/ (accessed July 22, 2024).
5 https://blogs.worldbank.org/water/water-security-way-out-conflict-climate-risk-trap-lake-chad-basin (accessed September 26, 2023).

References

Abdouraman, H. (2008). Le conflit frontalier Cameroun-Nigeria dans le lac Tchad: Les enjeux de l'île de Darak, disputée et partagée. *Cultures & Conflits*, *72*, 57–76. https://doi.org/10.4000/conflits.17311

Bachmann, J. (2008). The danger of 'undergoverned' spaces: The 'war on terror' and its effects on the Sahel region. In J. M. Eckert (Ed.), *The social life of anti-terrorism laws* (pp. 131–161). transcript Verlag.

Bankoff, G. (2001). Rendering the world unsafe: 'Vulnerability' as Western discourse. *Disasters*, *25*(1), 19–35. https://doi.org/10.1111/1467-7717.00159

Bøås, M. (2019). The Sahel crisis and the need for international support. Nordiska Afrikainstitutet.

Benjaminsen, T. A., & Hiernaux, P. (2019). From desiccation to global climate change: A history of the desertification narrative in the West African Sahel, 1900–2018. *Global Environment*, *12*(1), 206–236. https://doi.org/10.3197/ge.2019.120109

Bertoncin, M., & Pase, A. (2012). *Autour du Lac Tchad: Enjeux, tensions et conflits pour le contrôle de l'eau*. Harmattan.

Brandt, M., Mbow, C., Diouf, A. A., Verger, A., Samimi, C., & Fensholt, R. (2015). Ground- and satellite-based evidence of the biophysical mechanisms behind the greening Sahel. *Global Change Biology*, *21*(4), 1610–1620. https://doi.org/10.1111/gcb.12807

Buhaug, H. (2010). Climate not to blame for African civil wars. *Proceedings of the National Academy of Sciences*, *107*(38), 16477–16482. https://doi.org/10.1073/pnas.1005739107

Cherlet, M., Hutchinson, C., Reynolds, J., Hill, J., Sommer, S., & von Maltitz, G. (Eds.). (2018). *World atlas of desertification*. Publication Office of the European Union. https://doi.org/10.2760/06292

CIMA International. (2012). *Rapport final de l'étude de faisabilité du projet de transfert d'eau de l'Oubangui au Lac Tchad. Étude du cadrage environnemental et social*. Report.

Cooper, F., & Packard, R. M. (Eds.). (1997). *International development and the social sciences: Essays on the history and politics of knowledge*. University of California Press.

Czarniawska-Joerges, B., & Joerges, B. (1988). How to control things with words: Organizational talk and control. *Management Communication Quarterly*, *2*(2), 170–193. https://doi.org/10.1177/0893318988002002003

Daoust, G., & Selby, J. (2023). Understanding the politics of climate security policy discourse: The case of the Lake Chad Basin. *Geopolitics*, *28*(3), 1285–1322. https://doi.org/10.1080/14650045.2021.2014821

de Rycker, A., & Don, Z. M. (Eds.). (2013). *Discourse and crisis: Critical perspectives*. John Benjamins Publishers.
Dreyfus, H., & Rabinow, P. (1982). *Michel Foucault: Beyond structuralism and hermeneutics*. University of Chicago Press.
Duffield, M. (2007). *Development, security and unending war: Governing the world of peoples*. Polity Press.
Ferguson, J. (1990). *The anti-politics machine "development," depoliticization, and bureaucratic power in Lesotho*. Routledge.
Gangneron, F., Pierre, C., Robert, E., Kergoat, L., Grippa, M., Guichard, F., Hiernaux, P., & Leauthaud, C. (2022). Persistence and success of the Sahel desertification narrative. *Regional Environmental Change*, *22*(4), 1–11. https://doi.org/10.1007/s10113-022-01969-1
Giannini, A., Biasutti, M., & Verstraete, M. M. (2008). A climate model-based review of drought in the Sahel: Desertification, the re-greening and climate change. *Global and Planetary Change*, *64*(3–4), 119–128. https://doi.org/10.1016/j.gloplacha.2008.05.004
Glenzer, K. (2002). La Sécheresse: The social and institutional construction of a development problem in the Malian (Sudanese) Sahel, 1900–82. *Canadian Journal of African Studies/ Revue Canadienne des Études Africaines*, *36*(1), 1–34.
Goffner, D., Sinare, H., & Gordon, L. J. (2019), The Great Green Wall for the Sahara and the Sahel initiative as an opportunity to enhance resilience in Sahelian landscapes and livelihoods. *Regional Environmental Change*, *19*, 1417–1428. https://doi.org/10.1007/s10113-019-01481-z
Hajer, M. A. (1995). *The politics of environmental discourse: Ecological modernization and the policy process*. Clarendon Press.
Hall, P. A. (2020). *The political power of economic ideas: Keynesianism across nations*. Princeton University Press.
Haller, T., & Galvin, M. (2011). Challenges for participatory conservation in times of global change: Lessons from a comparative analysis and new developments. In U. Wiesmann & H. Hurni (Eds.), *Research for sustainable development: Foundations, experiences, and perspectives* (pp. 467–503). Geographica Bernensia.
Haller, T., Pase, A., Warner, J., Hashimshony-Yaffe, N., Kronenburg García, A., & Bertoncin, M. (2023). Mega-infrastructure projects in drylands: From enchantments to disenchantments. In A. Kronenburg García, T. Haller, H. van Dijk, C. Samimi, & J. Warner (Eds.), *Drylands facing change: Interventions, investments and identities* (pp. 112–131). Taylor & Francis.
Haller, T., & Weissman, S. (Eds.). (2024). *Disenchanted modernities: Mega-infrastructure projects, socio-ecological changes and local responses*. LIT Verlag.
Hickler, T., Eklundh, L., Seaquist, J. W., Smith, B., Ardö, J., Olsson, L., Sykes, M. T., & Sjöström, M. (2005). Precipitation controls Sahel greening trend. *Geophysical Research Letters*, *32*(21), 1–4. https://doi.org/10.1029/2005GL024370
Homer-Dixon, T. F. (1994). Environmental scarcities and violent conflict: Evidence from cases. *International Security*, *19*(1), 5–40. https://doi.org/10.2307/2539147
Kaweesi, E. S. (2023). The paradox of critical security and the African solutions to African problems. *European Journal of International Security*, 1–21. https://doi.org/10.1017/eis.2023.20
Khandekar, N., Arulingam, I., Joshi, D., Udalagama, U., Chakraborty, S., Ghosh, K., & Pacheco, P. (2023, June 15). What will it take to decolonize water science, policy, and practice? *Water Alternatives: the Water Dissensus Forum*. https://www.water-alternatives.org/index.php/blog/decolonize
Krätli, S. (2013, April). Global public policy narratives on the drylands and pastoralism. *Climate Resilient Dryland Development Briefing*, *2013*(1). https://www.jstor.org/stable/resrep01239?seq=1#metadata_info_tab_contents https://pubs.iied.org/pdfs/10040IIED.pdf
Krätli, S., & Toulmin, C. (2020). *Farmer-herder conflict in sub-Saharan Africa?* International Institute for Environment and Development (IIED).
Lake Chad Basin Commission (LCBC). (2000). *Terms of reference for the feasibility study and engineering design of project for water transfer from the Oubangui basin to Lake Chad basin*. Lake Chad Basin Commission.

Leach, M., Mearns, R., & Scoones, I. (1999). Environmental entitlements: Dynamics and institutions in community-based natural resource management. *World Development*, *27*(2), 225–247. https://doi.org/10.1016/S0305-750X(98)00141-7

Lemoalle, J., & Houdret, A. (1996). Atlas d'élevage du bassin du Lac Tchad = Livestock atlas of the Lake Chad basin. *De Zborowski Isolde (eE.)* (pp. 23–28). CIRAD-EMV.

Lemoalle, J., & Magrin, G. (Eds.). (2014). *Le développement du lac Tchad: Situation actuelle et futurs possibles*. IRD.

Lewis, J. (2010). Vulnerability, fear and denial and the social geography of risk. [Conference presentation]. *2010 Annual conference of the Disaster Management Institute of Southern Africa, Jeffrey's Bay, South Africa.*

Magrin, G. (2016). The disappearance of Lake Chad: History of a myth. *Journal of Political Ecology*, *23*(1), 204–222. https://doi.org/10.2458/v23i1.20191

Magrin, G., & Lemoalle, J. (2015). Les projets de transfert d'eau vers le lac Tchad. Des utopies initialese aux défis contemporaines. In G. Magrin, J. Lemoalle, & R. Pourtier (Eds.), *Atlas du lac Tchad* (pp. 156–158). Passages.

Magrin, G., & Mugelé, R. (2020). La boucle de l'Anthropocène au Sahel: Nature et sociétés face aux grands projets environnementaux (Grande Muraille Verte, Sauvegarde du lac Tchad). *Belgeo*, *3*(2020), 1–15. online: https://journals.openedition.org/belgeo/42872

Mimbale, J. (2021). The Transaqua Project: Narrative and Constructivist Analysis. Academia Letters.

Mitchell, T. (2005). The object of development: America's Egypt. In J. Crush (Ed.), *Power of development* (pp. 125–151). Routledge.

Molle, F. (2008). River-basin planning and management. The social life of a concept. *Geoforum*, *40*, 484–494. https://doi.org/10.1016/j.geoforum.2009.03.004

Morel, A. (2006). "Le désert avance…". In G. Courade (Ed.), *L'afrique des idées reçues* (pp. 106–111). Editions Belin.

Moseley, W. G., & Laris, P.. (2008). West African environmental narratives and development-volunteer praxis. *Geographical Review*, *98*(1), 59–81. https://doi.org/10.1111/j.1931-0846.2008.tb00288.x

Mugelé, R. (2018). La Grande muraille verte au Sahel: Entre ambitions globales et ancrage local. *Bulletin de l'association de géographes français*, *95*(2), 187–202. http://journals.openedition.org/bagf/3084; https://doi.org/10.4000/bagf.3084

Nagheeby, M., & Amezaga, J. (2023). Decolonising water diplomacy and conflict transformation: From security-peace to equity-identity, *Water Policy*, *25*(8), 835–850. https://doi.org/10.2166/wp.2023.043

Nett, K., & Rüttinger, L. (2016). *Insurgency, terrorism and organised crime in a warming climate. Analysing the links between climate change and non-state armed groups*. Adelphi.

Olivry, J. C., Chouret, A., Vuillaume, G., Lemoalle, J., & Bricquet, J. P. (1996). *Hydrologie du lac Tchad* (Vol. 12). Paris, France: Orstom.

Pearce, F. (2022, Olivry, J. C., Chouret, A., Vuillaume, G., Lemoalle, J., & Bricquet, J. P. (1996). *Hydrologie du lac Tchad* (Vol. 12). Paris, France: Orstom. February 8). It's not just climate: Are we ignoring other causes of disasters? *The Guardian*.

Peet, R., & Watts, M. (1993). Introduction: Development theory and environment in an age of market triumphalism. *Economic Geography*, *69*(3), 227–253. https://doi.org/10.2307/143449

Rahman, M. F., Lewis, D., Kuhl, L., Baldwin, A., Ruszczyk, H., Nadiruzzaman, M., & Mahid, Y. (2023). Managed urban retreat: The trouble with crisis narratives. *Urban Geography*, 1–10. https://doi.org/10.1080/02723638.2023.2228094

Reenberg, A. (2012). Insistent dryland narratives: Portraits of knowledge about human-environmental interactions in Sahelian environment policy documents. *West African Journal of Applied Ecology*, *20*(1), 97–111. https://www.ajol.info/index.php/wajae/article/view/86326

Ribot, J. (2004). *Waiting for democracy. The politics of choice in natural resource centralization*. World Resource Institute Report. http://www.wri.org/publication/waiting-democracy-politics-choice-natural-resource-decentralization

Rocheleau, D. E., Steinberg, P. E., & Benjamin, P. A. (1995). Environment, development, crisis, and crusade: Ukambani, Kenya, 1890–1990. *World Development*, *23*(6), 1037–1051. https://doi.org/10.1016/0305-750X(95)00016-6

Roe, E. M. (1995). Except-Africa: Postscript to a special section on development narratives. *World Development*, *23*(6), 1065–1069. https://doi.org/10.1016/0305-750X(95)00018-8

Roe, E., & van Eeten, M. J. (2004). Three—not two—major environmental counternarratives to globalization. *Global Environmental Politics*, *4*(4), 36–53. https://doi.org/10.1162/glep.2004.4.4.36

Sayan, R. C., Nagabhatla, N., & Ekwuribe, M. (2020). Soft power, discourse coalitions, and the proposed interbasin water transfer between Lake Chad and the river Congo. *Water Alternatives*, *13*(3), 752–778.

Scoones, I., & Nori, M. (2023). Living with and from uncertainty: Lessons from pastoralists for development. In I. Scoones (Ed.), *Pastoralism, Uncertainty and Development* (pp. 119–140).

Shore, C., & Wright, S. (Eds.). (2003). *Anthropology of policy: Perspectives on governance and power*. Routledge.

Starr, J. R., & Stoll, D. C. (Eds.). (1988). *The politics of scarcity: Water in the Middle East*. Westview Press.

Sullivan, S. (2000). Getting the science right, or introducing science in the first place. *Local 'factsfacts, global discourse – 'desertification' in north-west Namibia*. In P. Stott, & S. Sullivan (Eds.), *Political ecology: Science, myth and power* (pp. 15–44). Edward Arnold.

Swift, J. (1996). Desertification: Narratives, winners and losers. In M. Leach, & R. Mearns (Eds.), *Lie of the land: Challenging received wisdom on the African environment* (pp. 73–90). James Currey.

Tilho, J. (1928). Variations et disparition possible du Tchad. *Annales de Géographie*, *37*, 238–260. https://doi.org/10.3406/geo.1928.9299

Thomas, G. B. (1997). US environmental security policy: Broad concern or narrow interests. *The Journal of Environment & Development*, *6*(4), 397–425. https://doi.org/10.1177/107049659700600403

Treszakai, A. (2018). Water conflicts. Case study-Lake Chad conflict. Theme: Geopolitics and geostrategies. *Strategic Impact*, *2018*(3/4), 54–62. https://cssas.unap.ro/en/pdf_periodicals/si68-69.pdf

United Nations Convention to Combat Desertification (UNCCD). (2020). *The Great Green Wall: Implementationstatus and way ahead to 2030* United Nations Convention to Combat Desertification. https://catalogue.unccd.int/1551_GGW_Report_ENG_Final_040920.pdf

van Eeten, M. (1997). Sprookjes in rivierenland. Beleidsverhalen over wateroverlast en dijkversterking. *Beleid en Maatschappij*, *24*(1), 32–43.

Vázquez-Arroyo, A. Y. (2013). How not to learn from catastrophe: Habermas, critical theory and the 'catastrophization' of political life. *Political Theory*, *41(5*), 738–765. https://doi.org/10.1177/0090591713492776

Verhoeven, H. (2014). Gardens of Eden or hearts of darkness? The genealogy of discourses on environmental insecurity and climate wars in Africa. *Geopolitics*, *19*(4), 784–805. https://doi.org/10.1080/14650045.2014.896794

Warner, J. (2013). The politics of 'catastrophization. In D. Hilhorst (Ed.), *Disaster, conflict and society in crises: Everyday politics of crisis response* (pp. 76–94). Routledge.

Warner, J., Jomantas, S., Jones, E., Ansari, M. S., & De Vries, L. (2019). The fantasy of the Grand Inga hydroelectric project on the river Congo. *Water*, *11*(3), 407–14. https://doi.org/10.3390/w11030407

Warner, J., Kronenburg García, A., & Haller, T. (2023). Environmental crisis narratives in drylands. In A. Kronenburg García, T. Haller, H. van Dijk, C. Samimi, & J. Warner (Eds.), *Drylands facing change: Interventions, investments and identities* (p. 3250). Taylor & Francis.

Warner, J., & Thomas, V. (2015). Hydropolitics in the Harirud/Tejen river basin: Afghanistan as hydro-hegemon? *Water International*, *44*(4), 593–613. https://doi.org/10.1080/02508060.2015.1059164

3 Layered crises

Climate change, water insecurity and irrigation in the Sahel

Martina Angela Caretta, Mariasole Pepa and Edmond Totin

Introduction

Water insecurity has been and will be the first manifestation of climate change that the majority of the world's population has experienced or will experience (Caretta et al., 2022). Examples include increased heavy precipitations, glacial melting (Hock et al., 2019), frequency and magnitude of floods and droughts (Alfieri et al., 2017), and a deterioration of groundwater and water quality. Water security is defined as follows: the capacity to safeguard sustainable access to adequate quantities of acceptable quality water for sustaining livelihoods, human well-being and socio-economic development, for ensuring protection against water-borne pollution and water-related disasters, and for preserving ecosystems in a climate of peace and political stability (UN-Water, 2013, p. 1).

Water insecurity adversely affects over 4 billion people worldwide, and many African countries are at risk of water insecurity, given the high water demand for both irrigation and home consumption because of the rapid population growth in the continent.

Climate change-induced water insecurity impacts are unevenly distributed across the world, with Africa expected to bear the highest burden of water scarcity globally (Arnell et al., 2019). Climate change has led and will lead to an increase of floods and droughts, with detrimental impacts on an increasing number of people and potentially severe outcomes (Caretta et al., 2022). The magnitude and gravity of these consequences depend on the socio-economic vulnerability of local populations (Ferdous et al., 2019). Notably, while economic vulnerability due to drought has decreased globally, poor countries have experienced twice as high GDP losses caused by drought-related damages than high-income countries (Formetta & Feyen, 2019).

As it pertains to drylands, climate change has mixed effects. Depending on the region, the method used and the time period, both increases and decreases in vegetation productivity have been shown, with local and contextual implications for biodiversity, ecosystem services and water availability. However, a global trend

DOI: 10.4324/9781003429111-5

in dryland expansions has not emerged from the data; on the contrary, greening of drylands has been reported since the 1980s (Mirzabaev et al., 2022).

Globally, overall economic growth has been negatively affected by climate change, which, together with population growth, has resulted in additional water insecurity, in turn threatening food and energy security. In fact, 80% of consumptive water use occurs in agriculture, and 50% of global food production is irrigated (Food and Agriculture Organization [FAO], 2018). In the agricultural sector, water insecurity is projected to lead to potentially severe food insecurity for many people, with 6% of the combined African and Middle Eastern populations estimated to be adversely affected by production losses (Fitton et al., 2019). Dryland populations have experienced negative impacts on their livelihoods, health and well-being due to ecosystem degradation. Socio-economic vulnerability is determined by the unavailability of financial and educational resources, which are unequally distributed across populations due to race, gender and socio-economic class. The lack of services and infrastructure in their regions has constrained dryland populations to adapt to their local circumstances (Mirzabaev et al., 2022). Despite their limited access to resources, dryland populations have long relied on their local knowledge and Indigenous knowledge to combat climate change and implement place-based adaptation strategies (e.g., Caretta & Morgan, 2021; Rittelmeyer et al., 2024). Departing from an understanding of the ever-changing conditions in drylands and given the ongoing cascading and compounding climatic, land-based and geopolitical crises, in this chapter, we review climate change-induced water insecurity in the Sahel and explore the role that irrigation and local adaptation practices can play to lessen the burden of environmental degradation in the region's dryland communities.

Climate change in the Sahel

The Sahelian belt is often identified as the most vulnerable region in Africa because of its stressed climatic conditions and high sensitivity to changes in precipitation and potential evapotranspiration (Serdeczny et al., 2017). The region is characterised by a semi-arid climate, with the vitality of its natural ecosystems greatly reliant on water availability (Mirzabaev et al., 2022). A large number of the Sahelian population depend on rainfed agriculture and pastoralism for their livelihoods. Over the last 30 years, the region has experienced a significant warming trend, recurrent droughts and floods, which exacerbate the vulnerability of the population and ecological systems through unprecedented variations in precipitation and temperature (Trisos et al., 2022). Rainfall patterns in the Sahel have been characterised by the rapid alternation of dry spells and more intense rainfall events that significantly disrupt traditional farming practices, leading to decreased agricultural productivity and higher food insecurity (Ortiz-Bobea et al., 2021; Sultan et al., 2019). Increasing temperatures have intensified heat waves, causing a decrease in yields, livestock and fishery production, adversely affecting human health and well-being (Trisos et al., 2022). Droughts' frequency, intensity and duration contribute to water insecurity, increased pressure on natural resources and consequent tensions between farmers and pastoralists (Nunn & McGuirk, 2021), as well as negatively impact the rainfed agricultural sector (Lèye et al., 2021).

However, discerning the impacts of climate change is impossible without considering the multiple social, political, economic and environmental crises that have crossed the Sahelian space (Cepero et al., 2021). Indeed, climate change exacerbates an already fragile situation of recurrent crises and violence (Mbaye & Signé, 2022). In the context where water and land are the primary resources used for agriculture and pastoralism, the Sahelian population's main livelihoods, exploring the impacts of climate change, are crucial (Mbaye, 2020). Climate change needs to be situated in the broader context of multiple crises, such as the current war in Sudan and its impact on irrigation spaces (Chapter 9), armed group attacks in Sourou Valley, Burkina Faso (Chapter 11) and the Jihadist groups in Mali (Chapter 4). This confluence of crises has heightened the challenges of addressing climate change (Tesfaye, 2022), directly affecting irrigation spaces and constraining the ability to react and adapt accordingly.

The magnitude and the effects of climate change, whether direct or indirect, on conflicts in the area depend on future socio-economic and political events. Climate change is one of the multiple drivers of conflict. In the case of Mali, Benjaminsen et al. (2012) argue that other factors, such as corruption, mobility restrictions and political negligence, fuel pastoralist-farmer conflicts over resources rather than environmental degradation. It is indeed critical to explore the political and historical roots of conflicts, for instance, in natural resource governance, to understand the current dynamics in the Sahel and the role played by climate change in this complex picture (Benjaminsen & Svarstad, 2021).

Even if Africa's historical responsibility for anthropogenic climate change is negligible, the impacts that the continent has already experienced are substantially more serious than in many countries in the Global North that bear responsibility for high carbon dioxide (CO_2) emissions. The magnitude and gravity of these impacts are particularly related to the majority of the population's direct dependence on ecosystem-based livelihoods, such as agriculture, fishing and livestock rearing, which is especially the case of the Sahel. However, to understand these impacts, how to cope with them and prepare for future risks, appropriate, detailed and up-to-date scientific data should be available. Access to data represents one of the major challenges for the Sahel as it impairs the region's ability to adapt to climate change (Trisos et al., 2022). Sparse and intermittent weather stations seriously curtail climate data availability and accessibility, making it difficult to make future projections and anticipate risks. Survey data about the socio-economic conditions on the ground are also limited, restricting the possibility of understanding current impacts and predicting future ones. At the base of these scientific gaps is the extremely limited funding available for climate research in the whole African continent (Trisos et al., 2022). It is estimated that only 3.8% of global funding for climate research between 1990 and 2020 reached the African continent. This number is diametrically opposed to the high vulnerability that the continent faces due to climate change (Overland et al., 2022). Additionally, the Sahel is marred by multiple political crises, which make the area highly volatile and insecure, advising against direct data collection and the implementation of scientific investments and academic research in the area.

An underlying problem is the very definition of the area studies, in other words, the spatial classification that serves as a data container and thus a starting point for analytical tools. We should highlight the fact that the Sahel as a region is often categorised according to parameters that vary among different political and academic institutions. For example, on one hand, to report the vulnerability, adaptation and risks to climate change in the African continent, the Intergovernmental Panel on Climate Change 6th Assessment Report (IPCC AR6) used the African Union's categorisation of the continent into five regions: North, West, East, Central and South. This division effectively divides the Sahel into three regions: West (Mauritania, Mali, Ghana, Togo, Benin, Burkina Faso, Niger and Nigeria), Central (Chad) and North (Sudan). Thus, specific information about climate change in the Sahel is spread across different chapters of the IPCC report. On the other hand, the editors of this volume define the Sahel according to its seasonal and local rainfall patterns (200–900 mm of rain/year), including parts of all the above-mentioned countries (Chapter 5). These are just two examples of different categorisations of the Sahel according to its political, geographical or environmental characteristics that effectively challenge the possibility of comparing and complementing data gathering across these classifications. The difficulties in understanding and grasping the "breath of the Sahel" (Chapter 5) have contributed to the inability to design proper policy and development interventions that reflect the ever-changing geographies of these spaces. This failure stems from the lack of recognition and perhaps of interest in the intricate nature of the Sahelian space. These factors have resulted in inadequacies in designing climate change-mitigation strategies that are feasible (thus effectively considering the economic resource unavailability) and that reflect on-the-ground realities.

Irrigation as an adaptation (and maladaptation) measure

Irrigation has long been hailed as a way to avert yield failure in case of drought, protect crops in case of reduced rainfall and enhance crop productivity. These characteristics have become increasingly important in the contexts of rising global average temperatures and changing precipitation patterns (Siebert et al., 2014). Over the past five decades, the amount of irrigated fields has doubled, bringing the current irrigation coverage to approximately 40% of agricultural production worldwide, amounting to 20% of the global farmed surface (FAO, 2018; Rosa et al., 2020). Most of the world's irrigated lands are in Asia (69–73%), with Africa accounting for a minor percentage only (less than 5%), mostly in the Nile Basin, North Africa and South Africa (Meier et al., 2018).

In the context of its rising scarcity, it is becoming more and more important to save water. A wide variety of techniques increases water productivity—that is, more crops are produced per drop of water (Deligios et al., 2019). For instance, refurbishing leaking systems can save up to 76% of water while obtaining the same amount of yield (Jägermeyr et al., 2015). Likewise, irrigating less often for shorter periods, covering a smaller area or irrigating only during the crop development stage (i.e., deficit irrigation) can all be successful measures to ensure crop productivity while saving water (Malek & Verburg, 2018). Other water-saving technologies include the use of efficient pumps, micro-irrigation, laser levelling and improved agronomic practices, such as contouring or switching to more drought-resistant or short-term crops (Caretta & Börjeson, 2015; Traore et al., 2017).

Moreover, these water technologies are scarce in Sahelian countries due to a confluence of factors, ranging from constrained resources to economic and political instability. A case in point is Sudan, where even before the start of the war there, barriers to adopting water-saving technologies were evident. These technologies' accessibility to actors with financial means was impeded by logistical barriers at Port Sudan, the scarcity of spare parts and the imposition of US sanctions, thereby limiting the acquisition of innovative solutions. Meanwhile, for small-scale farmers, already grappling with securing basic agricultural inputs, such as seeds and fertilisers at the outset of the season, it seemed unrealistic to embrace advanced water-saving strategies like laser levelling or using efficient pumping systems (Chapter 10).

Irrigation has been recognised as providing several benefits, enhancing resilience and crop yield by reducing drought and heat stress and hence water insecurity during times of crop development and growth (Siebert et al., 2014; Zaveri & Lobell, 2019). It has been shown that irrigation is not only beneficial for crop growth itself but also plays a role in cooling local and regional climates (Cook et al., 2020; Thiery et al., 2020). Additionally, irrigation can reduce farmers' vulnerability by cushioning them from drought, thus reducing socio-economic and environmental risks and consequently poverty (Balasubramanya & Stifel, 2020). The resulting increased income has been reinvested in further improvements in irrigation. Specifically in the Sahel, the case of the Kenana Sugar Company in Sudan demonstrates that the income earned from irrigated sugarcane plantations can be reinvested in the diversification of production and technological innovation; however, the spill-over effects on the surrounding local communities can be limited and sometimes harmful (see Chapter 10).

Irrigation can also lead to maladaptation – given that increased efficiency can drive the uptake of larger farming areas, groundwater over-abstraction and aquifer salinisation (Foster et al., 2018; Pulido-Bosch et al., 2018). Many areas worldwide already face hard limits in water availability and thus are not in a position to expand irrigation any further (Albizua et al., 2019; Kummu et al., 2016). Particularly, given groundwater depletion due to over-abstraction, limitations to irrigation expansion exist in South and Central Asia, the Middle East and areas of North and Central America (Turner et al., 2019). From this viewpoint, it is clear why the Sahel is perceived by international investors as one of the last frontiers to acquire low-cost lands, framed as "available", next to significant reserves of water potentially usable for irrigation, particularly along the major rivers (see Chapters 1, 8 and 9). Since the early 2000s, Sahelian spaces have increasingly attracted foreign investors seeking vast tracts of land and access to water for promoting hyper-technological irrigation systems, thereby fostering the expansion of pivot-irrigation systems, commonly known as the "green circles" in the desert (Bertoncin et al., 2022). This situation contributes to added pressure on water availability and exacerbates tensions and conflicts over resources among various stakeholders, including local communities. Land grabbing and water grabbing pose further limits to water efficiency under severe climate change.

Local adaptation strategies in the Sahel

In the context of multiple and compounding crises, Sahelian communities have been able to adapt to increased water insecurity. Several well-rooted local techniques for water conservation and water harvesting are used in the Sahel, such as Zaï and half-moon (Partey et al., 2018). Zaï, a technique used in West Africa (Lèye et al., 2021), allows farmers to capture rain in pits that they had dug before the rainy season to conserve some water for the longer term, while concentrating organic matter and preserving soil moisture. Intercropping and mixing long- and short-term crops are also adaptation strategies that farmers have put in place in the Sahel for decades, effectively reducing the risk of food insecurity (Zougmoré et al., 2023).

Irrigation remains a promising climate change-adaptation strategy in the Sahel due to its potential to shield agricultural yields from the adverse impacts of water scarcity (Zorom et al., 2013). Irrigation enables off-season production and effectively mitigates climate impacts on farmers' livelihoods. Integrated water resource management and risk-based approaches are proven viable ways to enhance water availability and improve water security in arid and semi-arid regions like the Sahel (Stringer et al., 2021). These include in-situ water harvesting and storage technologies, such as small reservoirs, water conservation practices, watershed restoration and less water-intensive cropping, which have been demonstrated to enhance water availability for agriculture, particularly during dry periods (Abberton et al., 2021; Williams et al., 2021). For instance, in the Mandara Mountains in the north of Cameroon, where, during the dry season, there are several difficulties in water provision, several water harvesting techniques are practised. In particular, in the mid-1980s, micro-dams known as *bief* were constructed by non-government organisations (NGOs). Bief were not designed to create permanent reservoirs on the surface but to slow down the rate of runoff and retain some of the flood water, encouraging its infiltration and thus favouring groundwater recharge (Bertoncin

& Pase, 2008). Even if these micro-dams were close to local practices, the development agency that engaged in spreading this technique faced the disruption of the once well-established connections between water management and political power in the villages.

Another example can be drawn from Pepa's conversation with the General Directorate of Agriculture Extension and Technology Transfer Administration (held in Khartoum in February 2023), an important service under the Ministry of Agriculture and Forestry. Their meeting revealed the significant impact of climate change on livelihoods in Sudan, particularly in the rainfed sector. Their discussion shed light on the diverse ramifications of climate change in the region, including instances of drought, severe floods, erratic rainfall patterns, dust storms and desertification. These environmental challenges have notably hampered agricultural productivity, exacerbated soil degradation and intensified conflicts over resources, consequently driving rural-to-urban migration. The traditional rainfed agricultural sector, characterised by small-scale family-operated farms (5–20 *feddans*, 1 *feddan* is equal to 0.42 ha), constitutes the majority of farmers in Sudan. Its operations rely on manual labour and traditional farming methods, with limited access to both markets and modern agricultural technologies. This sector has been heavily impacted by environmental degradation. During the COVID-19 pandemic, agricultural extension members went around the country to document and gather information pertaining to local techniques and traditional knowledge aimed at climate change mitigation and adaptation. This effort yielded a wealth of simple methods rooted in local knowledge, such as storing water in *Bao Peppa's* trees and collecting rainwater (known locally as *Fula*). Many traditional techniques for conserving water and combating desertification are practised in different areas of Sudan. Among the documented techniques were small-scale water harvesting (*Sheerak*) and the use of dry plant fences. Predominantly practised in the western and central-western regions of Sudan, *Sheerak* involves farmers anticipating rainfall and collecting rainwater into small squares (1–2 m) defined by bare soil ridges. This method serves as a localised form of natural micro-water harvesting, crucial for preserving soil moisture that is essential for seasonal crop cultivation. A similar technique known as *dingiiji* or *diguettes* is applied in *karal* soil, vertisols, clayey soils, in the extreme north of Cameroon to cultivate sorghum, locally named *muskuwaari* (Seignobos, 2000). Furthermore, in marginal rainfed areas of North Kordofan state, farmers protecting small-scale plots employ dry plant fences to mitigate the effects of desertification.

Financial and decision-making support for locally based adaptation strategies, partly grounded on Indigenous knowledge, could be crucial to increase the local populations' adaptive capacity in the water sector by means of different techniques, ranging from behaviour-based (e.g., water-saving) to technical solutions (e.g., drip irrigation). Supporting these strategies will contribute to implementing adaptation strategies that reflect local populations' everyday realities, aligned with their needs.

Conclusion

Climate change acts as a compounding factor that exacerbates the existing vulnerabilities in the Sahel region and creates new challenges for water security and irrigation management. Changes in the amplitude, frequency and timing of extreme water-related events are expected to continuously affect lake levels, rates of river discharge and runoff and groundwater recharge in sub-Saharan Africa (Darko et al., 2019). With populations in the Sahel highly dependent on agriculture and pastoralism for their livelihoods (Ben Mohamed, 2011), addressing water insecurity through effective and climate-smart pathways becomes crucial to ensure sustainable production, safeguard livelihoods and promote people's resilience.

Globally, irrigation plays a crucial role in safeguarding agricultural yields against the risks posed by water insecurity and serves as a chief climate-adaptation strategy. By providing a reliable water supply for agricultural activities, irrigation mitigates the challenges arising from changing rainfall patterns and contributes to enhanced resilience in the face of climate change. However, addressing the complex challenges related to water insecurity, inadequate infrastructure and governance issues is critical for the successful implementation of irrigation projects. By incorporating sustainable water management practices, more efficient irrigation technologies and institutional support, water security, agricultural productivity and overall resilience in the context of climate change can be enhanced. However, as it pertains to the Sahel, structural limitations (e.g., lack of resources, difficulty in accessing irrigation space and receiving spare parts) cannot be ignored.

Addressing climate change and water insecurity thus means taking into account and working through pre-existing complexities and ongoing compounding and cascading crises, which cannot be disentangled. Therefore, to maximise the benefits of irrigation and enhance resilience to climate change in the Sahel, a comprehensive approach encompassing various strategies is required. Thus, it is important to reconsider the local agricultural techniques' role in water harvesting and conservation, especially when it is impossible to access other kinds of equipment during times of crisis. Moreover, integrated water resource management practices, including traditional and modern techniques, such as rainwater harvesting, small-scale water storage and conjunctive use of surface and groundwater, can help optimise water availability and improve the resilience of irrigation systems. For these adaptation measures to be implemented, there needs to be a shift in the climate change policy narrative in the Sahel so that small-holder farmers and pastoralists are heard and favoured instead of disregarded and marginalised (Benjaminsen & Hiernaux, 2019). In this sense, strengthening governance frameworks and institutional capacity and promoting farmer-led initiatives are essential means for the successful implementation and sustainable management of irrigation projects. Moreover, consolidating local institutions and community participation and promoting cross-sectoral collaboration represent fundamental ways to enable the conditions for a promising adaptation to water challenges in the Sahel. To facilitate such changes of narrative and action by local governments, different types of knowledge should be combined. On one hand, monitoring and measuring environmental change ought to be prioritised to obtain a more comprehensive and up-to-date picture of the status and advancement of climate change in the Sahel. On the other hand, Indigenous knowledge and local innovations must be increasingly appreciated by national and international governmental and academic actors. Drawing from these two different ontological perspectives will allow the development of more effective and socially acceptable adaptation measures, thus increasing the likelihood of properly addressing climate change-induced water insecurity and lessening future climatic risks.

References

Abberton, M., Abdoulaye, T., Arinloye, D. A., Asiedu, R., Ayantunde, A., Bayala, J., Cofie, O., Lamien, N., Potaka, K., Jalloh, J., Tabo, A., Tenkouano, R., Tepa-Yotto, A., Whitbread, G., Worou, A., Zougmore, O. N., & Zwart, R. (2021). *Priority interventions for transformational change in the Sahel* (CGIAR Research Program on Climate Change, Agriculture and Food Security Working Paper).

Albizua, A., Corbera, E., & Pascual, U. (2019). Farmers' vulnerability to global change in Navarre, Spain: Large-scale irrigation as maladaptation. *Regional Environmental Change*, *19*(4), 1147–1158. https://doi.org/10.1007/s10113-019-01462-2

Alfieri, L., Bisselink, B., Dottori, F., Naumann, G., de Roo, A., Salamon, P., Wyser, K., & Feyen, L. (2017). Global projections of river flood risk in a warmer world. *Earth's Future*, *5*(2), 171–182. https://doi.org/10.1002/2016EF000485

Arnell, N. W., Lowe, J. A., Bernie, D., Nicholls, R. J., Brown, S., Challinor, A. J., & Osborn, T. J. (2019). The global and regional impacts of climate change under representative concentration pathway forcings and shared socioeconomic pathway socioeconomic scenarios. *Environmental Research Letters*, *14*(8), 084046. https://doi.org/10.1088/1748-9326/ab35a6

Balasubramanya, S., & Stifel, D. (2020). Viewpoint: Water, agriculture & poverty in an era of climate change: Why do we know so little? *Food Policy*, *93*, Article 101905. https://doi.org/10.1016/j.foodpol.2020.101905

Ben Mohamed, A. (2011). Climate change risks in Sahelian Africa. *Regional Environmental Change*, *11*, 109–117. https://doi.org/10.1007/s10113-010-0172-y

Benjaminsen, T. A., & Hiernaux, P. (2019). From desiccation to global climate change: A history of the desertification narrative in the West African Sahel, 1900–2018. *Global Environment*, *12*(1), 206–236. https://doi.org/10.3197/ge.2019.120109

Benjaminsen, T. A., Alinon, K., Buhaug, H., & Buseth, J. T. (2012). Does climate change drive land-use conflicts in the Sahel? *Journal of Peace Research*, *49*(1), 97–111. https://doi.org/10.1177/0022343311427343

Benjaminsen, T. A., & Svarstad, H. (2021). Climate change, scarcity and conflicts in the Sahel. In T. A. Benjaminsen, & H. Svarstad (Eds.), *Political ecology* (pp. 183–205). Palgrave Macmillan. https://doi.org/10.1007/978-3-030-56036-2_8

Bertoncin, M., & Pase, A. (2008). *Attorno al lago Ciad. Sguardi diversi sullo sviluppo*. Harmattan Italia.

Bertoncin, M., Pase, A., & Turrini, S. (2022). Water, land and Arab investments in irrigation projects: Continuity and innovation in Sudan. In J.-N. Bach (Ed.), *Routledge handbook on Horn of Africa* (pp. 657–669). Routledge. https://doi.org/10.4324/9780429426957

Caretta, M. A., & Börjeson, L. (2015). Local gender contract and adaptive capacity in smallholder irrigation farming: A case study from the Kenyan drylands. *Gender. Place Culture*, *22*(5), 644–661. https://doi.org/10.1080/0966369X.2014.885888

Caretta, M. A., & Morgan, R. A. (2021). Special issue on Indigenous knowledge for water-related climate adaptation. *Climate and Development*, *13*(9), 761–765. https://doi.org/10.1080/17565529.2021.1993627

Caretta, M. A., Mukherji, A., Arfanuzzaman, M., Betts, R. A., Gelfan, A., Hirabayashi, Y., Lissner, T. K., Liu, J., Lopez Gunn, E., Morgan, R., Mwanga, S., & Supratid, S. (2022). Water. In H.-O. Pörtner, D. C. Roberts, M. Tignor, E. S. Poloczanska, K. Mintenbeck, A. Alegría, M. Craig, S. Langsdorf, S. Löschke, V. Möller, A. Okem, & B. Rama (Eds.), *Climate change 2022: Impacts, adaptation, and vulnerability. Contribution of working group II to the sixth assessment report of the intergovernmental panel on climate change* (pp. 551–712). Cambridge University Press.

Cepero, O. P., Desmidt, S., Detges, A., Tondel, F., Van Ackern, P., Foong, A., & Volkholz, J. (2021). Climate change, development and security in the Central Sahel. CASCADES. https://www.cascades.eu/wp-content/uploads/2021/06/Climate-Change-Development-and-Security-in-the-Central-Sahel.pdf

Cook, B. I., McDermid, S. S., Puma, M. J., Williams, A. P., Seager, R., Kelley, M., …, & Aleinov, I. (2020). Divergent regional climate consequences of maintaining current irrigation rates in the 21st century. *Journal of Geophysical Research: Atmospheres*, *125*(14), e2019JD031814. https://doi.org/10.1029/2019JD031814

Darko, D., Kpessa-Whyte, M., Obuobie, E., Siakwah, P., Torto, O., & Tsikata, D. (2019). The context and politics of decision making on large dams in Ghana: An overview. *FutureDAMS working paper 002*. Manchester: The University of Manchester.

Deligios, P. A., Chergia, A. P., Sanna, G., Solinas, S., Todde, G., Narvarte, L., & Ledda, L. (2019). Climate change adaptation and water saving by innovative irrigation management applied on open field globe artichoke. *Science of the Total Environment*, *649*, 461–472. https://doi.org/10.1016/j.scitotenv.2018.08.349

Ferdous, M. D., Ruknul, A. W., Brandimarte, L., Slager, K., Zwarteveen, M., & Di Baldassarre, G. (2019). The costs of living with floods in the Jamuna floodplain in Bangladesh. *Water*, *11*(6), 1238. https://doi.org/10.3390/w11061238

Fitton, N., Alexander, P., Arnell, N., Bajzelj, B., Calvin, K., Doelman, J., Gerber, J. S., Havlik, P., Hasegawa, T., Herrero, M., Krisztin, T., van Meijl, H., Powell, T., Sands, R., Stehfest, E., West, P. C., & Smith, P. (2019). The vulnerabilities of agricultural land and food production to future water scarcity. *Global Environmental Change*, *58*(July), Article 101944. https://doi.org/10.1016/j.gloenvcha.2019.101944

Food and Agriculture Organization (FAO). (2018). *World food and agriculture – Statistical pocketbook 2018*. Retrieved December 18, 2020, from http://www.fao.org/3/CA1796EN/ca1796en.pdf

Formetta, G., & Feyen, L. (2019). Empirical evidence of declining global vulnerability to climate-related hazards. *Global Environmental. Change*, *57*, Article 101920. https://doi.org/10.1016/j.gloenvcha.2019.05.004

Foster, S., Pulido-Bosch, A., Vallejos, Á, Molina, L., Llop, A., & MacDonald, A. M. (2018). Impact of irrigated agriculture on groundwater-recharge salinity: A major sustainability concern in semi-arid regions. *Hydrogeology Journal*, *26*(8), 2781–2791. https://doi.org/10.1007/s10040-018-1830-2

Hock, R., Rasul, G., Adler, C., Cáceres, B., Gruber, S., Hirabayashi, Y., Jackson, M., Kääb, A., Kang, S., Kutuzov, S., Milner, A., Molau, U., Morin, S., Orlove, B., & Steltzer, H. (2019). High mountain areas. In H.-O. Pörtner, D. C. Roberts, V. Masson-Delmotte, P. Zhai, M. Tignor, E. Poloczanska, K. Mintenbeck, A. Alegría, M. Nicolai, A. Okem, J. Petzold, B. Rama, & N. M. Weyer (Eds.), *IPCC special report on the ocean and cryosphere in a changing climate* (pp. 131–202). Cambridge University Press.

Jägermeyr, J., Gerten, D., Heinke, J., Schaphoff, S., Kummu, M., & Lucht, W. (2015). Water savings potentials of irrigation systems: Global simulation of processes and linkages. *Hydrology and Earth System Sciences*, *19*(7), 3073–3091. https://doi.org/10.5194/hess-19-3073-2015

Kummu, M., Guillaume, J. H., de Moel, H., Eisner, S., Flörke, M., Porkka, M., …, & Ward, P. J. (2016). The world's road to water scarcity: Shortage and stress in the 20th century and pathways towards sustainability. *Scientific Reports*, *6*(1), 1–16. https://doi.org/10.1038/srep38495

Lèye, B., Zouré, C. O., Yonaba, R., & Karambiri, H. (2021). Water resources in the Sahel and adaptation of agriculture to climate change: Burkina Faso. In S. Diop, P. Scheren, & A. Niang (Eds.), *Climate change and water resources in Africa* (pp. 309–331). Springer. https://doi.org/10.1007/978-3-030-61225-2_14

Malek, Ž., & Verburg, P. H. (2018). Adaptation of land management in the Mediterranean under scenarios of irrigation water use and availability. *Mitigation and Adaptation Strategies for Global Change*, *23*, 821–837. https://doi.org/10.1007/s11027-017-9761-0

Mbaye, A. A. (2020). Climate change, livelihoods, and conflict in the Sahel. *Georgetown Journal of International Affairs*, *21*(1), 12–20. https://doi.org/10.1353/gia.2020.0020

Mbaye, A. A., & Signé, L. (2022). *Climate change, development, and conflict-fragility nexus in the Sahel*. Brookings.

Meier, J., Zabel, F., & Mauser, W. (2018). A global approach to estimate irrigated areas–a comparison between different data and statistics. *Hydrology and Earth System Sciences*, *22*(2), 1119–1133. https://doi.org/10.5194/hess-22-1119-2018

Mirzabaev, A., Stringer, L. C., Benjaminsen, T. A., Gonzalez, P., Harris, R., Jafari, M., Stevens, N., Tirado, C. M., & Zakieldeen, S. (2022). Cross-chapter paper 3: Deserts, semiarid areas and desertification. In H.-O. Pörtner, D. C. Roberts, M. Tignor, E. S. Poloczanska, K. Mintenbeck, A. Alegría, M. Craig, S. Langsdorf, S. Löschke, V. Möller, A. Okem, & B. Rama (Eds.), *Climate change 2022: Impacts, adaptation and vulnerability. Contribution of working group II to the sixth assessment report of the intergovernmental panel on climate change* (pp. 2195–2231). Cambridge University Press. https://doi.org/10.1017/9781009325844.020

Nji, A., & Fonteh, M. F. (2002). Water harvesting: Its potential in the greening and poverty reduction of northern Cameroon. *Journal of the Cameroon Academy of Sciences*, *2*, 33–48.

Nunn, N., & McGuirk, E. (2021). How climate shocks trigger inter-group conflicts: Evidence from Africa's transhumant pastoralists. Retrieved December 15 2023, from https://www.nber.org/system/files/working_papers/w28243/w28243.pdf

Ortiz-Bobea, A., Ault, T. R., Carrillo, C. M., Chambers, R. G., & Lobell, D. B. (2021). Anthropogenic climate change has slowed global agricultural productivity growth. *Nature Climate Change*, *11*(4), 306–312. https://doi.org/10.1038/s41558-021-01000-1

Overland, I., Fossum Sagbakken, H., Isataeva, A., Kolodzinskaia, G., Simpson, N. P., Trisos, C., & Vakulchuk, R. (2022). Funding flows for climate change research on Africa: Where do they come from and where do they go?. *Climate and Development*, *14*(8), 705–724. https://doi.org/10.1080/17565529.2021.1976609

Partey, S. T., Zougmoré, R. B., Ouédraogo, M., & Campbell, B. M. (2018). Developing climate-smart agriculture to face climate variability in West Africa: Challenges and lessons learnt. *Journal of Cleaner Production*, *187*, 285–295. https://doi.org/10.1016/j.jclepro.2018.03.199

Pulido-Bosch, A., Rigol-Sanchez, J. P., Vallejos, A., Andreu, J. M., Ceron, J. C., Molina-Sanchez, L., & Sola, F. (2018). Impacts of agricultural irrigation on groundwater salinity. *Environmental Earth Sciences*, *77*, 1–14. https://doi.org/10.1007/s12665-018-7386-6

Rittelmeyer, P., Caretta, M. A., Dowler, C., Vora, S., Seigerman, C. K., Reddy, E. U. B., NR, L., Parajuli, J., Srinivasan, J., Priya, R., & Mukherji, A. (2024). Adaptation to water-induced disaster: Exploring local knowledge and Indigenous knowledge-led strategies. *AlterNative: An International Journal of Indigenous Peoples*, *20*(1), 96–108. https://doi.org/10.1177/11771801241235332

Rosa, L., Chiarelli, D. D., Sangiorgio, M., Beltran-Peña, A. A., Rulli, M. C., D'Odorico, P., & Fung, I. (2020). Potential for sustainable irrigation expansion in a 3°C warmer climate. *Proceedings of the National Academy of Sciences*, *117*(47), 29526–29534.

Seignobos, C. (2000). Sorghos et civilisations agraires. In C. Seignobos, & O. Iyébi-Mandjek (Eds.), *Atlas de la province Extreme-Nord Cameroun* (pp. 82–87). IRD.
Serdeczny, O., Adams, S., Baarsch, F., Coumou, D., Robinson, A., Hare, W., Schaeffer, M., Perrette, M., & Reinhardt, J. (2017). Climate change impacts in Sub-Saharan Africa: From physical changes to their social repercussions. *Regional Environmental Change, 17*, 1585–1600. https://doi.org/10.1007/s10113-015-0910-2
Siebert, S., Ewert, F., Rezaei, E. E., Kage, H., & Graß, R. (2014). Impact of heat stress on crop yield—On the importance of considering canopy temperature. *Environmental Research Letters*, *9*(4), 044012. https://doi.org/10.1088/1748-9326/9/4/044012
Stringer, L. C., Mirzabaev, A., Benjaminsen, T. A., Harris, R. M., Jafari, M., Lissner, T. K., Stevens, N., & Tirado-von Der Pahlen, C. (2021). Climate change impacts on water security in global drylands. *One Earth*, *4*(6), 851–864. https://doi.org/10.1016/j.oneear.2021.05.010
Sultan, B., Defrance, D., & Iizumi, T. (2019). Evidence of crop production losses in West Africa due to historical global warming in two crop models. *Scientific Reports*, *9*(1), Article 12834. https://doi.org/10.1038/s41598-019-49167-0
Tesfaye, B. (2022). *Climate change and conflict in the Sahel* (Discussion Paper Series on Managing Global Disorder No. 11). Council on Foreign Relations.
Thiery, W., Visser, A. J., Fischer, E. M., Hauser, M., Hirsch, A. L., Lawrence, D. M., …, & Seneviratne, S. I. (2020). Warming of hot extremes alleviated by expanding irrigation. *Nature Communications*, *11*(1), 290. https://doi.org/10.1038/s41467-019-14075-4
Traore, K., Sidibe, D. K., Coulibaly, H., & Bayala, J. (2017). Optimizing yield of improved varieties of millet and sorghum under highly variable rainfall conditions using contour ridges in Cinzana, Mali. *Agriculture & Food Security*, *6*, 1–13.
Trisos, C. H., Adelekan, I. O., Totin, E., Ayanlade, A., Efitre, J., Gemeda, A., Kalaba, K., Lennard, C., Masao, C., Mgaya, Y., Ngaruiya, G., Olago, D., Simpson, N. P., & Zakieldeen, S. (2022). Africa. In H.-O. Pörtner, D. C. Roberts, M. Tignor, E. S. Poloczanska, K. Mintenbeck, A. Alegría, M. Craig, S. Langsdorf, S. Löschke, V. Möller, A. Okem, & B. Rama (Eds.), *Climate change 2022: Impacts, adaptation and vulnerability. Contribution of working group II to the sixth assessment report of the intergovernmental panel on climate change* (pp. 1285–1455). Cambridge University Press. https://doi.org/10.1017/9781009325844.011
Turner, S. W., Hejazi, M., Calvin, K., Kyle, P., & Kim, S. (2019). A pathway of global food supply adaptation in a world with increasingly constrained groundwater. *Science of the Total Environment*, *673*, 165–176. https://doi.org/10.1016/j.scitotenv.2019.04.070
UN-Water (2013). *Water security & the Global Water Agenda*. The UN-Water Analytical Brief 53.
Williams, P. A., Simpson, N. P., Totin, E., North, M. A., & Trisos, C. H. (2021). Feasibility assessment of climate change adaptation options across Africa: An evidence-based review. *Environmental Research Letters*, *16*(7), 073004. https://doi.org/10.1088/1748-9326/ac092d
Zaveri, E., & Lobell, B. (2019). The role of irrigation in changing wheat yields and heat sensitivity in India. *Nature Communications*, *10*(1), 4144. https://doi.org/10.1038/s41467-019-12183-9
Zorom, M., Barbier, B., Mertz, O., & Servat, E. (2013). Diversification and adaptation strategies to climate variability: A farm typology for the Sahel. *Agricultural Systems*, *116*, 7–15. https://doi.org/10.1016/j.agsy.2012.11.004.
Zougmoré, R., Segnon, A. C., & Thornton, P. (2023). Harnessing Indigenous knowledge and practices for effective adaptation in the Sahel. *Current Opinion in Environmental Sustainability*, *65*, Article 101389. https://doi.org/10.1016/j.cosust.2023.101389

4 The "Glocal" nature of Jihadist insurgencies?

Conflict, space, land and resources in the Sahel[1]

Edoardo Baldaro

Introduction

For many international security experts and practitioners, over the past decade, the Sahel[2] has become one of the epicentres of the so-called Global Jihad. According to estimations, in 2023, the region recorded more than 11,000 militant Islamist-linked fatalities (Africa Center for Strategic Studies, 2024). At the time of writing—March 2024—large portions of the territories of Mali, Niger and Burkina Faso are more or less effectively under the control of insurgent groups who have declared their allegiance to al-Qaeda or to the Islamic State and are implementing veritable systems of "jihadi governance" (Bøås & Strazzari, 2020). The growing insecurity, coupled with the inability of Sahelian states and Western interveners, such as France and the European Union (EU), to contain and counter the spread of these insurgencies, has produced a detrimental impact at the political level as well. On one hand, between 2020 and 2023, militaries seized power in all of the three countries mentioned above, justifying their coup d'états with the need to reestablish security and state control over their entire national territories. On the other hand, mounting tensions with previous international security partners and the UN have led to unprecedented geopolitical competition. More specifically, starting in 2021, Mali, followed by Niger and Burkina Faso, have loosened or even broken their diplomatic and military ties with France, the US and other Western actors. Military juntas have started looking for alternative political and security partnerships, a situation that has been intercepted and exploited, especially by Russia, which has strongly increased its military and diplomatic presence in the area over the past few years. This change in international alliances has not produced any improvement in the stabilisation of the Sahel or in the protection of local populations up to now (Baldaro & Strazzari, 2023). On the contrary, a large body of evidence suggests that jihadist activities are expanding towards West African coastal states, with an increasing number of attacks recorded in the northern regions of Cote d'Ivoire, Togo, Benin and Ghana.

Under the influence of the "Global War on Terror" security and policy framework, many observers have explained the emergence and the development of these jihadist insurgencies in the Sahel by employing the analytical lenses of the so-called

DOI: 10.4324/9781003429111-6

ungoverned spaces. According to this approach, the Sahelian states' weakness and incapacity to assert their presence in the peripheral areas of their territories, coupled with enduring conditions of extreme poverty and limited socio-economic opportunities, have created a sort of political and governance void, ready to be filled by non-state actors of a different nature (Dowd & Raleigh, 2013). In the case of the Sahel, a region that was previously stable and peaceful would have consequently been turned into a conflict zone by the obscure and exogenous forces of the Global Jihad. In the academic debate, this is the interpretation supported by the "globalists", namely those who consider the different jihadist insurgencies around the globe as expressions of a unique and coordinated transnational Islamic fight (Kilcullen, 2005) and who also perceive jihadist groups in the Sahel as a direct filiation of a global Islamic insurrection whose origins are located in the Middle East (Zenn, 2017).

On the other extreme of this debate are the "localists" who view the affiliation to al-Qaeda or Daesh by Sahelian insurgent groups mainly as a performative posture thought to reinforce their symbolic legitimacy, while their agendas and motivations remain embedded in the local contexts and political struggles (Thurston, 2020). According to this analysis, violence and conflict in the Sahel have little in common with a presumed global Islamist fight, being rooted in local dynamics and grievances where religion only plays an ancillary and instrumental role. The followers of this approach have advanced different explanations for the current condition of violence in the Sahel. Some have focused on the ecological conditions and the tensions generated within the—mainly agrarian and pastoralist—system of production in the area to explain the concurrent and violent mobilisation of conflicting communities (Benjaminsen & Ba, 2019). Others have identified the abuses committed by local regimes, usually against specific communities and ethnic groups, as the main causes of conflict (Raineri, 2022). Further works have framed the insurgencies in the Sahel as an expression of a widening urban–rural political and economic cleavage (Van Dijk & de Bruijn, 2022) or again as a sort of "youth uprising" against a social order perceived as immutable and favouring older generations and leaders (United Nations Development Programme [UNDP], 2017).

As some scholars have started to suggest, these varying interpretations should not be regarded as contrasting one another since they identify different layers of explanation within a complex and entangled crisis (van Dijk & de Bruijn, 2022). In a similar way, the case of the Sahel suggests that even the "global vs. local" opposition could be nuanced, considering how the jihadist groups have changed and evolved over time, in a dialectical relation to both the international imaginary of the Global Jihad and the multilayered dynamics of local conflicts (Bencherif, 2021).

Building on these considerations, in this chapter, I discuss the onset and development of the jihadist presence in the Sahel, in order to offer some insights about how jihadist fighters have shaped and worsened the security crisis in the area, as well as how they have turned spaces, lands and resources into tools of contention. Started as an external penetration of foreign actors coming from Algeria and loosely linked to al-Qaeda, jihadism in the Sahel has evolved over time: the sahelisation process has transformed both the membership and the agenda of the insurgent groups. Accordingly, my analysis is organised by distinguishing among three different phases and focusing on slightly different socio-geographical and ecological contexts. In the "Beginning of jihadist insurgency in the Sahel" section, I focus on the period from the initial establishment of the jihadist groups in the transborder area between Eastern Mauritania and Northern Mali (early 2000s) to the 2012 crisis in Mali, corresponding to the decade during which al-Qaeda in the Islamic Maghreb (AQIM) became embedded in the Sahelian political and conflict dynamics. In this phase, the essential characteristics of the Sahelian context at the beginning of the crisis are depicted. In the "2012 Rebellion in Northern Mali and its aftermath" section, I discuss the rebellion that took place in Northern Mali in 2012–2013, in which jihadist fighters played a fundamental role, and its aftermath. In fact, it was at this event, for the first time, that the jihadist groups intercepted and appropriated the grievances and, to a certain extent, the political agenda of (part of) the populations of Northern Mali and particularly the Tuareg community. This takeover has been translated into an embryonic form of jihadi governance, which has eventually mixed *sharia*-based rules with approaches to the management of property and mobility, more in line with the customs and interests of their local constituencies. "Expansion beyond Northern Mali: Katiba Macina in the Inner Niger Delta" section moves the focus to the Inner Niger Delta in Central Mali: this is the area where since 2015 jihadist territorial control is probably more developed. In this latter case, the regulation of the access to and use of land has become a central feature of jihadist rulers, who are fully embedded in one of the more fertile—and conflictual—areas of the Sahel.

Beginning of jihadist insurgency in the Sahel

The first Salafist armed group that established its presence between Northern Mali and Eastern Mauritania during the early 2000s was almost exclusively formed by Algerian fighters who left their country of origin during the ending phase of the Algerian civil war (1991–2002). When the defeat in the civil war started to appear inevitable, in fact, the leaders of the Group Salafiste pour la Prédication et le Combat (Salafist Group for Preaching and Combat [GSPC]) decided to "move south" and to reorganise the bases of the organisation outside Algeria (Harmon, 2016).

In the northern band of the Sahel, particularly in Northern Mali, the members of the GSPC found a favourable environment. This part of the Sahel mostly extends over a semi-arid and arid desert area highly affected by climatic variability, where mobility and circulation represent the main assets for securing a livelihood and where a huge smuggling economy connected with Algeria had developed since the mid-1970s (McDougall & Scheele, 2012). Mainly inhabited by mobile and semi-nomadic communities that based their economic survival and social structures on pastoralism and transboundary trade, at the beginning of the 2000s Northern Mali was a place quite abandoned by the country's central government in terms of both public services and military presence. Indeed, aiming at concentrating the state's (limited) resources in the South, while trying to contain Northern (and Tuareg in particular) nationalism and independentism, the regime in Bamako had opted for limiting its presence in those lands. It rather favoured the emergence of an unstable and fragmented system of governance built around various traditional leaders, economic and/or armed entrepreneurs and criminal actors (Baldaro, 2018). This fluid and unsettled socio-political and economic context finally offered the jihadists the ideal conditions for securing their presence and their influence.

The first episode that attested to the establishment of the GSPC in Northern Mali occurred in February 2003, when the group kidnapped 32 European tourists in Southern Algeria and held them hostage for around six months in Northern Mali. Ransom was finally paid, and the hostages were freed in August of the same year (International Crisis Group [ICG], 2012). Starting from that time, while the GSPC exponentially augmented its activities in Mali and Mauritania, it also began to pursue a sort of hybrid strategy, aimed at reinforcing its connection with al-Qaeda's galaxy while simultaneously strengthening its anchoring in the local context. In fact, in 2007, the GSPC made its allegiance to al-Qaeda public, with the group rebranding its name to AQIM. While this move implied a change in the group's tactics—that introduced kamikaze attacks and strongly escalated its offensive against local security forces—AQIM also became more embedded in its new context, creating strong linkages with local communities and criminal entrepreneurs over time.

The transformation of AQIM into a locally grounded armed actor took place in two main dimensions. On one hand, to guarantee the group's economic survival and to accumulate the necessary capital for developing their activities, jihadists created a veritable "kidnapping industry" across Mali, Mauritania and Algeria, mostly targeting Western citizens and interests. After the hostage crisis of 2003, which probably allowed the group to collect around US$5 million from the German government, it has been calculated that between 2003 and 2012, AQIM gained up to US$150 million from ransoms (Nünlist, 2013). Part of this money was redistributed among the local population or used for buying the loyalty of local leaders or corrupting members of the security forces, Malian in particular (Chauzal & Van Damme, 2015). During this same period, jihadist fighters also built contingent forms of collaboration—mainly based on the protection of the convoys—with regional criminal entrepreneurs involved in drug and arms trafficking (Lacher, 2012). These initiatives allowed AQIM to become an influential actor in the Northern Mali socio-economic landscape since its activities in Mauritania were highly reduced over time. New allegiances with local communities were established by marrying women, mostly of Tuareg and Arab origins, while the group also started to recruit fighters among the local youth (Harmon, 2016). Due to these initiatives, at the beginning of the 2010s, AQIM had become one of the most prominent political armed actors in the area.

2012 Rebellion in Northern Mali and its aftermath

The rebellion in Northern Mali started in January 2012 and is commonly considered the veritable turning point of the security crisis in the Sahel. From that point in time, what began as an independentist rebellion led by Tuareg fighters, now expanded from Mali to its neighbours, Niger and Burkina Faso, turning into a constellation of autonomous but interconnected insurgencies, having jihadist groups linked to al-Qaeda or the Islamic State at their centre.

The 2012 uprising was the fourth Tuareg insurrection in the history of Mali. Indeed, post-independence, Mali has identified in the "Tuareg question" one of the main challenges to its state-building process since its inception. When defining the institutional architecture of the new Malian state, the government of Modibo Keità (1960–1968) implemented a policy of administrative centralisation, cultural homogenisation and forced sedentarisation, which exacerbated Tuareg's grievances and refusal to be integrated into the Malian institutional system (Baldaro & Raineri, 2020). The relations between Bamako and its North rapidly evolved towards violence, as demonstrated by the outbreak of the first Tuareg rebellion in 1963. A further event that reinforced Tuareg opposition vis-à-vis Mali's central authorities emerged from the catastrophic droughts that hit the Sahel in the 1970s and the 1980s. While these extreme climatic events almost destroyed the pastoral economy of Northern Mali and caused the deaths or departure of hundreds of thousands of Tuareg, the military regime of Moussa Traoré purposely mismanaged humanitarian aid to weaken the northern communities (Lecocq, 2010).

These different episodes definitively crystallised the political dynamic and narrative, which described Tuareg as a threat to the nation in Bamako, while presenting the Malian central authorities as abusive and predatory in the eyes of the Tuareg. At the same time, the successive rebellions in the 1990s and the mid-2000s, during which violent intra-Tuareg clashes were recorded, have also shown how northern communities are not unitarian actors but are rather shaped by class, clan, and ethnic struggles and personal divisions that have been exploited by the jihadist groups as well. During the 2000s, jihadist actors proved to be particularly effective in intercepting the political grievances, economic interests and power struggles among the Tuareg—and, to a lesser extent, the Arab—communities of Northern Mali. Consequently, when the Tuareg fighters of the National Movement for the Liberation of the Azawad (MNLA as its French acronym) started preparing for the 2012 insurgency, three different jihadist formations supported their move. In the months before the beginning of the insurrection, in fact, two groups other than AQIM had made their appearance in the area. On one hand, the Movement for Oneness and Jihad in West Africa (MOJWA) was born as AQIM's splinter group, formed by fighters who have had a strong rapprochement with Northern Mali's Arab communities. On the other hand, a Tuareg-jihadist group called Ansar Dine (Defenders of the Faith) was also created. Founded by Iyad ag Ghali, a former rebel and leader of one of the most important clans of the Tuareg confederation,

Ansar Dine encapsulated a further step in the process of sahelisation of the jihadists fighters in the area, as the group was driven by a political project that was simultaneously Islamist and ethno-nationalist (Lebovich, 2019).

After an initial collaboration that allowed the Tuareg-jihadist alliance to succeed and conquer the quasi-totality of Northern Mali, in June 2012, jihadist formations expelled MNLA from the main cities it occupied during the previous months and started implementing their first experiment in jihadist governance. In Kidal, Gao and Timbuktu, local "governors" of AQIM, Ansar Dine and MOJWA defined an initial set of juridical norms and governing rules in accordance with their own literal interpretation of the *sharia* law. Notwithstanding some internal oppositions, the jihadists did not hesitate in implementing the most violent aspects of (their version of) *sharia*, such as prohibiting music, requiring women to be fully veiled, establishing Islamic courts and a police force, or employing corporal punishments of "criminals". They also showed a certain capacity to adapt their system of Islamic governance to the specific characteristics and presumed needs of local populations and occupied territories.

In a certain sense, the structure of the spaces and borders pursued by jihadist rulers represented an attempt to create a viable synthesis between two hardly compatible projects of social and territorial organisation. On one hand, concepts such as nomadism, connectivity, statelessness and borderless spaces captured the socio-economic arrangements that historically defined power and livelihood in Northern Mali (Raineri & Strazzari, 2015). Cross-border exchanges and transactions have been the sources of cultural identities and economic interests across Saharan borderlands, where categories such as land as private property would have hardly planted their seeds. On the other hand, the creation of territorial states strongly affected the organisation of local societies, reshaping intercommunity allegiances and economic functions (McDougall & Scheele, 2012). Within this complex framework, jihadist rulers tried to propose their own vision of a nomadic state.

They eliminated custom fees and informal trespassing taxations—particularly on pasturelands—while starting to create an embryonic form of a welfare state. These initiatives were justified as actions taken in accordance with Islamic law, but they also appeared as ways to gain support among economically dynamic but hierarchically subordinate communities, particularly traders, traffickers, smugglers and pastoralists (Raineri & Strazzari, 2015). Hence, while explicitly rejecting the system codified by the Western state, jihadist rulers contributed to redefining the relation between authority and space and forging a new social pact based on the re-elaboration of traditional practices, Islamic-inspired norms and modern tools of governance.

The French military operation *Serval* halted the insurgents' advance towards Bamako and reconquered most of the territory of Northern Mali during the first half of 2013. Even if the failure of the 2012–2013 rebellion represented a severe setback for jihadists in the Sahel, these groups were not completely defeated, and in successive years, they applied the strategy of "submersion" in rural areas in order to reorganise their ranks and capacities. This approach produced two main consequences. On one hand, the attempt to exploit the tribalised armed politics (Bencherif & Campana, 2017) generated by (national and international) counterinsurgency policies led to the creation of different jihadist sub-groups located in specific geographical areas and/ or ethnic constituencies, as had been the cases of the Katiba Macina in Central Mali or Ansarul Islam in Northern Burkina Faso. On the other hand, the emergence of an "alternative" to al-Qaeda in the Islamist galaxy led to the creation of the Islamic State in the Greater Sahara (ISGS, meantime rebranded as the Islamic State Sahel Province or IS Sahel) in 2015. In response to this move, the different groups linked to al-Qaeda merged in 2017, forming Jama'at Nasr al-Islam wal Muslimin (JNIM).

Nevertheless, the abuses committed by Sahelian states and national security forces in the name of the fight against terrorism have been key to the survival and success of the jihadist groups after 2013 (Thurston, 2020). In fact, in the years that followed the rebellion in Northern Mali, Sahelian political elites and institutions opted for criminalising civilian dissent, while presenting in the public discourse the most marginalised and unsettled communities such as the Tuareg and the Fulani as sympathetic with the jihadist fight. With this approach, the governments of Mali, Niger and Burkina Faso finally allowed the jihadist insurgents to present themselves as protectors and liberators of the abused populations and to turn their fundamentalist claims into radical instruments of contestation against abusive states (Cissé, 2020). This is also one of the main factors explaining the expansion of the security crisis from Northern Mali to the rest of the Sahel.

Expansion beyond Northern Mali: Katiba Macina in the Inner Niger Delta

Most of the observers consider 2015 as the second turning point of the security crisis in the Sahel (Van Dijk & de Bruijn, 2022). It is at that time when Amadou Kouffa, a Fulani Islamist preacher and leader of the jihadist group Katiba Macina, first established himself in the Inner Niger Delta in Central Mali. Other than representing the first successful move beyond Northern Mali—accomplished by jihadist insurgents—the arrival of Katiba Macina in the area was also instrumental in mobilising (part of) the Fulani community against Sahelian regimes, a key element for understanding the spread of the insurgency in Niger, Burkina Faso and beyond (Rupesinghe & Bøås, 2019).

At least two different factors created a fertile ground for this armed insurgency to take hold in the Inner Niger Delta (Cissé, 2020). First, Katiba Macina took advantage of the animosity of some Fulani and their search for protection vis-à-vis the Malian state. Fulani mostly comprises a pastoral and semi-nomadic community present in most of West Africa and the Sahelian belt. While historically marginalised from Mali's political and economic centres of power, which are mainly occupied by southern elites and ethnic groups such as the Bambara, the relations between this community and the post-independent state did not turn violent for decades. Things started to change during the 1990s, when Fulani first experienced armed abuses from both Tuareg rebels and state security forces, which worsened after the National Pact was signed in 1992, believed to address and resolve the root causes of the Tuareg insurgency. In the eyes of the Fulani, while the North (formally) obtained political autonomy, economic investments and impunity for the leaders of the revolt, the Centre did not receive any compensation or recognition by central authorities, something that strengthened the Fulani's sense of injustice and will to take revenge vis-à-vis the Malian state (ICG, 2016). The re-explosion of the conflict during the Malian civil war in 2012–2013 and the violence that followed (Fulani civilians and chiefs were targeted [again] by rebel groups and the Malian army) pushed members of the Fulani community to take up arms and subsequently recognise in Katiba Macina potential protectors against the abusers of the state.

The second element to be taken into account concerns the specific socio-ecological and economic conditions of the Inner Niger Delta and how they shape conflict dynamics. Due to the seasonal floods of the Niger River, the Inner Niger Delta is one of the most fertile areas of the Sahel, but it is also one of the most affected by climate change and reduced precipitation. This space is principally inhabited by Fulani herders and Dogon farmers.[3] Given the specific ecological context of the Inner Niger Delta, inter- and intra-ethnic relations of power and rule tend to overlap with socio-economic factors and the roles occupied in the local production chain. For this same reason, most of the disputes observed in the area are about the exploitation and use of land and water resources, as well as access to them. Along this line, many observers have highlighted how violence in the area should be primarily viewed as an agro-pastoral conflict opposing (Fulani) herders to (Dogon) farmers, who compete for control and rule the land (ICG, 2016). In the Fulani community, there are also profound differences—and tensions—between those who own lands and herds and those who have to pay for access to pastures. Indeed, Fulani society is not monolithic; extremely hierarchical and organised in castes, the customary rules and traditions that govern intra-community and caste relations are not uncontested. While tensions among herders and farmers and between upper and lower Fulani castes already existed in the past, worsening climatic conditions and the recent demographic growth in the area have also had an impact. Coupled with a (perceived) peak in the level of corruption and the clientele behaviours of those in charge of the system of land management and conflict resolution—both the state's representatives and the customary (upper caste) leaders—all have negatively affected and furthered the use of violence in the Inner Niger Delta (Benjaminsen & Ba, 2019, 2021).

Feeding on these existing tensions, the appearance of the Katiba Macina played a fundamental role in driving the escalation of violence in the area. It favoured a vicious cycle of inter-ethnic violence, which ultimately strengthened the jihadist group. During the initial phase, Katiba Macina penetrated the Inner Niger Delta with a strategy based principally on violence and fear. The group attacked mayors, judges, customary leaders and anyone suspected of collaborating with its opponents. Then, from 2016 onwards and (partly) as a reaction to these attacks, various ethnic self-defence militias, organised principally by armed individuals from the Dogon and—to a lesser extent—Bambara communities, appeared in the area, usually with the tacit support of the central government. The widespread violence and various episodes of ethnic cleansing finally defined two main areas of influence, with the Inner Niger Delta under the control of Katiba Macina, and the provinces of Bandiagara, Bankass, Koro and Douentza mostly occupied by Dogon militias (Baldaro & Diall, 2020).

As Katiba Macina began to secure its position of military power, it also refined its governing approach with the objective of obtaining support, through co-optation and/or association, from at least part of the local population. Trying to adapt to and exploit the opportunities offered by the context, Katiba Macina opted for not completely subverting the customary rules applied in the area. On one hand, the group effectively imposed *sharia*, forcing women to veil themselves, banning traditional celebrations or imposing the Islamic judges as the only authorities who can rule over disputes in the area (Raineri, 2018). On the other hand, the group did not call into question the customary system governing land management. Following an administrative scheme that was elaborated under the Macina Empire in the 19th century but whose functioning up to the present time was fixed under the French colonisation and reaffirmed during the initial years of postcolonial Mali, part of the grazing lands of the Inner Niger Delta belongs to the Djowros, the statutory Fulani upper class, who regularly collect tributes from other herders. In a similar vein, public lands are ruled by a sort of right of precedence and exploitation possessed by the Djowros. In this regard, Katiba Macina decided to limit the tributes to be paid to the Djowros but did not contest the basic functioning of the system (Benjaminsen & Ba, 2019), thus leaving intact the intra-Fulani tensions about access to pastures. Moreover, in their further attempt to win the support of the ruling classes in the area under their control, Katiba Macina eliminated all state taxes and customs, substituting them with *zakat* (effectively a tribute paid to fund the jihadist effort), and built a limited welfare system.

The governing system put into place by Katiba Macina along with their choice to work with the Fulani ruling classes and traditional authorities have also been at the centre of the conflict that exploded between JNIM/Katiba Macina and ISGS, starting from early 2020. Once again, the conflict between JNIM and ISGS/IS Sahel could be framed as a competition whose roots remain in the dynamics of the Global Jihad as much as it is driven by local struggles and land disputes. Reproducing the conflict that characterises the divisions between al-Qaeda and the Islamic State at the highest level, ISGS accused Katiba Macina of blasphemy because of its presumed failure to enforce Islamic law. At the same time, Katiba Macina relied on and reinforced the Fulani ruling classes, particularly the Djowros. In the name of a "purer" interpretation of the Quran, the ISGS tried to mobilise the Fulani lower classes by putting at the centre of its agenda the complete collectivisation of land, particularly pastures. For this reason, most of the combatants who have joined the Islamic State in Mali and beyond come from the Fulani lower classes who practise pastoralism and have to pay for access to grazing lands during the annual transhumance (Baldaro & Diall, 2020).

As an indirect consequence of the sahelisation of the insurgent jihadist groups, the fight for control over the Inner Niger Delta has become a struggle to impose a specific system of governance deeply anchored in the local dynamics of power and socio-ecological conditions. This turn of events also explains the relative success of ISGS in obtaining the support of part of the population, particularly the Fulani subaltern classes. These cleavages are generating self-reinforcing mechanisms of violent contention, which now threaten to expand to other communities and ethnic groups.

Conclusion

Twenty years of expansion of the jihadist activities in the Sahel have produced a deep impact on the region. While the nature, agendas and reach of the jihadist groups have evolved over time, their struggle has transformed the socio-political and security landscape of the region, which is now considered one of the most unstable and conflict-prone areas of the African continent. As I have attempted to show in this chapter, the jihadist phenomenon in the Sahel should be treated neither as an external penetration nor solely as a local struggle; rather, it represents one of those "glocal" insurgencies that are now fuelling political grievances and conflicts in different parts of the world. As a consequence, space and land disputes, as well as struggles with access to and exploitation of resources, have been captured within the broader dynamics of conflict and contention, which have eventually created the conditions for expanding and deepening the security crisis in the Sahel region. Given these conditions, the current events in the Sahel—particularly the different coup d'états in Mali, Niger and Burkina Faso—are negatively impacting the already unstable regional context, potentially opening new avenues for manoeuvre by armed entrepreneurs and violent insurgents, who are already emerging as the real "winners" of the security crisis in the Sahel.

Notes

1 Research for this work was supported by Fonds National De La Recherche Scientifique—FNRS—"Postdoctoral Researcher" grant n. 40000345.
2 When dealing with security issues and jihadist insurgencies, international observers and policymakers, as well as many scholars focusing on security studies, tend to delimit the "Sahel" to the territories of five states, namely Mauritania, Mali, Niger, Burkina Faso and Chad. Unless indicated otherwise, this same territorial definition is adopted in this chapter.
3 These categories should not be regarded as fixed or reified since both Dogon and Fulani societies are more complex than can be expressed in this contribution.

References

Africa Center for Strategic Studies. (2024, January 29). Deaths linked to militant Islamist violence in Africa continue to spiral. *Infographic*.

Baldaro, E. (2018). A dangerous method: How Mali lost control of the north and learned to stop worrying. *Small Wars and Insurgencies*, *29*(3), 579–603. https://doi.org/10.1080/09592318.2018.1455323

Baldaro, E., & Diall, Y. S. (2020). The end of the Sahelian exception: Al-Qaeda and Islamic State clash in central Mali. *The International Spectator*, *55*(4), 69–83. https://doi.org/10.1080/03932729.2020.1833566

Baldaro, E., & Raineri, L. (2020). Azawad: A parastate between nomads and Mujahidins? *Nationalities Papers*, *48*(1), 100–115. https://doi.org/10.1017/nps.2019.62

Baldaro, E., & Strazzari, F. (2023). Strategic misalignment: European security and P/CVE engagement in the Sahel. *Mediterranean Politics*, 1–11. https://doi.org/10.1080/13629395.2023.2289795

Bencherif, A. (2021). Unpacking "glocal" jihad: From the birth to the "sahelisation" of Al-Qaeda in the Islamic Maghreb. *Critical Studies on Terrorism*, *14*(3), 335–353. https://doi.org/10.1080/17539153.2021.1958171

Bencherif, A., & Campana, A. (2017). Alliances of convenience: Assessing the dynamics of the Malian insurgency. *Mediterranean Politics*, *22*(1), 115–134.

Benjaminsen, T. A., & Ba, B. (2019). Why do pastoralists in Mali join jihadist groups? A political ecological explanation. *The Journal of Peasant Studies*, *46*(1), 1–20. https://doi.org/10.1080/03066150.2018.1474457

Benjaminsen, T. A., & Ba, B. (2021). Fulani-Dogon killings in Mali: Farmer-herder conflicts as insurgency and counterinsurgency. *African Security*, *14*(1), 4–26. https://doi.org/10.1080/19392206.2021.1925035

Bøås, M., & Strazzari, F. (2020). Governance, fragility and insurgency in the Sahel: A hybrid political order in the making. *The International Spectator*, *55*(4), 1–17. https://doi.org/10.1080/03932729.2020.1835324

Chauzal, G., & Van Damme, T. (2015). *The roots of Mali's conflict: Moving beyond the 2012 crisis*. Conflict Research Unit Report. The Clingendael Institute.

Cissé, M. G. (2020). *Understanding Fulani perspectives on the Sahel crisis*. Africa Center for Strategic Studies.

Dowd, C., & Raleigh, C. (2013). The myth of global Islamic terrorism and local conflict in Mali and the Sahel. *African Affairs*, *112*(448), 498–509. https://doi.org/10.1093/afraf/adt039

Harmon, S. A. (2016). *Terror and insurgency in the Sahara-Sahel region: Corruption, contraband, jihad and the Mali war of 2012–2013*. Routledge.

International Crisis Group (ICG). (2012). *Mali: Avoiding escalation*. Africa Report n. 109.

International Crisis Group(ICG). (2016). *Central Mali: An uprising in the making*? Africa Report n. 238.

Kilcullen, D. J. (2005). Countering global insurgency. *Journal of Strategic Studies*, *28*(4), 597–617. https://doi.org/10.1080/01402390500300956

Lacher, W. (2012). *Organized crime and conflict in the Sahel-Sahara region*. The Carnegie Papers – Middle East. Carnegie Endowment for International Peace.

Lebovich, A. (2019). *Mapping armed groups in Mali and the Sahel*. Report. European Council on Foreign Relations.

Lecocq, B. (2010). *Disputed desert: Decolonization, competing nationalisms and Tuareg rebellions in Mali*. Brill.

McDougall, J., & Scheele, J. (Eds.). (2012). *Saharan frontiers: Space and mobility in Northwest Africa*. Indiana University Press.

Nünlist, C. (2013). *Kidnapping for ransom as a source of terrorism funding*. CSS Analysis in Security Policy n. 141. Center for Security Studies (CSS).

Raineri, L. (2018). *If victims become perpetrators: Factors contributing to vulnerability and resilience to violent extremism in the Central Sahel*. International Alert Report. International Alert.

Raineri, L. (2022). Explaining the rise of jihadism in Africa: The crucial case of the Islamic State of the Greater Sahara. *Terrorism and Political Violence*, *34*(8), 1632–1646. https://doi.org/10.1080/09546553.2020.1828078

Raineri, L., & Strazzari, F. (2015). State, secession, and jihad: The micropolitical economy of conflict in Northern Mali. *African Security*, *8*(4), 249–271. https://doi.org/10.1080/19392206.2015.1100501

Rupesinghe, N., & Bøås, M. (2019). *Local drivers of violent extremism in Central Mali*. Norwegian Institute of International Affairs.

Thurston, A. (2020). *Jihadists of North Africa and the Sahel: Local politics and rebel groups*. Cambridge University Press.

United Nations Development Programme (UNDP). (2017). *Journey to extremism in Africa: Drivers, incentives and the tipping point for recruitment*. United Nations Development Programme Regional Bureau for Africa.

van Dijk, H., & de Bruijn, M. (2022). Religious movements in the drylands: Ethnicity, jihadism, and violent extremism. In A. Kronenburg Garcia, T. Haller, H. van Dijk, C. Samimi, & J. Warner (Eds.), *Drylands facing change* (pp. 155–173). Routledge.

Zenn, J. (2017). Demystifying al-Qaida in Nigeria: Cases from Boko Haram's founding, launch of jihad and suicide bombings. *Perspectives on Terrorism*, *11*(6), 173–189.

Part II

Cartographic Sketches of the Sahel

5 Defining the Sahel

B-ordering an area study[1]

Andrea Pase, Federico Gianoli, Ludovica Crocitto, Luca De Felice, Michael Cherlet, Marina Bertoncin, Angela Kronenburg García, Mariasole Pepa and Carla Braga

Introduction

We chose the word "Sahel" as the title of our chapter because it specifically identifies our study area. It is the main protagonist. This choice raises two critical questions. The first concerns the meaning of "area studies" and their implicit assumptions. The second concerns the question of whether it makes sense to identify an area study as "the Sahel". In this first section of the chapter, we address these two dimensions, which are in many ways radical. After that, we delve into specifics, looking for the geographic limits of the Sahel, those proposed in the literature and by practitioners, as well as those that we can try to identify and map. Drawing a limit performs a double function: it delimits, that is, it defines a context, whether of a political or social analysis and yet, at the same time, proposes an ordering, that is, defining what makes sense to identify, what counts and what does not. Suggestive from this point of view is the idea of the "b/ordering space", proposed by van Hotum et al. (2005), which precisely juxtaposes the border in its operational effectiveness and the ordering of the social, as the construction of contexts of meaning. This applies to us as it does to all those who have proposed limits for the Sahel. In delimiting an area in a map what is done is not only to construct a framework that allows us to aggregate and analyse information but also to propose a certain reading of that space as having a social meaning. One is therefore faced with something far from trivial: the boundary of an area study not only defines a container of data and thus allows one to operate on it but also contributes to creating an identity that takes on a symbolic value. This is the awareness required to consciously approach the delimitation of a field of study in a critical way.

Did you say "area studies"?

In recent years, a debate has involved the so-called area studies (Sidaway, 2013; Sidaway et al., 2016); a heated discussion that highlights their epistemological nature, leaving room for indeterminacy or even ambiguity. First, area studies are very contextualised research — one area or another in its specificity, in its unique, particular historical path. Inevitably, therefore, they incorporate a high degree of difference, which makes these studies quite different from context to context. They

DOI: 10.4324/9781003429111-8

also have a hybrid status. On one hand, different disciplines converge in the same spatial context to acquire an alleged "whole" scientific understanding: area studies are therefore composed of different layers of knowledge, obtained using very diverse methods. On the other hand, they are frequently linked to the necessity on the part of certain actors to equip themselves with interpretation tools to make policy choices and drive practical interventions. It is no coincidence that in American universities, area studies have found a privileged context of development (Sidaway, 2013). The identification and management of policies by a superpower interested and involved in all global chessboards inevitably requires the ability to produce knowledge and interpretive frameworks that are expected to be effective. In fact, area studies are supposed to orient the strategic choices and tactical decisions in US international politics. In the same way, inevitably, the other major players involved in the global geopolitical games are called upon to produce visions of the world that are articulated on a regional scale; this is valid, for example, for China as well as for the European Union (Pichon, 2020). Area studies have faced many criticisms for their colonial origins and the ever-present risk of exoticism (Orientalism, according to Said, 1978). They are also criticised because they often implicitly reproduce those academic power dynamics that are always placed at the centre of the Western and even more Anglo-American research centres (Sidaway, 2013; Sidaway et al., 2016). However, area studies have significant potential because a considerable degree of creativity can be brought into play. They could provide a platform for constructing innovative knowledge, one that overturns taken-for-granted approaches, imagining different frameworks through multiple reconfigurations of space, not anchored to claims of permanent identities but oriented towards dynamism and hybridisation. In this sense, we can consider Scott's (2009) proposal, as much cited as criticised, of a "new" region called Zomia, a mountainous area of anarchic resistance in the heart of the Indochinese mountains, in contrast to the empires of the plains. Identifying, naming and delimiting one region rather than another is, therefore, always both a political affair and a scientific one, whether politicians or scientists lean towards the acceptance of inherited regions or aim at the invention of new regional configurations (or new representations of such spaces, as is the case in this chapter), to go beyond self-evident concepts and to escape the various "territorial traps" (Agnew, 1994). A theoretical-methodological deepening of area studies can only pass through direct experience, a confrontation with a specific area, which is what we have started to do.

Does the Sahel exist?

We now arrive at the second question. Does it make sense to identify an area as "the Sahel"? A comprehensive answer exceeds the scope of this chapter, but it does not mean that the relevant issues can be evaded, although the reflections proposed will inevitably be partial. A French geographer, D. Retaillé (2018), challenges the very possibility of identifying the Sahel (with the definite article and an initial capital letter). He studied the transformation of meaning that has affected the Arabic word *sahel*, whose original meaning was *shore* or *edge* (Retaillé, 2018; see also Agnew & Chappell, 1999). In the African context, the term was initially used to indicate areas adjoining the Sahara Desert, both to the North (e.g., the Algerian or Tunisian Sahels) and to the South. The geographers and, more generally, the French savants in the colonial era were the ones who carried out a significant semantic transposition from many *sahels* to "the Sahel", that is, a sub-Saharan region delimited or at least definable with scientific presuppositions (particularly climatic-botanical). This conceptual operation is anything but neutral; in Retaillé's view, it has been useful in organising the knowledge necessary to identify and then build the cornerstones of colonial territorial control. Starting from Retaillé's observations, the French anthropologist J.P. Amselle moves in the same direction when writing about the "colonial invention of the Sahel" (2022, pp. 13–20). During colonialism, the Sahel became the place of confrontation between "sedentary farmers" and "nomadic herders", entities who are generalised and hypostatised through the attribution of these different characteristics to well-separated ethnic groups. The colonisers can rely on sedentary farmers, more easily integrated into the projects for the exploitation of Sahelian resources and more easily controllable; in contrast, the desert, the pre-desert steppes and the nomads are dangerous and tend to be hostile. This definition of the Sahel has solidified over time, entering maps, atlases and geographical descriptions and eventually reactivated, especially in emergency contexts, such as the great Sahelian droughts of the 1970s and the 1980s, or the current phase of political instability and explosion of terrorist and repressive violence. In short, the identification of the Sahel appears always *functional to* colonial dominance, the management of humanitarian emergencies and the present political-security situations. Retaillé and Walther (2019) instead highlight the value of the movement (both spatial and social) between the desert and the savannas, emphasising the fluidity in the use of resources (herders who sometimes cultivate, farmers who rear livestock, sedentary people who move their locations, nomadic elites who are actually sedentary in the oases, etc.) and the exaggerated weight attributed to "ethnic frontiers". The region (sahel) should rather be read as a hinge, not a delimited area; it is a point of connection, a crossroad or a constellation of places, itself tied to other hinges, both to the North and the South. The Sahel should encompass the Sahara rather than be separated from it; it should be understood as a mobile space, not anchored, delimited, defined or, perhaps worse, a *definitive* space. According to Retaillé, understanding of the sahel (lowercase) can only be geohistorical, and the attempt to comprehend it through a bioclimatic partition is futile (2018, p. 37). One can absolutely agree with many points of Retaillé's analysis: the extreme mobility that characterises the territorial context, the importance of historical legacies and the connection with the desert. However, there remains one problem: why "a sahel"? This choice (lowercase and indefinite) can be read as a diminution, a downscaling. Let us try to reverse the situation; how would it sound if an African or an American author spoke of "a europe" or "a mediterranean"?

Retaillé (2018, p. 52) also disagrees with the adherence to that cartographic imperative, so keenly felt by geographers, which seeks to produce "a clearly framed and defined fixed image". From his viewpoint, rather than a tool of scientific analysis, cartography is an instrument of control and propaganda. Here as well, one can only agree on the necessary critical scrutiny of the production and use of cartography, especially in a spatial-temporal context where cartographic techniques come from external sources and were initially conveyed by colonial power (Pase, 2011, 2024). However, this does not prevent Retaillé from being then forced to define and delimit the area of analysis; for example, he confines it to West Africa and excludes the Nile axis, returning several times to justify this choice, which inevitably rips important relationships that have moved along the parallels (consider just the routes of pilgrims to Mecca). In fact, this choice substantially limits the analysis to Francophone Africa. In another passage, Retaillé and Walther state that Senegal is "a purely Sahelian country"; thus, even for them, a Sahelian identity exists (2019, p. 18). While adopting Retaillé and Walther's caution in identifying and representing the Sahel (see also OECD/Sahel and West Africa Club, 2014), we deem it necessary to still reason on its denomination and delimitation, although choosing the path proposed (among others) by Braudel (1995) for history and by Dematteis (1985, 2021) for geography — multiply the metaphors to describe the world. For Braudel, there is not one but many Mediterraneans, not one Germany but at least three and likewise in the case of the Atlantics. Our choice follows this direction: not one Sahel but many Sahels, giving plurality to the denomination and delimitation, which is still needed to organise discourses, propose interpretations and formulate operational hypotheses. After all, the term *sahel* itself originated as a metaphor. Accepting the challenge of mobility in space and transformation in time means engaging in data processing and cartography drawing, attempting to thus enhance the capabilities of interpretation and representation.

Towards a new cartographic approach to the Sahel

Identifying a new way of mapping the Sahel through the increasing potentialities derived from remote sensing and the availability of big data is a complex challenge from both theoretical and methodological standpoints. At the theoretical level, there is a significant junction concerning the incessant change of Sahelian spaces, starting from a physical-geographical matrix (primarily climatic-botanical), and the flexible forms of resources used by Sahelian societies. However, this mobility is intercepted by the rigidity imposed by political-administrative divisions and modern infrastructure in the territory (e.g., consider the dams and the regulation of floods in the Sahelian rivers; see Chapter 7). How can cartography record this mobility and take into account the current (perhaps better, presumed) constraints on it? The second problem concerns regional limits; it is necessary to identify, despite mobility, a sensible cut-out that allows indicating what is inside and what is outside the Sahel. This operation is inevitably delicate, politically we might say, in the sense that it constructs an identity (which is always built from the definition of limits, as Remotti (1996) reminds us); therefore, we must be lucidly aware of the meanings it conveys. Nonetheless, these limits are somehow drawn by someone and become operational limits, containers of projects. Moreover, without a definition of limits, it is impossible to reason with data to establish correlations and propose comparisons. Limits are needed; in the first place, it is not easy to identify the better-motivated and potentially effective ones. However, after having decided the locations of the borders, not anchoring to them is even more challenging, with full awareness of the *limits of limits* — their irreducible temporariness and relativity — thus avoiding the traps of an essentialist vision.

At the methodological level, two factors emerge as the most significant, further highlighting the complexity of this area of study. The first is the abundance of data, derived as much from remote sensing as from databases, to some extent georeferenced. Such is the profusion of information available today (for our area of interest, we refer to the Africa Platform proposed by the European Union) that the selection of what is relevant is neither a minor nor an innocent operation; it cannot be done automatically. The choice of data and analysis methodologies allows or prevents illuminating factual dimensions and territorial and sociopolitical tensions. The second relevant factor is that the macro-view from above and the quantitative perspective of big data, to be effectively useful and well-calibrated, need fieldwork or a micro-view from below, qualitative and adhering to specificities, which allows verification, on one hand, and an expansion of the meanings of the information itself, on the other hand. With satellites and fieldwork, computing capacity and notebooks, if the challenge of connecting such distant dimensions of research is won, a stereoscopic view can perhaps be obtained, capable of providing three-dimensionality, conferring corporeality to territorial analyses. However, if this step is mistaken, there is a risk of strabismus that does not enable focusing on the problems; therefore, rather than increasing sharpness, it produces blurred images.

While attempting a new cartographic approach to the Sahel is certainly a complex task, it is also highly significant, in our opinion, from both scientific and operational perspectives. The scientific importance of this operation of territorial analysis and cartographic representation lies in the opportunity offered by the Sahel to reason on the frame of research, geographyically and beyond. There is the issue of the choice of the scale of analysis and representation because, as we know, the scale creates the phenomenon. Then again, we have the crucial issue of territorial traps (Agnew, 1994, 2010, 2015), that is, the pre-constituted contexts of analysis that are derived either from political dimensions (the "container" of states) or from traditions of space partitioning, such as the very idea of a continent or of different regional areas. The Sahel precisely offers an opportunity because we have in front of us a cut-out consolidated in scientific literature, with internal political partitions, but at the same time, it is an imprecise cut-out, or better, it has many different definitions. As mentioned, it is also in constant movement. The Sahelian states present a particular path, inherited by colonialism. In many situations, they are in such a crisis as to question qualifying them as "failed" states when they are not even "collapsed" (Rotberg, 2003). Instead, the operational value lies in building informational and analytical capital useful for interpreting or at least for diversifying and multiplying the representations of one of the most fragile, disconnected and difficult areas of the world. This region, precisely because of its multifactorial crises, requires further cognitive effort to penetrate the opacity that currently characterises its image.

A cognitive challenge of such significance requires multiple contributions of skills and experiences, inevitably leading to the convergence of diverse institutional research contexts. In this case, they comprise a university research group, linked to the 30-year tradition of fieldwork in the Sahel by the geographers of the University of Padua (Bertoncin & Faggi, 2006; Bertoncin & Pase, 2012, 2017; Bertoncin et al., 1995, 2019), and several units linked to the Joint Research Centre (JRC) of the European Commission, particularly the team of the World Atlas of Desertification (Cherlet et al., 2018) and the Africa Knowledge Platform.

We continue by discussing the different limits of the Sahel in scientific literature, outlining what we believe to be an innovative proposal. The databases, the methodological choices and the processing procedures followed by the research group are then identified. Finally, the main results obtained are presented, particularly the delimitation of the Sahel as an area of study; the assessment of the movement of the Sahelian region over time, through the analysis of rainfall; and the critical discussion of the political partitions that segment the regional context.

Between nature and politics: Climatic limits and political borders

The name *sahel*, meaning edge or shore, suggests a spatial placement, with the great desert as its primary reference. If this original definition appears unbalanced because it indicates a northern limit yet remains indeterminate towards the South, at the same time it already also introduces an essential theme: the edge of the shore is not a sharp limit but rather a zonal one and is subject to movement, like the tides and marine currents along the coast. Where then is the Sahel — or rather, how has it been identified by scholars who have studied the area?

The cartographies available in scientific literature, as well as in popular science, in the grey literature of international organisations and on the internet, trace the boundaries of the Sahel in very different ways, offering multiple interpretations of this area squeezed between the Sahara and the humid Sudanese climate zone. Many maps prioritise the political dimension based on the borders of the Sahelian states, out of the need to provide operational delimitations and to facilitate the identification of interlocutors for discussing the implementation of intervention plans and aid by organisations, such as the United Nations, the World Bank and the FAO. However, this approach may expand the Sahelian zone to the Gulf of Guinea or to the heart of the Sahara, or to include the entire Ethiopian Highlands or to exclude areas that are definitely Sahelian because the state where they belong is not listed among the Sahelian ones (in the map in Fig. 5.1, the case of the Far North of Cameroon is an example). This path is evidently limited but reveals a point of great interest — the importance of the political dimension in defining area studies (Sidaway, 2013; Sidaway et al., 2016) and, in our specific case, the Sahel.

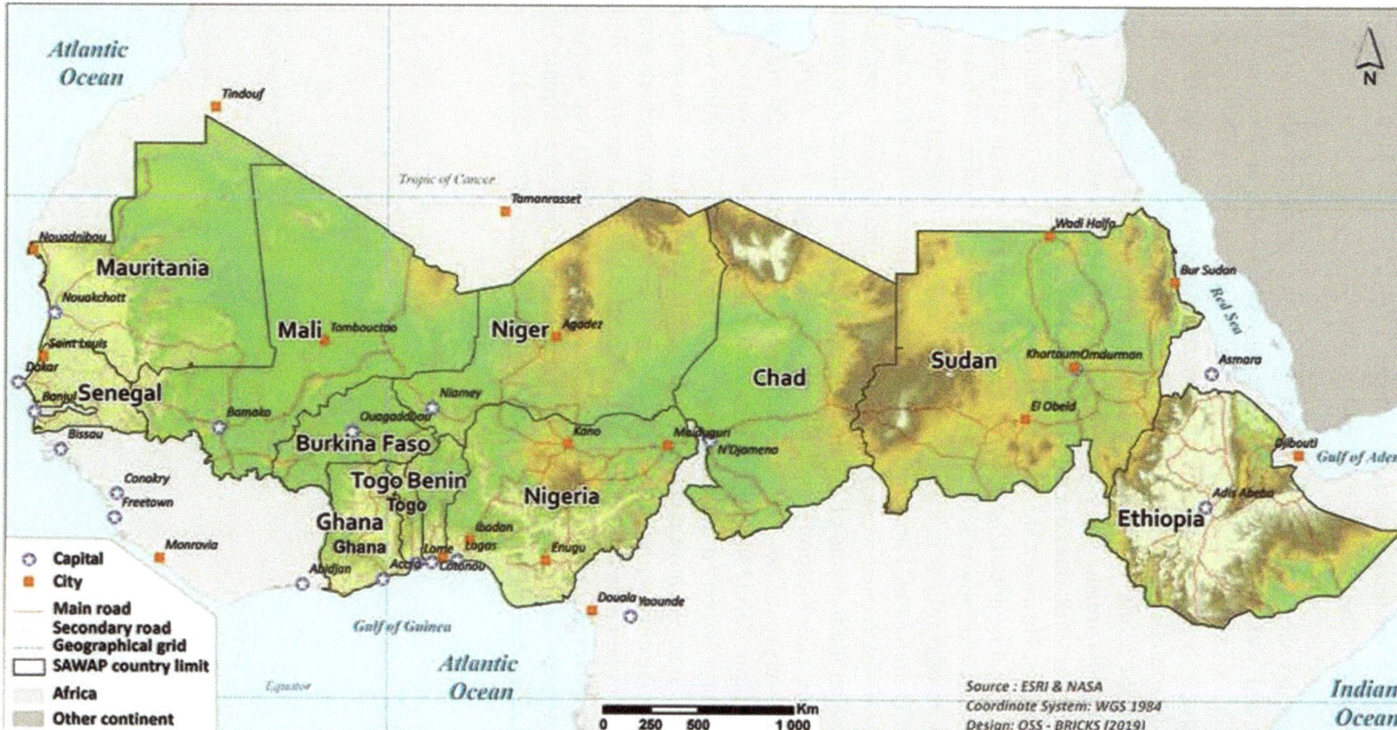

Figure 5.1 Political boundaries of the Sahel (Sahara and Sahel Observatory [OSS], 2019).

The political delimitation cannot be expelled; however, dealing with it through the acceptance of the territorial traps (Agnew, 1994, 2010, 2015) of state spaces is limiting if not misleading. This issue will be revisited.

Another method used to define the Sahel is the criterion of latitude; for example, Sivakumar and Wallace (1991) suggest the latitudes 10–15° N (based in growing season isolines of 60 and 150 days), while Nicholson and Palao (1993) propose the latitudes 10–20° N (for a detailed discussion, see: Agnew & Chappel, 1999). Using this method certainly highlights a characteristic of the Sahel, its extension along the parallels, but the result is inevitably too general and static for an area characterised by undeniable dynamism (as mentioned). The difference between the two cited proposals is quite stark: one covers twice the area of the other, encompassing a large part of the desert. However, the most commonly used reference to define the Sahel is pluviometric; it is indeed the alternation of dry and wet seasons and the gradient of decreasing rainfall towards the North that defines the Sahelian field. Among the many authors who follow this approach, Agnew (1982), Davy et al. (1976), Giri (1983) and Hulme (1992) are cited in this chapter. Specifically (from the earliest to the latest of these cited works), Davy et al. (1976) identify the "morphological Sahel" between the isohyets of 100 and 700 mm/year (the authors also summarise other previous classifications in Tables II.1a and 1b, pp. 7–8). Agnew (1982) chooses the isohyets of 200 mm/year for the northern limit and 700 mm/year for the southern limit. Giri (1983) identifies the Sahel between the isohyets of 150 mm/year in the North and 650 mm/year in the South. Hulme (1992) uses the isohyets of 100–600 mm/year. We have created the map in Fig. 5.2 by taking into account the definitions of the cited authors, both those who have chosen the parallels and those who have identified pluviometric limits. In the latter case, the thresholds identified by the authors were maintained, but the course of the isohyet was recalculated on the same series of data that we used in our proposal (i.e., the Climate Hazards Group InfraRed Precipitation with Station [CHIRPS] data, 1981–2023).

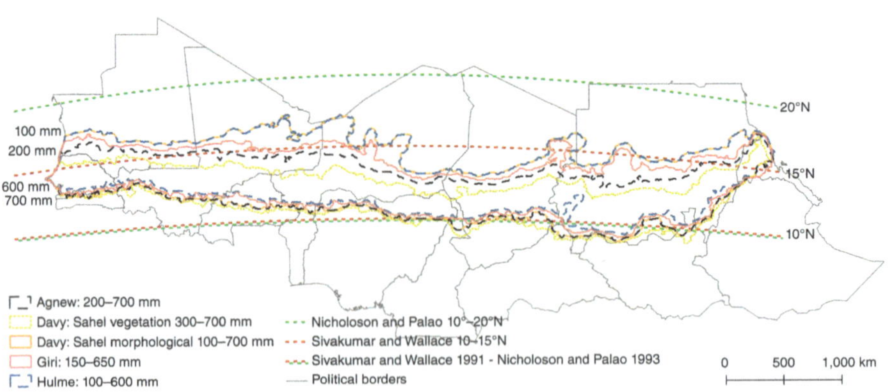

Figure 5.2 Borders of the Sahel.

Further subdivisions of the Sahelian area are then identified in two studies. Davy et al. (1976) describe a "Vegetational Sahel" (300–700 mm/year). Giri (1983) distinguishes a "Nomadic Sahel" (150–400 mm/year) and a "Sedentary Sahel" (400–650 mm/year) and then introduces a transitional zone towards the truly Sudanese climate (650–850 mm/year). This need to differentiate zones clearly indicates that both Davy et al. and Giri recognise the importance of distinction within the Sahelian belt, an issue that was already reported in their cited literature. Based on these various definitions, we have sought to identify a delineation of the Sahel that not only takes these studies into account but also takes a further step towards capturing the intrinsic mobility of the Sahelian zone. Retaillé (2018, pp. 50–51) contests the use of isohyets drawn from annual precipitation averages because they "flatten" and simplify zonal and temporal differentiations. Indeed, relying solely on averages over a 30-year period or longer, and only on a regional scale, plunges us into a night where "all cats are grey". It becomes necessary to distinguish between dry and wet periods to account for the continuous spatial and temporal mobility of rains. However, aside from the pluviometric limits, political limits must also be considered from the very beginning because these mark the Sahel, too. We cannot simply take them for granted as such; they need careful consideration as well.

From the intersection of these first *two Sahels* (the third, the hydrological one, is added in Chapter 7), a cartographic representation starts to emerge, perhaps capable of better capturing the complexity of this area study. In the following section, we present the sources and methods we used.

Sahel on the move: Sources and methods

The pluviometric Sahel

To define the first Sahel, the pluviometric one, it is evidently necessary to start from climatic data and, more precisely, from those related to rains. We remember that the rainy season is linked to the longitudinal movements of the Intertropical Convergence Zone: the precipitation period roughly begins in June and ends in September, peaking between the end of July and August, and progressively becomes shorter towards the North (De Planhol & Rognon, 1970; Fensholt et al., 2013). For the identification of thresholds, we chose to rely on the already mentioned proposal by the geographer Giri (1983), in its most extensive version, therefore including the transitional zone towards the Sudanese climates. Thus, identifying the isohyets of 150 and 850 mm of annual rainfall as useful for defining an expanded Sahel, we updated the representation, obtaining the pluviometric averages using the CHIRPS data. The CHIRPS dataset is in a raster format with a spatial resolution of 0.05°, approximately 5 km/pixel at the equator. This dataset consists of a historical series of daily, weekly and monthly precipitation from 1981 to the present, at the global level. The main feature of this dataset is that in addition to the use of models, it relies on field observations collected by weather stations for validation and data integration (Funk et al., 2015). For the analysis of the dataset, we used Google Earth Engine (GEE), a cloud-based geospatial analysis platform, which allows the analysis of historical series of satellite images and raster and vector data on a global scale. Using the distributed computing power of Google servers enables efficient processing of geospatial data (Gorelik et al., 2017). Google servers have already loaded terabytes of satellite data (Landsat, Sentinel, Modis, etc.), as well as products (CHIRPS, NDVI, Land Cover, etc.). This formidable tool makes it possible to analyse and represent historical data series in a very short time, thus obtaining information that was unthinkable to process just a few years ago. GEE makes remote sensing accessible to everyone without needing special hardware systems, but users should have basic knowledge of programming languages, such as JavaScript or Python, necessary for writing the commands that GEE will then process through the network nodes.

For the definition of the isohyets, the code is available in the repository (https://github.com/fgianoli/PRIN_UNIPD). The code is written in JavaScript, a scripting language commonly used to define actions on webpages. In this case, it is used to send analysis commands to GEE with the definition of data variables and functions that need to be applied. The written code unfolds in four phases: first, the data are defined. Second, the data are filtered in the time period of interest for the analysis. The third phase involves conducting the actual analyses of extracting pluviometric data with the calculation of isohyets. The final step of coding allows exporting the results of the analysis to be opened in a common GIS software, such as QGIS.

To avoid the flattening effect produced by a series over time, we decided to identify the five driest and the five wettest years of the entire data series. With the described code, the following data were calculated:

- Isohyets 150–850 mm/year, 1981–1984 (the driest five-year period of the time series) and
- Isohyets 150–850 mm/year, 2018–2023 (the wettest five-year period of the time series).

For the definition of the pluviometric Sahel limits, the dataset of global watershed basins defined by the World Wildlife Fund for Nature (WWF), precisely level 8, was also used. The WWF watershed basins are products obtained from the NASA Shuttle Radar Topography Mission in 2000 (Lehner, 2008). This dataset is derived from global elevation data at a resolution of 15 arc-seconds, approximately 500 m at the equator, tracking where water flows (if and when it falls, which is not obvious, especially on the desert side of the Sahel). The data have various levels of detail, from level 1, the most generalised, to level 12, the most detailed. Level 8 of aggregation was chosen because it provides a precise scale of analysis in conformity with other data platforms such as the Freshwater Ecosystems Explorer (https://www.sdg661.app). This geospatial platform allows statistics related to watershed basins, by country, in support of

Sustainable Development Goal (SDG) 6.6.1, developed by the United Nations Environment Programme in collaboration with Google and the JRC of the European Commission. The SDGs are the targets that the UN hopes to achieve by 2030. There are 17 SDGs, and each point is articulated and specified in subpoints. In particular, SDG 6 concerns access to sources of clean water for all, while target 6.6.1 aims at monitoring, over time, the spatial extension of surface waters and ecosystems related to them. The isohyets obtained from CHIRPS were thus intersected with the WWF watershed basins to provide a consistent hydrological definition of the North and the South limits of the study area (150–850 mm/year).

The political Sahel

The pluviometric Sahel defines a region, climatic and biogeographic and, in some respects, unitary. It is divided into territorial boundaries, essentially resulting from the scramble for Africa by European colonial powers between the late 19th and the early 20th centuries (Gann & Duignan, 1969; Gavin & Betley, 1973; Griffiths, 1986; Pase, 2011). In this chapter, our intention is to reflect on the different delimitations of the Sahel; for this reason, it is necessary to take state borders into account in order to understand, in a politically fragile context, what is, situation by situation, their actual *strength* (i.e., their real impact on the lives of the populations, their capacity to *separate*, i.e., to mark differences). Not all borders have the same value and operability, as a generic political map (see Fig. 5.1) might suggest. The impacts of the borders on Sahelian Africa vary in space and time, depending on the different states' abilities to exercise control over their borders and the cross-border areas. Smuggling, illegal flows of people and goods (from weapons to drugs) and the entry of terrorist groups are just some of the aspects at play.

The Global Administrative Unit Layer dataset (2015) defined by the FAO was used as a reference for the delineation of the borders, collecting data on administrative boundaries globally, from national to provincial borders.

The logic followed by our analysis first involved conceptually splitting the single boundary line into two, thus considering it as the sum of two different elements, to which it is then possible to attribute the different strengths derived from the situation in the two neighbouring states. If a border is traditionally represented by a line that belongs to both states, in order to depict the borders in a differentiated way, we decided to draw the border with two separate and adjacent lines, one for each state. This methodological choice is fundamental for a more detailed analysis of the different roles and weights of the specific border sections. The analytical structure is based on the initial attribution of a value of 10 to each border section of each country, and then points are added or subtracted, depending on the different conditions listed below. First, each border section is characterised in relation to the distance from its capital, calculating the distances with an SQL script in a PostgreSQL/PostGIS database, reported in the same repository. (https://github.com/fgianoli/PRIN_UNIPD/blob/main/04_distances_from_capital.sql).

Table 5.1 Values attributed to border sections according to distance from the capital city.

Distance	*Value*
0–100 km	*+2*
101–300 km	*0*
301–600 km	*−2*
>600 km	*−4*

The distances considered were 0–100, 101–300, 301–600 and over 600 km (Table 5.1). Proximity to the capital is considered a factor that increases the strength of the border. In contrast, as the distance from the capital increased, especially in the Sahelian context with its poor road infrastructure and difficult connections, we decided to decrease the value of the border sections.

Another spatial dimension deemed appropriate to consider is the presence of two particularly problematic areas that have a specific relation with the borders. These are the intersections of three states, specifically the tripoints Burkina–Niger–Mali and Nigeria–Niger–Chad. A circle with a radius of 150 km was drawn from these points, attributing a value of −1 to the border sections falling within because they are particularly exposed to violent events, often with transboundary origins and dynamics.

The subsequent calculation involves weighing the borders of individual states based on four parameters, utilising the first approximation of the different weights of the Sahelian states: population size (Table 5.2), gross domestic product per capita (Table 5.3), political stability (regime changes/coups/problematic power transitions over the past 20 years: Table 5.4) and episodes of violence (Table 5.5). Threshold values were defined for each of the considered parameters, which were used to attribute an overall differentiated score to the borders of each Sahelian state. The identification of threshold values depends on the evaluation, entrusted to the skills and experiences of the working group. To quantify the weights of the different border sections, the methodology used is fundamentally qualitative.

Table 5.2 Strength as a function of population size.

Class	*Weight*
<200 million inhabitants (Nigeria)	*+4*
>25 million inhabitants	*+2*
10–25 million inhabitants	*+1*
<10 million inhabitants	*0*

Source: https://data.worldbank.org/indicator/SP.POP.TOTL; accessed July 1, 2023

Table 5.3 Strength as a function of GDP per capita (in US dollars, IMF [2023] data, World Economic Outlook, accessed November 7, 2023).

Class	*Weight*
>1500	+3
1001–1500	+2
500–1000	+1
<500	0

For the population, whose size is significant for many reasons (e.g., in determining the size of the internal market and the attractiveness of demographic concentrations (metropolitan areas)), we used data from the World Bank (2022; https://data.worldbank.org/indicator/SP.POP.TOTL; accessed July 1, 2023). The states were divided into three classes, plus a specific situation for Nigeria, the demographic giant of the area with its 218 million inhabitants (Table 5.2).

Data from the International Monetary Fund (IMF, 2023; https://www.imf.org/external/datamapper/NGDPDPC@WEO/OEMDC/ADVEC/WEOWORLD) were used for weighing the differences of states based on the GDP per capita. Four classes were identified (Table 5.3). Notably, most Sahelian countries are found in the lower positions of the global ranking.

Another indicator identifies regime changes/coups d'état or otherwise problematic political transitions, based on an analysis of political events over the past 20 years

Table 5.4 Weakness as a function of number of regime changes/political traumas, 2003–2023.

Class	*Weight*
None	0
1	−1
>1	−2

Table 5.5 Weakness as a function of the number of violent episodes recorded in the last five years, November 2018–October 2023 (ACLED).

Class	*Weight*
0–100 episodes	*0*
101–1000 episodes	*−1*
1001–2000 episodes	*−2*
>2000	*−3*

Table 5.6 Values attributed by the Fragile State Index (FFP [2023], accessed November 7, 2023).

Class	*Weight*
Warning	*0*
Elevated warning	*−1*
High warning	*−2*
Alert	*−3*
High alert	*−4*

in each country within the study area (Table 5.4). These transitions highlight the instability of the different political contexts.

The last element considered is the number of episodes of violence (various types: killings, bombings, insurrections, etc.), as reported by Armed Conflict Location & Event Data (ACLED), an NGO that collects data, analyses and activities concerning armed conflicts on a global scale (Table 5.5). The ACLED database compiles (in Africa, since 1997) all available information on conflicts, noting the dates, locations, motivations and groups involved. The sources of this data are local media, government reports and humanitarian agency reports. The data are updated in real time and can be viewed and downloaded directly from the ACLED portal (https://acleddata.com).

We decided it was appropriate to conduct a further investigation of a state's political strength/weakness. We used the Fragile States Index 2023. This is a synthetic index produced by the Fund for Peace (FFP; https://fundforpeace.org/fsi/), which ranks all the states of the world using economic, political, cohesion and social indicators that classify states according to their strength or fragility. The higher the value, the more fragile the state is considered. Sahelian states fall into the Warning and especially the Alert categories used by the FFP; thus, they are all characterised by high or extreme fragility. For our purposes, five different classes are defined (Table 5.6).

The weight thus obtained was compared with the sum of the weights of the two indicators of political instability and violence that we assessed.

Sahel on the move: results

Having clarified the sources and methods followed in our work, we are able to present the cartographic results and the analytical interpretations derived from them.

The pluviometric Sahel

The pluviometric Sahel was identified by setting the northern limit at the isohyet of 150 mm/year from the wettest five-year period and the southern limit at the isohyet

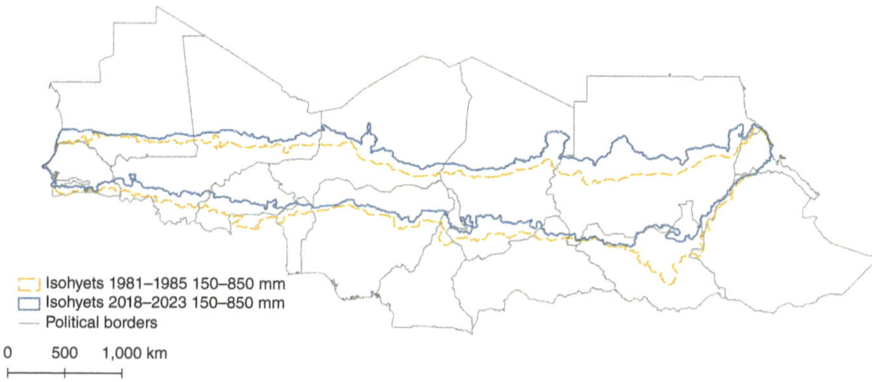

Figure 5.3 Sahel on the move, between the wettest and the driest periods.

of 850 mm/year from the driest five-year period. In this way, starting from Giri's (1983) extended definition (the actual Sahel and the transitional zone towards the Sudanese climate), we have identified the maximum extension reached, under different rainfall conditions, by the Sahel in the period considered (Fig. 5.3). This is the maximum expansion of the Sahelian region in the alternation of wet and dry conditions.

The two isohyets were then reprocessed based on the definition of level 8 of the WWF hydrography, extending the area of interest to include all the basins involved, for that scale of detail (Fig. 5.4). The cut-out thus identified then becomes our study area. From now on, when we refer to the Sahel, we mean this extension, which aims to be inclusive of the different climatic conditions that the area has encountered since the CHIRPS data series began (1981).

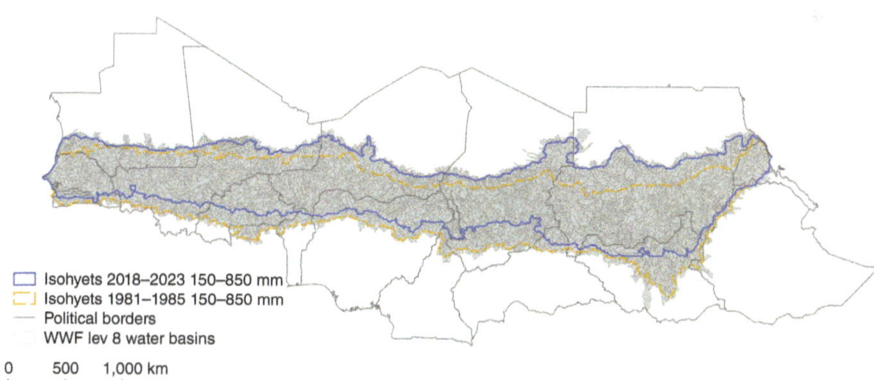

Figure 5.4 Area study, defined to the North by the 150 mm/year isohyet (2018–2023) and to the South by the 850 mm/year isohyet (1981–1985), reported to WWF level 8 hydrographic limits.

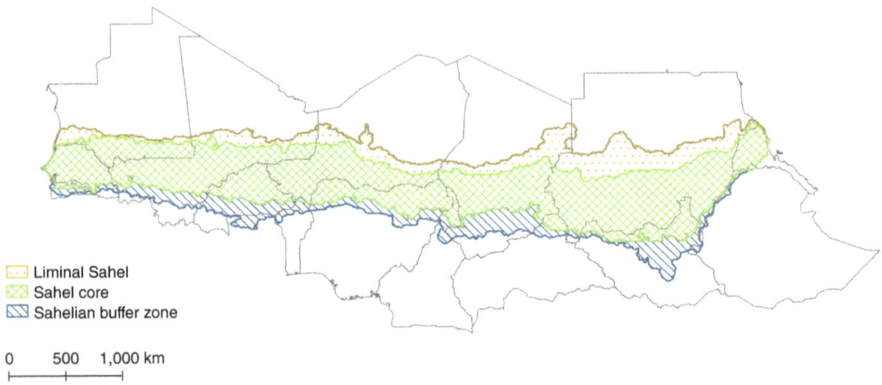

Figure 5.5 The core of the Sahel, the liminal Sahel and the buffer zone.

The map in Fig. 5.5 presents an internal subdivision of the Sahel, based on two isohyets: 150 mm/year during the driest five-year period in the North and 850 mm/year during the wettest five-year period in the South. In this way, we identify three distinct zones: the core of the Sahel (between the two isohyets just described) or the area that is always Sahel, both in drought phases and in wetter phases; the liminal Sahel, between 150 mm/year in drought conditions and 150 mm/year in wet conditions, which is active only during the wetter phases; and finally, the Sahelian buffer zone, between 850 mm/year in wet conditions and 850 mm/year in dry conditions, which is the area where the Sahel recedes during drought conditions and where it has different and more favourable pluviometric conditions in wetter years.

The political Sahel

As noted, it is necessary — however delicate — to take into account the political delimitations that carve out the study area, which are often used to define the Sahel's location. In our analysis, as indicated in the Sources and Methods section, we used two spatial distinctions: the distance from the capital and the proximity to two particularly exposed tripoints. Furthermore, we identified four parameters to attribute different strengths to the borders of different countries (Table 5.7). The values of political instability and violence were added together, with a control index, whose value was then compared with the Fragile States Index. The comparison essentially confirms the attributed relative weight: when a divergence is recorded, if possible, the average of the two values is taken; otherwise, the lesser value is chosen, in consideration of the deeply deteriorated context of the Sahelian countries.

Each border segment is weighted by the sum of the value obtained from Table 5.7 and the values determined by the spatial variables. This formula makes it possible to construct a representation (Fig. 5.6) that is capable of visually rendering the complex situation of the borders in the Sahel, highlighting where they appear weaker or

Table 5.7 Weighting of the relative "strengths" of the different countries involved.

	Population (2022)	*GDP per capita (2023)*	*Political instability*	*Violence*	*Control index*	*Fragile States Index (2021)*	*Weighted index*	*Total weight*	*Index (starting from 10)*
Senegal	+1	+3	0	−1	*−1*	−1	*−1*	*+3*	*13*
Mauritania	0	+3	−2	0	*−2*	−2	*−2*	*+1*	*11*
Mali	+1	+1	−2	−3	*−5*	−3	*−4*	*−2*	*8*
Burkina Faso	+1	+1	−2	−3	*−5*	−3	*−4*	*−2*	*8*
Nigeria	+4	+3	0	−3	*−3*	−3	*−3*	*+4*	*14*
Niger	+2	+1	−2	−3	*−5*	−3	*−4*	*−1*	*9*
Chad	+1	+1	−1	−1	*−2*	−4	*−3*	*−1*	*9*
Cameroon	+2	+3	0	−3	*−3*	−3	*−3*	*+2*	*12*
Sudan	+2	+1	−2	−3	*−5*	−4	*−5*	*−2*	*8*
Eritrea	0	+1	0	0	*0*	−3	*−2*	*−1*	*9*
South Sudan	+1	0	−1	−3	*−4*	−4	*−4*	*−3*	*7*
The Gambia	0	+1	−1	0	*−1*	−1	*−1*	*0*	*10*
Togo	0	+1	−1	−1	*−2*	−2	*−2*	*−1*	*9*
Ghana	+2	+3	0	−1	*−1*	0	*−1*	*+4*	*14*
Benin	+1	+2	0	−1	*−1*	−1	*−1*	*+2*	*12*
Central African Republic	0	+1	−2	−2	*−4*	−4	*−4*	*−3*	*7*

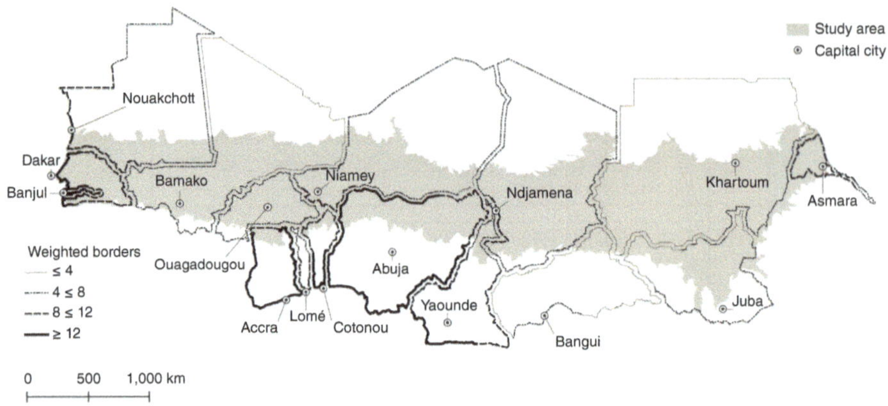

Figure 5.6 Sahel's political, weighted borders.

even in fact irrelevant, and where, on the contrary, they are more consolidated. The difference from a normal political map of the region (see Fig. 5.1) is evident. The role that the weakness of state borders can play is critical. On one hand, it can favour the formation of cross-border poles more or less alternative to the established territorial arrangements (Bennafla, 1999, 2002; Igue, 1995). On the other hand, it can push the spread of centrifugal forces and destabilise processes (Fournet-Guérin & Magrin, 2018; Grégoire, 2019).

Conclusion

At the end of this chapter, we propose a first map that synthesises the *b/ordering* of the Sahel: identification of the core of the pluviometric Sahel and weighting of political borders (Fig. 5.7). This cartographic representation seems to be the

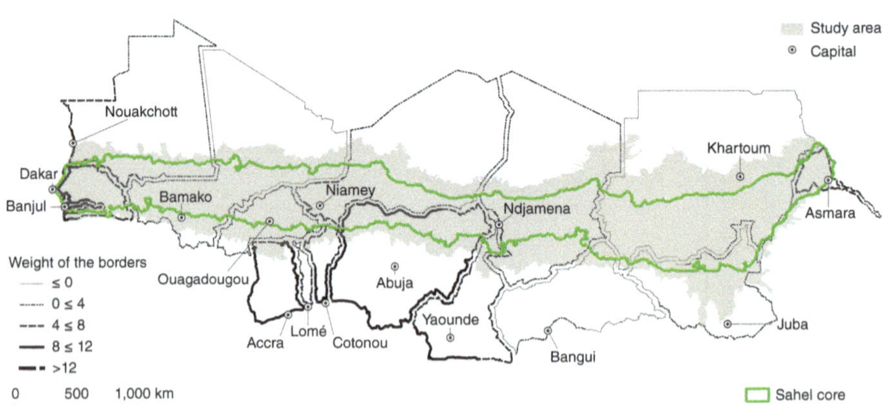

Figure 5.7 A first cartographic synthesis: Pluviometric and political Sahel.

first possible answer to Retaillé's (2018) challenge regarding the substantial "impossibility" of defining the Sahel and thus proposing plural and meaningful delimitations.

The map (Fig. 5.7) collects and intersects two different Sahels, highlighting both the mobility of the area in space and the existence of a Sahelian core and emphasising the changes in the value of the state borders, depending on the political-economic evolution of the different countries. As it has been constructed, the representation can be continuously updated, drawing on information made available over time by remote sensing and the different databases used. We have tried to go beyond the average precipitation and the fixedness of borders. In investigating each of these concepts, we have valued the factor of transformation over time. In this way, we seem to have at least alleviated the "rigidity" that the cartographic imperative imposed on the Sahel, in Retaillé's judgement (2018), without renouncing a definitional effort in both naming and delimitation. While we certainly cannot claim to have answered the question, "What is the Sahel?", at least we can assert to have reasoned around it.

Note

1 An early version of this chapter appeared as an article in *Rivista Geografica Italiana* (Pase et al., 2022). The present chapter partly takes up the article but with significant changes in the structure of the text and with updated data for all tables and maps.

References

Agnew, C. T. (1982). Water availability and development of rainfed agriculture in S.W. Niger. *Transactions of the Institute of British Geographers*, *7*(4), 419–457.

Agnew, J. (1994). The territorial trap: The geographical assumptions of international relations theory. *Review of International Political Economy*, *1*, 53–80. https://doi.org/10.1080/09692299408434268

Agnew, J. (2010). Still trapped in territory? *Geopolitics*, *15*(4), 779–784. https://doi.org/10.1080/14650041003717558

Agnew, J. (2015). Revisiting the territorial trap. *Nordia Geographical Publications*, *44*(4), 43–48. Retrieved June 3, 2024, from https://nordia.journal.fi/article/view/64824

Agnew, C. T., & Chappell, A. (1999). Drought in the Sahel. *GeoJournal*, *48*, 299–311. https://doi.org/10.1023/A:1007059403077

Alcaro, R., & Pirozzi, N. (Eds.). (2014). *Transatlantic security from the Sahel to the Horn of Africa*. Edizioni Nuova Cultura. Istituto Affari Internazionali.

Amselle, J.-L. (2022). *L'invention du Sahel*. Croquant.

Bennafla, K. (1999). La fin des territoires nationaux? État et commerce frontalier en Afrique centrale. *Politique Africaine*, *73*, 24–49.

Bennafla, K. (2002). *Le commerce frontalier en Afrique centrale: Acteurs, espaces, pratiques*. Khartala.

Bertoncin, M., Bicciato, F., Bonollo, L., Croce, D., Faggi, P., Mariani, L., Minoia, P., & Pase, A. (1995). Irrigazione, stato e territorio in Sudan: Il gioco della posta in gioco. *Terra d'Africa* 8–40.

Bertoncin, M., & Faggi, P. (Eds.). (2006). *Cosa resta nel piatto? Fallimenti e promesse dell'agricoltura irrigua nella valle del Senegal*. Torino L'Harmattan.

Bertoncin, M., & Pase, A. (2012). *Autour du Lac Tchad: Enjeux, tensions et conflits pour le contrôle de l'eau*. Harmattan.

Bertoncin, M., & Pase, A. (2017). Interpreting mega-development projects as territorial traps: The case of irrigation schemes on the shores of Lake Chad (State of Borno, Nigeria). *Geographica Helvetica*, *72*, 243–254. https://doi.org/10.5194/gh-72-243-2017

Bertoncin, M., Pase, A., Quatrida, D., & Turrini, S. (2019). At the junction between state, nature and capital: Irrigation mega-projects in Sudan. *Geoforum*, *106*, 24–37. https://doi.org/10.1016/j.geoforum.2019.07.002

Braudel, F. (1995). *The Mediterranean and the Mediterranean world in the age of Philip II* (Vol. I). University of California Press.

Cherlet, M., Hutchinson, C., Reynolds, J., Hill, J., Sommer, S., & von Maltitz, G. (Eds.). (2018). *World atlas of desertification*. Publication Office of the European Union.

Davy, E. G., Mattei, F., & Solomon, S. I. (1976). *An evaluation of the climate and water resources for development of agriculture in the Sudan-Sahelian zone of West Africa*. WMO Special Environment Report No. 9. Retrieved June 3, 2024, from https://library.wmo.int/doc_num.php?explnum_id=8268

Dematteis, G. (1985). *Le metafore della Terra. La geografia umana tra mito e scienza*. Feltrinelli.

Dematteis, G. (2021). *La geografia come immaginazione. Tra piacere della scoperta e ricerca di futuri possibili*. Donzelli.

De Planhol, X., & Rognon, P. (1970). *Les zones tropicales arides et subtropicales*. Colin.

Fensholt, R., Rasmussen, K., Kaspersen, P., Huber, S., Horion, S., & Swinnen, E. (2013). Assessing land degradation/recovery in the African Sahel from long-term Earth observation-based primary productivity and precipitation relationships. *Remote Sensing*, *5*(2), 664–686. https://doi.org/10.3390/rs5020664

Fournet-Guérin, C., & Magrin, G. (2018). L'Afrique, du Sahel et du Sahara à la Méditerranée: Intégrations, circulations et fragmentations. *Bulletin de l'Association de Géographes Français, Géographies*, *95*(2), 165–174.

Funk, C., Peterson, P., Landsfeld, M., Pedreros, D., Verdin, J., Shukla, S., Husak, G., Rowland, J., Harrison, L., Hoell, A., & Michaelsenet, J. (2015). The climate hazards infrared precipitation with stations – A new environmental record for monitoring extremes. *Scientific Data*, *2*, 150066. https://doi.org/10.1038/sdata.2015.66

Gann, L. H., & Duignan, P. (1969). Reflections on imperialism and the scramble for Africa. In L. H. Gann & P. Duignan (Eds.), *Colonialism in Africa 1870–1960* (Vol. 1, pp. 100–129). Cambridge University Press.

Gavin, R. J., & Betley, J. A. (1973). *The scramble for Africa*. Ibadan University Press.

Giri, J. (1983). *Le Sahel demain catastrophe ou renaissance?* Karthala.

Gorelik, N., Hancher, M., Dixon, M., Ilyushchenko, S., Thau, D., & Moore, R. (2017). Google Earth Engine: Planetary-scale geospatial analysis for everyone. *Remote Sensing of Environment*, *202*, 18–27. https://doi.org/10.1016/j.rse.2017.06.031

Grégoire, E. (2019). Le Sahel et le Sahara entre crises et résilience. *Hérodote*, *172*, 5–22.

Griffiths, I. (1986). The scramble for Africa: Inherited political boundaries. *The Geographical Journal*, *152*(2), 204–216.

Hulme, M. (1992). Rainfall changes in Africa. *International Journal of Climatology*, *12*(7), 685–699. https://doi.org/10.1002/joc.3370120703

Igue, J. O. (1995). *Le territoire et l'État en Afrique. Les dimensions spatiales du développement*. Karthala.

Lehner, B., Verdin, K., & Jarvis, A. (2008). New global hydrography derived from spaceborne elevation data. *EOS, Transactions American Geophysical Union*, *89*(10), 93–94. https://doi.org/10.1029/2008EO100001

Nicholson, S. E., & Palao, I. M. (1993). A re-evaluation of rainfall variability in the Sahel. *International Journal of Climatology*, *13*, 371–389. http://dx.doi.org/10.1002/joc.3370130403

OECD/Sahel and West Africa Club. (2014). *An atlas of the Sahara-Sahel. Geography, economics and security*. West African Studies. OECD Publishing. http://dx.doi.org/10.1787/9789264222359-en

Pase, A. (2011). *Linee sulla terra. Confini politici e limiti fondiari in Africa subsahariana*. Carocci.

Pase, A. (2024). Emptying and filling. Maps of inland Africa. In T. Rossetto & L. Lo Presti (Eds.), *Routledge handbook of cartographic humanities* (pp. 235–245). Routledge. http://dx.doi.org/10.424/97810027578

Pase, A., Gianoli, F., De Felice, L., Bertoncin, M., Cherlet, M., & Kronenburg Garcìa, A. (2022). Il respiro del Sahel. Rappresentazioni di uno spazio in movimento. *Rivista Geografica Italiana*, *CXXIX*(1), 32–59.

Pichon, E. (2020). *Understanding the EU strategy for the Sahel*. Briefing EU Policies – Insight. EPRS, PE 652.050. Retrieved June 3, 2024, from https://www.europarl.europa.eu/RegData/etudes/BRIE/2020/652050/EPRS_BRI(2020)652050_EN.pdf

Remotti, F. (1996). *Contro l'identità*. Laterza.

Retaillé, D. (2018). Sahel, sahel. *Information Géographique*, *1*, 34–81.

Retaillé, D., & Walther, O. (2019). *Mapping the Sahelian space*. Retrieved June 3, 2024, from https://arxiv.org/abs/1906.02223

Rotberg, R. I. (2003). Failed state, collapsed states, weak states: Causes and indicators. In R. I. Rotberg (Ed.), *State failure and state weakness in a time of terror* (pp. 1–25). Brookings Institution.

Sahara and Sahel Observatory (OSS). (2019). *Sahel and West Africa – Atlas of land cover maps – Strengthening resilience through services related to innovation, communication and knowledge – BRICKS (Benin, Burkina Faso, Chad, Ethiopia, Ghana, Mali, Mauritania, Niger, Nigeria, Senegal, Sudan and Togo)*. Retrieved June 28, 2021, from chrome-extension://efaidnbmnnnibpcajpcglclefindmkaj/https://www.oss-online.org/sites/default/files/2023-03/OSS-BRICKS-Atlas_En.pdf

Said, E. W. (1978). *Orientalism*. Pantheon Books.

Scott, J. C. (2009). *The art of not being governed: An anarchist history of upland Southeast Asia*. Yale University Press.

Sidaway, J. D. (2013). Geography, globalization, and the problematic of area studies. *Annals of the Association of American Geographers*, *103*(4), 984–1002. http://www.jstor.org/stable/23485682

Sidaway, J. D., Ho, E. L., Rigg, J. D., & Woon, C. Y. (2016). Area studies and geography: Trajectories and manifesto. *Environment and Planning D: Society and Space*, *34*(5), 777–790. https://doi.org/10.1177/0263775816656520

Sivakumar, M. V. K., & Wallace, J. S. (1991) Soil water balance in the Sudano-Sahelian zone: Need, relevance and objectives of the workshop. In M. V. K. Sivakumar, J. S. Wallace, C. Renard, & C. Giroux (Eds.), *Soil water balance in the Sudano-Sahelian zone* (pp. 3–10). IAHS Publication 199, Wallingford.

van Houtum, H., Kramsch, O., & Zierhofer, W. (Eds.). (2005). *B/Ordering space*. Ashgate.

6 Land degradation and the Convergence of Evidence in the Sahel

Federico Gianoli, Mélanie Weynants and Andreas Brink

Introduction

In this chapter, we aim to assess the hotspots of land degradation in the Sahel based on Earth Observation (EO) data. This approach has been undertaken by numerous authors employing various methodologies, such as indicators of Based Primary Productivity and Precipitation Relationships (Fensholt et al., 2013). However, our approach differs because we are convinced that land degradation cannot be captured by a single indicator but is a multifaceted process resulting from a combination of driving factors. Some of these factors can be surveyed through EO using satellites such as Sentinel or Landsat. By combining indicators of these factors, we aim to determine whether they collectively reveal a common narrative.

We first frame land degradation in the context of the UN Sustainable Development Goals' (SDGs') framework and draw on the Convergence of Evidence (CoE), a methodology introduced in the third edition of the *World Atlas of Desertification (WAD)* (Cherlet et al., 2018). Our analysis includes a range of indicators, both biophysical, such as land productivity and soil erosion, and anthropogenic, such as population growth or irrigated areas. The results are examined in the context of the three Sahel boundaries, as defined in Chapter 5.

SDG 15 aims to "protect, restore, and promote sustainable use of terrestrial ecosystems, sustainably manage forests, combat desertification, and halt and reverse land degradation and halt biodiversity loss" (United Nations Department of Economic and Social Affairs, n.d.). SDG 15.3.1 specifically measures the "proportion of land that is degraded over total land area." This indicator is critical for tracking progress towards achieving a land degradation-neutral world by 2030. The significance of SDG 15.3.1 lies in its ability to provide quantitative measures of land degradation, which include changes in land cover, land productivity and carbon stocks. These measures help identify areas at risk of degradation or those undergoing restoration. This indicator is essential for governments and organisations to implement and monitor restoration projects and to prioritise interventions that can prevent or reverse degradation. By focusing on the sustainable management of land resources, SDG 15.3.1 supports ecological health, climate change mitigation

DOI: 10.4324/9781003429111-9

and resilience in ecosystems, which in turn contribute to broader environmental and development goals (Sims et al., 2021).

However, SDG 15.3.1 does not capture the full complexity of land degradation, which encompasses a range of processes affecting global land resources and whose evaluation poses notable difficulties. The *WAD* introduced the CoE as a tool for calling attention to regions where human-driven processes called global land change issues (LCIs) impact the land capacity to sustain thriving ecosystems, potentially leading to land degradation (Cherlet et al., 2018). Instead of mapping degradation directly, the CoE method examines the co-occurrence of these processes and their spatial arrangement. The presence of such convergence at any location suggests stress on land resources, indicating potential or existing land degradation (Reynolds, 2021). This methodology also aligns with Rasmussen's idea that land degradation, as defined by the United Nations Convention to Combat Desertification (UNCCD; www.unccd.int), cannot be studied from a single perspective. Instead, it is the result of multiple human and biophysical factors that must be examined together, along with the techniques used to investigate them (Rasmussen et al., 2016).

In this study, we tailor the CoE framework to the Sahelian region by selecting LCIs that are relevant to the area and using more precise input data to identify those areas where the land is more stressed and where we expect land degradation to prevail.

Convergence of Evidence

The CoE map is constructed using two basic types of global data: land cover and global change issues (GCIs) (Gianoli et al., 2023). For the land cover component of the analysis, six land cover classes have been identified to represent the total global land mass, although the stratification could alternatively be performed using another relevant dataset, such as those comprising biomes, ecoregions or climate zones.

The Copernicus Dynamic Global Land Cover (GLC) data provide detailed and high-resolution information on the types of surfaces covering the Earth's surface. This dataset is part of the Copernicus Land Monitoring Service, funded and coordinated by the European Union. The GLC dataset is available at a resolution of 100 m and provides comprehensive and up-to-date land cover maps for the years 2015–2019, which are essential for various environmental and land management applications (Zenodo, n.d.). For the CoE analysis, the GLC data are reclassified into six broad classes to simplify the land cover types, as shown in Fig. 6.1, according to the classes described in Table 6.1.

Once the land cover layer with its classes is established, the next step is to identify data indicative of land degradation processes in the area. This involves finding datasets that contain both human-induced and biophysical indicators, which

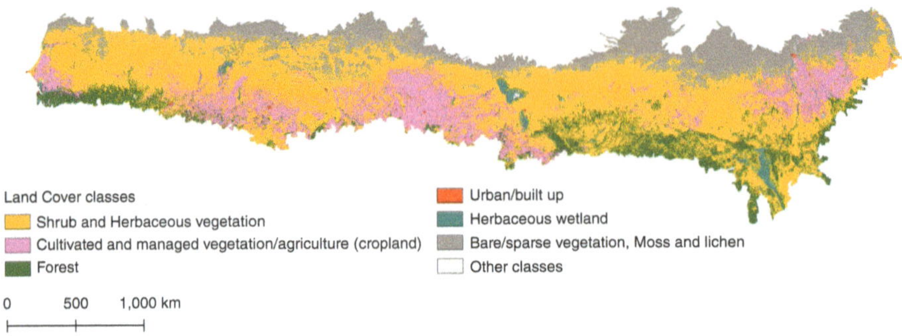

Figure 6.1 Land cover classes, merged for stratification process.

can drive land degradation, either individually or collectively. The selection of these indicators is essential for a comprehensive assessment of land degradation factors. Human-induced indicators may include data on population density, urban expansion, agricultural practices and industrial activities. Biophysical indicators may be measures of soil erosion, water availability, vegetation cover and climatic conditions. In recent years, sub-Saharan Africa has experienced significant changes in land cover and land use due to both human and natural factors. Population growth, economic development and globalisation have continuously eroded the region's natural ecosystems and resources. Natural hazards such as floods, landslides, droughts and climate change have further intensified this degradation (Brink & Eva, 2009).

Integrating these diverse datasets makes it possible to increase our understanding of which different processes interact and how they may contribute to land degradation. This approach not only helps in identifying current degradation hotspots but also aids in predicting potential future risks, enabling more effective land management and conservation strategies.

Table 6.1 Merged land cover classes used for the stratification. (GLC classes refer to the Copernicus Global Land Cover legend.)

Classes	*Description of land cover classes*
1	Shrubs and herbaceous vegetation
2	Cultivated and managed vegetation/agriculture (cropland)
3	Forest
4	Urban/built up
5	Herbaceous wetland
6	Bare/sparse vegetation, moss and lichen
0	Other

We have identified nine indicators, further referred to as LCIs, owing to their relevance and timely availability:

1 **Population density** measures the number of people in a specific area. A higher density can exert significant pressure on land resources. This indicator is taken from the Global Human Settlement Layer (GHSL), produced by the European Commission's Joint Research Centre (JRC) under the Copernicus programme. The GHSL dataset provides detailed information on the distribution and density of human settlements across the globe. It is a valuable resource for assessing urbanisation trends, population distribution and the impact of human activities on the environment (Schiavina et al., 2019).
2 **Population change** monitors the changes in population density over time since a rapid population increase can lead to unsustainable use of land resources. This indicator is computed using the GHSL of the year 2020 and the projection for 2030.
3 **Drought risk** provides information on the probability and impacts of drought conditions, which can severely affect soil productivity and water availability. The indicator used in this study is obtained from the World Resource Institute (WRI) Aqueduct Project. Drought risk assesses the likelihood of drought occurrences, the exposure of populations and assets and their vulnerability to negative impacts. Higher values signify a greater risk of drought (Kuzma et al., 2023).
4 **Revised Universal Soil Loss Equation (RUSLE)** is a widely used empirical model designed to estimate the long-term average annual rate of soil erosion caused by rainfall and associated surface runoff. It factors in various components, including rainfall intensity, soil type, topography, crop system and management practices. It helps in predicting soil loss and planning conservation measures to mitigate erosion (Ghosal & Das Bhattacharya, 2020; Petito et al., 2022). The indicator is computed using Google Earth Engine, and the code is available at https://github.com/fgianoli/PRIN_UNIPD.
5 **Built-up areas** indicate the extent of urbanised areas since soil sealing due to construction is a form of land degradation. The indicator comes from the GHSL.
6 **Water stress** evaluates the ratio of water withdrawal to available renewable water resources, indicating areas where competition for water is highest. Water is crucial for the advancement of human societies and necessary for both a healthy environment and a flourishing economy. It is vital for food production, electricity generation and manufacturing, among other uses (Kuzma et al., 2023).
7 **Groundwater table decline** is selected because groundwater resources are crucial for both ecosystems and human livelihoods. However, excessive withdrawal of groundwater can lead to a decline in groundwater levels, which in turn can cause seawater intrusion, land subsidence, depletion of streamflow and drying up of wells (Jasechko et al., 2024). The source of this indicator is WRI's *Water Risk Atlas*.

8 **Tree loss** records the reduction in tree cover, which can lead to land degradation through loss of vegetation and increased soil erosion. The indicator is derived from the Hansen Forest Loss, which tracks the annual loss of tree cover worldwide. Using high-resolution satellite imagery, the dataset identifies changes in forest cover, providing detailed information on where and how much forest loss has occurred. These data are crucial for monitoring deforestation, understanding environmental changes and informing conservation efforts. The Hansen Forest Loss dataset is widely used by researchers, policymakers and environmental organisations to track and address forest degradation and deforestation globally (Ceccherini et al., 2020; Hansen et al., 2013).

9 **Land productivity dynamics (LPD)** measures changes in land productivity over time, reflecting the health and capacity of the land to support vegetation. Analysing vegetation is essential for identifying land degradation patterns. Farmers, pastoralists, agro-pastoralists and forest users all depend on vegetation resources and are affected by changes in their availability. Vegetation dynamics are influenced by a combination of natural and human factors, including climate change, atmospheric carbon dioxide levels, grazing pressure, bush fires and the expansion or contraction of agricultural activities (Mbow et al., 2014). The LPD dataset from the JRC combines (a) a 15-year baseline observation period from 2000 to 2015 and (b) a 16-year reporting period from 2004 to 2019. These maps provide information on the direction, intensity and persistence of trends and changes in above-ground biomass (surface biomass) generated by photosynthetically active vegetation cover, which is broadly equivalent to the gross primary production of the global land surface (Rotllan-Puig et al., 2021). Having LPD data is essential for identifying land degradation hotspots because the information provides detailed insights into the trends, intensity and persistence of changes in vegetation biomass. This knowledge helps in monitoring long-term land productivity trends and pinpointing areas with persistent declines.

These indicators collectively provide a comprehensive overview of the factors contributing to land degradation, enabling targeted and effective land management strategies.

An important note is that while other studies, such as the *WAD* (Cherlet et al., 2018) and the CoE in Europe (Gianoli et al., 2023), have given significant weight to the effects of fires in identifying land degradation hotspots, this indicator has not been considered for the Sahel region. Being aware of agricultural practices (in the Sahel) that involve burning (Aboul-Hadi et al., 2023) is important because these fires constitute a crucial part of local farming, used to rejuvenate the land after grazing and enhance future productivity. However, these often numerous and small fires complicate the identification of land degradation hotspots (Nielsen et al., 2003). While agricultural fires comprise a useful indicator in many studies, in our opinion, their frequency and spread can obscure other causes of land degradation, making it challenging to isolate and analyse the most critical areas of degradation.

Stratification

The stratification process is essential for accurately assessing and addressing land degradation by establishing specific thresholds for each LCI based on the characteristics of different land cover types. The method involves a number of critical steps to ensure that the analysis is both accurate and meaningful.

The process begins with the results and data from the previous section of this chapter, specifically the classification of land cover data into six categories. This classification provides a foundational framework for subsequent analysis. Once the land cover is classified, relevant data are collected for various LCIs, which can significantly impact land degradation. Each indicator is clipped to the Sahel study area, as defined in Chapter 5, and resampled to the same projection, spatial resolution (EPSG:54009 Mollewide, 1 km/pixel) and grid alignment to ensure consistency.

For each land cover class, the median value of each LCI is calculated. This median serves as a threshold to identify areas where the LCI poses a potential concern. Specifically, if an LCI value at a location exceeds the median threshold, it suggests conditions that may induce land degradation; values below the median indicate lower risks. These thresholds are used to assign binary values to each pixel, with 1 indicating concern and 0 indicating no concern. This binary classification simplifies data, making it easier to identify patterns and trends. Spatial binary layers for each LCI are then created and overlaid on the land cover map to identify areas where multiple LCIs converge, highlighting regions under significant pressure from various land degradation drivers.

To compute the stratification, we developed a tool using R and QGIS. This tool extracts values for each land cover class and calculates the median value, which serves as the threshold. The tool then performs a Boolean reclassification based on the median value. This step is applied only to quantitative indicators, such as population density or RUSLE, not to qualitative ones, such as LPD or water stress, where only the most impactful classes are selected and reclassified with the value of 1.

The reclassified indicators can be analysed individually or in combination by summing them, as shown in Figure 6.2, which presents the combination of all stratified layers.

The analysis of these layers helps identify hotspots where land resources are most at risk, requiring closer scrutiny and remedial action. This stratified approach ensures that the assessment is tailored to the unique characteristics of different land types, leading to more effective land management and conservation strategies. This structured stratification process allows a nuanced understanding of land degradation, providing critical insights for policymakers, researchers and land managers aiming to protect and restore fragile ecosystems.

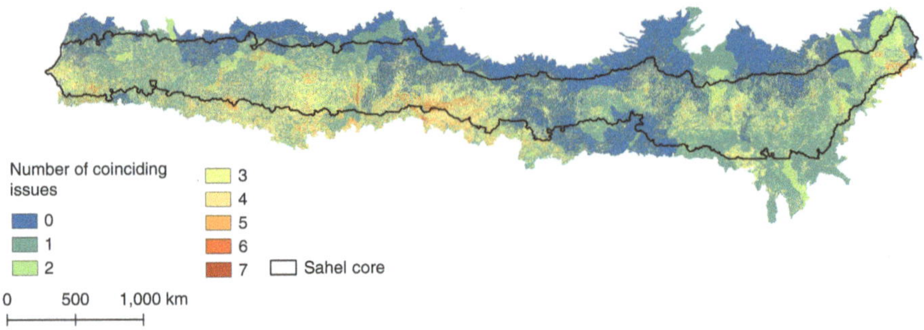

Figure 6.2 Convergence of Evidence map with nine land change issues.

Statistical outputs

We analyse the overall convergence results in three distinct areas: the liminal Sahel in the north, the core area and the Sahelian buffer zone in the south. This boundary delimitation is defined by precipitation patterns, creating three distinct areas: a central core region and two zones, one to the north and one to the south. These boundaries shift, expanding and contracting based on rainfall, effectively illustrating the "breath" of the Sahel; see Chapter 5 (Pase et al., 2022). Figure 6.3 shows that the number of coinciding issues increases from north to south. This trend can be interpreted as an increase in human activity and land management dynamics moving away from the desert borders. In the southern regions, there is greater water availability, and agriculture predominates over the nomadic pastoralism typical of the north. Additionally, looking at the map, it is noticeable that there is a convergence of GCIs in the western part of the Sahel, while in the central-eastern part, the number of co-occurring GCIs decreases.

Figure 6.4 illustrates the highly variable spatial distribution of the individual GCIs. Most of the study area appears to be primarily affected by issues related to human

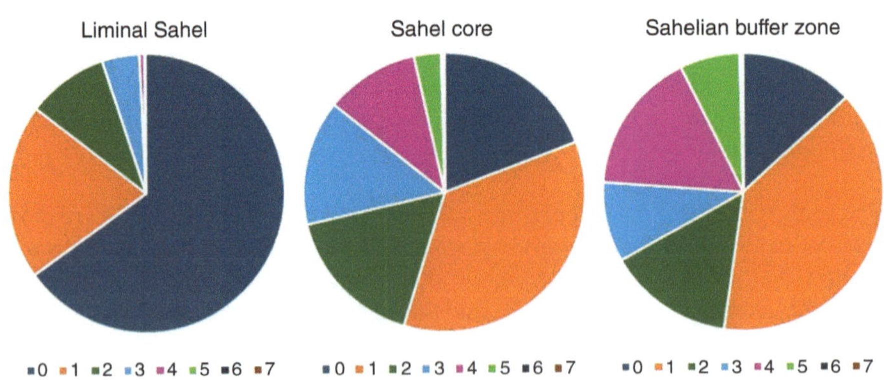

Figure 6.3 Percentage of GCI coverage for each area of the Sahel.

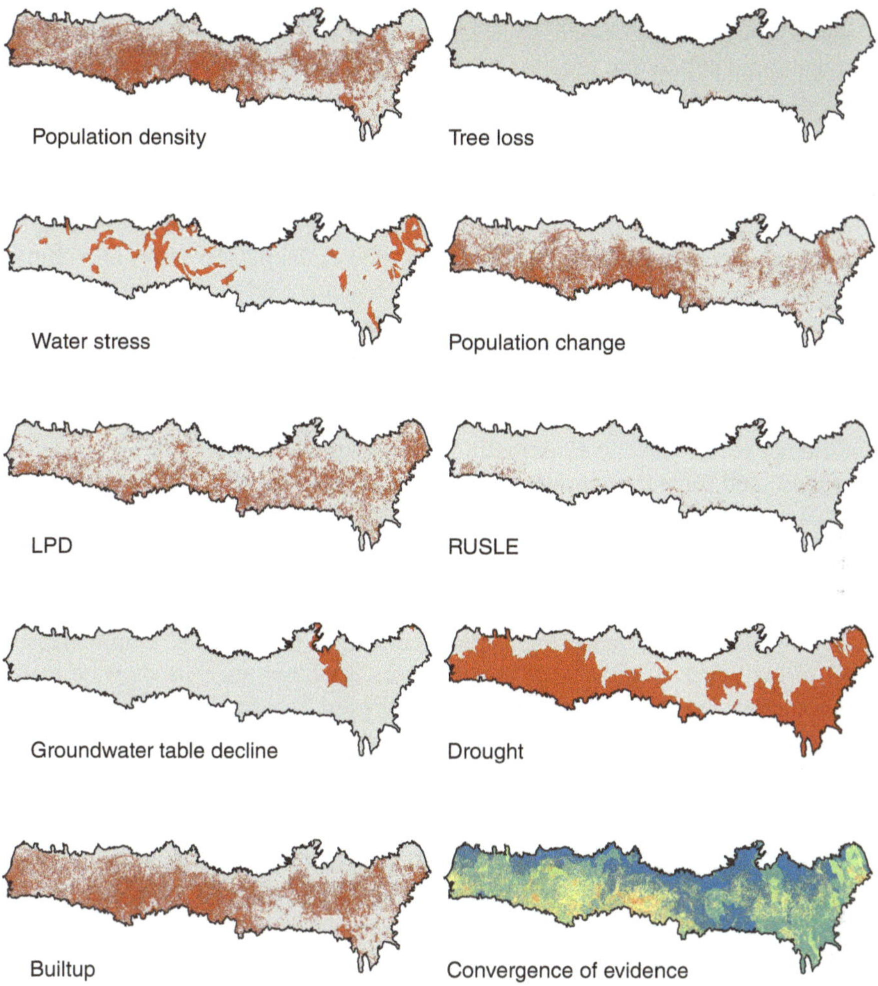

Figure 6.4 Spatial distribution of the indicators considered for the calculation of the Convergence of Evidence.

activities and substantial population growth, which, in turn, depletes the local resources and drives more changes in land use.

Despite numerous projects aimed at increasing vegetative cover and limiting the effects of erosion caused by wind, which is a major phenomenon and can affect soil fertility (Pierre et al., 2023), and water, such as those indicated by the RUSLE, challenges remain. One prominent initiative is the Great Green Wall, the objective of which is to increase the amount of arable land in the Sahel (Goffner et al., 2019); for further considerations about the Great Green Wall, refer to Chapter 2. Recent analyses show that vegetation cover has stabilised due to increased rainfall, even with continued agricultural expansion. However, the observed greening of the Sahel since the early 21st century does not change the fact that rapid population growth is decreasing the natural resources available per capita (Cesaro et al., 2023).

Conclusion

The Sahel, a region marked by widespread land degradation, is experiencing the compounded negative effects of population growth, human activities and climate variability, leading to recurrent droughts and a steady decline in natural resources and land productivity. Agriculture, livestock and forestry, which sustain over 80% of the population, are increasingly threatened by the overexploitation of natural resources and unsustainable water, agricultural and silvo-pastoral land management practices. These challenges have direct and lasting harmful impacts on food, water and energy security, while exacerbating social inequalities, conflicts over land and resources, and forced migration. Despite the critical situation in the region, there are viable solutions to reverse these trends (United Nations Convention to Combat Desertification [UNCCD], 2019).

In this chapter, we have applied the CoE methodology to identify the areas where various indicators of land degradation converge, providing a comprehensive view of the Sahel. According to the indicators used, our analysis shows a higher occurrence of land degradation moving southward, where anthropogenic factors appear to be more impactful.

This finding aligns with the conclusions of other authors, who have noted that while degradation is certainly a reality in specific localities or with respect to certain environmental components in the Sahel, there is no evidence of a catastrophic regional environmental crisis (Behnke & Mortimore, 2016), at least in the northern boundaries, in the liminal zone adjacent to the Sahara Desert. Additionally, claims of land degradation, derived from onsite observations of reduced tree cover and species disappearance, do not necessarily contradict the findings that no land degradation is occurring, based on increased vegetation productivity observed through the time series of EO data (Rasmussen et al., 2016).

It is important to clarify that in the northern part of the Sahel, the risk of desertification—understood as a severe and nearly irreversible process of land degradation—is a condition that is present continuously over time. However, the patterns of overlap are fewer because this area is less subject to anthropogenic pressures. This is coupled with a very different use of the land, as mentioned in other chapters of this book (particularly Chapters 12, 13 and 15), leading to conflicts between the pastoralist and farmer populations over resource use (Shettima & Tar, 2008). This analysis is further confirmed by the increasing east–west pattern, with a peak in the central part of the Sahel corresponding to Nigeria, which has a significantly denser population compared to neighbouring countries (Adesola et al., 2024).

Following the Sahel Drought of 1972–1974, the second significant drought cycle in 1984–1985 and a decline in average annual rainfall by 25% or more from the 1960s to the 1990s, the region has become and continues to be an archetype of dryland challenges (Hermann & Sop, 2016). However, starting in the mid-1990s, the situation began to change. The prevailing narrative has lagged behind the realities on the ground (Mbow et al., 2021). This shift in environmental crisis narratives in the Sahel is described in Chapter 2.

Studies have emphasised the unsustainability of livestock farming practices due to insufficient biomass production, land use changes and other factors. As the land becomes less productive, competition for limited resources intensifies, heightening tensions and contributing to regional instability (Rahimi et al., 2021). Moreover, the UNCCD explicitly connects land degradation with poverty, migration and conflict. It contends that if land degradation is not addressed, it will lead to increased extremism, radicalisation and resource-driven crises as people struggle to secure their livelihoods (Toulmin & Brock, 2016).

To fully understand the growing demand for land resources, it is important to situate this issue in a broader and more complex context that includes higher-scale political and economic dynamics, such as land and water grabbing and resource appropriation by elites. These factors exacerbate environmental stressors, intensifying competition for resources and deepening existing inequalities. This state of affairs underscores the urgent need for sustainable land management practices, not only to mitigate conflicts but also to ensure the long-term viability of both agricultural and pastoral livelihoods in the region.

The issue of land degradation is surrounded by different levels of complexity. The nexus of land degradation, climate change and conflicts is certainly not straightforward. Researchers and policymakers should approach it critically and refrain from proposing an oversimplifying narrative. Indeed, this would bear the risk of justifying external interventions, as demonstrated in Chapter 2.

In conclusion, the CoE methodology applied to our study area has allowed us to identify patterns and hotspots associated with potential land degradation. A more detailed analysis, incorporating additional biophysical stress factors, such as wind erosion, agricultural burning practices and intensive agricultural methods (as discussed in Chapter 8), alongside in situ case studies, can offer a more comprehensive understanding of land degradation. This recommended in-depth investigation is especially important in the Sahel, a region characterised by the complex interplay of multiple critical factors. Moreover, it is essential to acknowledge how Sahelian communities address these overlapping crises in innovative ways, offering valuable lessons (see Chapters 1 and 17).

Acknowledgements

We express our gratitude to Ludovica Crocitto for her work on calculating the Revised Universal Soil Loss Equation (RUSLE) index for the Sahel, to Andrea Pase for reviewing this chapter and serving as an invaluable source of inspiration, to Michael Cherlet for designing and promoting the Convergence of Evidence approach and to Joaquin Marquez Perez for his revisions.

References

Aboul-Hadi, S., Omoul-kaïrou, N., & Soulé, M. (2023). Effects of bush fires on biodiversity in West Africa Sahel: A review. *Research in Ecology*, *5*, 40–50. https://doi.org/10.30564/re.v5i2.5140

Adesola, R. O., Opuni, E., Idris, I., Okesanya, O. J., Igwe, O., Abdulazeez, M. D., & Lucero-Prisno, D. E. (2024). Navigating Nigeria's health landscape: Population growth and its health implications. *Environmental Health Insights*, *18*. https://doi.org/10.1177/11786302241250211

Behnke, R., & Mortimore, M. (2016). Introduction: The end of desertification? In R. Behnke & M. Mortimore (Eds.), *The end of desertification?: Disputing environmental change in the drylands* (pp. 1–34). Springer Earth System Sciences. Springer. https://doi.org/10.1007/978-3-642-16014-1_1

Brink, A. B., & Eva, H. D. (2009). Monitoring 25 years of land cover change dynamics in Africa: A sample-based remote sensing approach. *Applied Geography*, *29*, 501–512. https://doi.org/10.1016/j.apgeog.2008.10.004

Ceccherini, G., Duveiller, G., Grassi, G., Lemoine, G., Avitabile, V., Pilli, R., & Cescatti, A. (2020). Abrupt increase in harvested forest area over Europe after 2015. *Nature*, *583*, 72–77. https://doi.org/10.1038/s41586-020-2438-y

Cesaro, J.-D., Mbaye, T., Ba, B., Ba, M. F., Delaye, E., Akodewou, A., & Taugourdeau, S. (2023). Reforestation and sylvopastoral systems in Sahelian drylands: Evaluating return on investment from provisioning ecosystem services, Senegal. *Frontiers in Environmental Science*, *11*. https://doi.org/10.3389/fenvs.2023.1073124

Cherlet, M., Hutchinson, C., Reynolds, J., Hill, J., Sommer, S., & Von Maltitz, G. (2018). *World atlas of desertification: Rethinking land degradation and sustainable land management*. Publications Office.

Fensholt, R., Rasmussen, K., Kaspersen, P., Huber, S., Horion, S., & Swinnen, E. (2013). Assessing land degradation/recovery in the African Sahel from long-term Earth observation-based primary productivity and precipitation relationships. *Remote Sensing*, *5*, 664–686. https://doi.org/10.3390/rs5020664

Ghosal, K., & Das Bhattacharya, S. (2020). A review of RUSLE model. *Journal of the Indian Society of Remote Sensing*, *48*, 689–707. https://doi.org/10.1007/s12524-019-01097-0

Gianoli, F., Weynants, M., & Cherlet, M. (2023). Land degradation in the European Union – Where does the evidence converge? *Land Degradation & Development*. https://doi.org/10.1002/ldr.4606

Goffner, D., Sinare, H., & Gordon, L. J. (2019). The Great Green Wall for the Sahara and the Sahel initiative as an opportunity to enhance resilience in Sahelian landscapes and livelihoods. *Regional Environmental Change*, *19*, 1417–1428. https://doi.org/10.1007/s10113-019-01481-z

Hansen, M. C., Potapov, P. V., Moore, R., Hancher, M., Turubanova, S. A., Tyukavina, A., Thau, D., Stehman, S. V., Goetz, S. J., Loveland, T. R., Kommareddy, A., Egorov, A., Chini, L., Justice, C. O., & Townshend, J. R. G. (2013). High-resolution global maps of 21st-century forest cover change. *Science*. https://doi.org/10.1126/science.1244693

Hermann, S., Sop, T. K., (2016). The Map Is not the Territory: How Satellite Remote Sensing and Ground Evidence Have Re-shaped the Image of Sahelian Desertification. In R. Behnke

& M. Mortimore (Eds.), *The end of desertification?: Disputing environmental change in the drylands* (pp. 117–145). Springer. https://doi.org/10.1007/978-3-642-16014-1_2

Jasechko, S., Seybold, H., Perrone, D., Fan, Y., Shamsudduha, M., Taylor, R. G., Fallatah, O., & Kirchner, J. W. (2024). Rapid groundwater decline and some cases of recovery in aquifers globally. *Nature*, *625*, 715–721. https://doi.org/10.1038/s41586-023-06879-8

Kuzma, S., Bierkens, M.F.P., Lakshman, S., Luo, T., Saccoccia, L., Sutanudjaja, E.H. & Van Beek, R. (2023). *Aqueduct 4.0: Updated decision-relevant global water risk indicators. Technical Note.* World Resources Institute. doi.org/10.46830/writn.23.00061.

Mbow, C., Fensholt, R., Nielsen, T. T., & Rasmussen, K. (2014). Advances in monitoring vegetation and land use dynamics in the Sahel. *Geografisk Tidsskrift-Danish Journal of Geography*, *114*, 84–91. https://doi.org/10.1080/00167223.2014.886515

Mbow, C., Halle, M., El Fadel, R., & Thiaw, I. (2021). Land resources opportunities for a growing prosperity in the Sahel. The dryland social-ecological systems in changing environments. *Current Opinion in Environmental Sustainability*, *48*, 85–92. https://doi.org/10.1016/j.cosust.2020.11.005

Nielsen, T. T., Rasmussen, K., Mbow, C., & Touré, A. (2003). The fire regime of Senegal and its determinants. *Geografisk Tidsskrift-Danish Journal of Geography.*, *103*, 43–53. https://doi.org/10.1080/00167223.2003.10649478

Pase, A., Gianoli, F., De Felice, L., Bertoncin, M., Cherlet, M., & García, A. K. (2022). The breath of the Sahel. Representations of a space in motion. *Rivista Geografica Italiana.*, *129*, 32–59. https://doi.org/10.3280/rgioa1-2022oa13366

Petito, M., Cantalamessa, S., Pagnani, G., Degiorgio, F., Parisse, B., & Pisante, M. (2022). Impact of conservation agriculture on soil erosion in the annual cropland of the Apulia Region (Southern Italy). Based on the RUSLE-GIS-GEE framework. *Agronomy*, *12*, Article 281. https://doi.org/10.3390/agronomy12020281

Pierre, C., Rajot, J. L., Faye, I., Dorego, G. S., Bouet, C., Marticorena, B., Bergametti, G., Ka, A., Amar, B., Tall, A., Diagne, N., & Feron, A. (2023). A contrasting seasonality of wind erosivity and wind erosion between Central and Western Sahel. *Aeolian Research*, *62*, Article 100879. https://doi.org/10.1016/j.aeolia.2023.100879

Rahimi, J., Haas, E., Grote, R., Kraus, D., Smerald, A., Laux, P., Goopy, J., & Butterbach-Bahl, K. (2021). Beyond livestock carrying capacity in the Sahelian and Sudanian zones of West Africa. *Scientific Reports*, *11*, Article 22094. https://doi.org/10.1038/s41598-021-01706-4

Rasmussen, K., D'haen, S., Fensholt, R., Fog, B., Horion, S., Nielsen, J. O., Rasmussen, L. V., & Reenberg, A. (2016). Environmental change in the Sahel: Reconciling contrasting evidence and interpretations. *Regional Environmental Change*, *16*, 673–680. https://doi.org/10.1007/s10113-015-0778-1

Reynolds, J. F. (2021). Desertification is a prisoner of history: An essay on why young scientists should care. *Ecosistemas*, *30*, Article 2302. https://doi.org/10.7818/ECOS.2302

Rotllan-Puig, X., Ivits, E., & Cherlet, M. (2021). LPDynR: A new tool to calculate the land productivity dynamics indicator. *Ecological Indicators*, *133*, Article 108386. https://doi.org/10.1016/j.ecolind.2021.108386

Schiavina, M., Freire, S., & MacManus, K. (2019). *GHS-POP R2019A – GHS population grid multitemporal (1975-1990-2000-2015).* https://doi.org/10.2905/0C6B9751-A71F-4062-830B-43C9F432370F

Shettima, A. G., & Tar, U. (2008). Farmer-pastoralist conflict in West Africa: Exploring the causes and consequences. *Information, Society and Justice*, *1*(2), 163–184.

Sims, N. C., Newnham, G. J., England, J. R., Guerschman, J., Cox, S. J. D., Roxburgh, S. H., Viscarra Rossel, R. A., Fritz, S., & Wheeler, I. (2021). *Good practice guidance.* SDG Indicator 15.3.1.

Toulmin, C., & Brock, K. (2016). Desertification in the Sahel: Local practice meets global narrative. In R. Behnke & M. Mortimore (Eds.), *The end of desertification?:*

Disputing environmental change in the drylands (pp. 37–63). Springer. https://doi.org/10.1007/978-3-642-16014-1_2

United Nations Convention to Combat Desertification (UNCCD). (2019). *The global land outlook. West Africa thematic report. Land degradation neutrality: Benefits for human security*. Retrieved July 31, 2024, from https://www.unccd.int/resources/publications/global-land-outlook-west-africa-thematic-report-land-degradation-neutrality

United Nations Department of Economic and Social Affairs. (n.d.). *Sustainable Development Goal 15*. Retrieved July 30, 2024, from https://sdgs.un.org/goals/goal15

Zenodo. (n.d.). *Copernicus Global Land Service: Land cover 100 m: Version 3 Globe 2015–2019: Product user manual*. Retrieved February 3, 2022, from https://zenodo.org/record/3938963#.Yfuz_-rMJHY

7 Troubled waters

Luca De Felice, Federico Gianoli and Andrea Pase

Introduction

The crucial nature of water in the Sahel lies at the heart of this book. In this chapter, through our analysis of remotely sensed data, georeferenced information and use of GIS tools, we focus on a number of interrelated and fundamental aspects. The first is the hydrological behaviour of the great Sahelian rivers, particularly the inundation of contiguous lands linked to the annual floods. The second concerns human intervention to regulate this behaviour and (attempt to) control its dynamics. The last aspect involves the level of anthropogenic pressure on the major rivers themselves, even if only roughly visualised through a single indicator. In many respects, these are troubled waters. They are complicated in themselves because of the considerable variability and unpredictability of floods, which since antiquity have inspired the design of forecasting instruments, of which the Nilometer in Cairo is the most significant example (Ardagh, 1889; Popper, as cited in D. S. R., 1955; Torrey, 1952). River waters are tricky due to their importance in entering semi-arid regions of the Sahel or even the middle of the desert and bringing with them the gift of moisture at times when it would not be there (far from the rainy season) or in places that would never have it otherwise. The diverse and often conflicting interests that this moisture arouses make these troubled waters the focus of a crowded arena of narratives and practices. The large-scale irrigation schemes whose genesis and evolution are discussed in Chapter 1 depend on these waters. The fate of these gigantic terraforming operations makes water management problematic and always on the edge.

In the first section, we define the contours of the hydrological Sahel and analyse some relevant issues, such as the functioning of the region's main water bodies in relation to the seasonal floods. In the second section, we discuss the main hydraulic interventions on the basin scale, that is, dams and reservoirs. Finally, in the third section, we attempt to construct a graphical rendering of the increasing densification of human communities by following the course of major rivers.

DOI: 10.4324/9781003429111-10

The hydrological Sahel

The Sahel encompasses some major hydrological basins, notably Senegal, Volta, Niger, Chad and Nile, from west to east. The Senegal basin spans Guinea, Mali, Mauritania and Senegal. The river, originating in the Fouta Djallon highlands in Guinea and joined by its main tributary Falémé upstream of the city of Bakel, flows north and west through Mali and Mauritania before reaching the Atlantic Ocean in Senegal. The Volta basin spans six countries: Ghana, Burkina Faso, Togo, Benin, Ivory Coast and Mali. Most of the basin is located in Ghana and Burkina Faso. The basin is named after the Volta River, which is formed by the confluence of the Black Volta, White Volta and Red Volta Rivers. These tributaries originate in the highlands of Burkina Faso and flow southwards. The Niger River basin, which is the largest in West Africa, covers a significant portion of the Sahel. Niger River, vital for many countries in the region, such as Mali, Niger and Nigeria, has several tributaries; however, some of them contribute little or nothing to the main Niger River and are considered endorheic basins (Descroix et al., 2009; Rameshwaran et al., 2021). The Lake Chad basin is also an important hydrological basin in the Sahel; it is the world's largest interior drainage basin, covering an approximate area of 2,500,000 km^2 (Ndehedehe et al., 2016), but the basin's active hydrological extent is primarily concentrated on its southern part, covering an area of 967,000 km^2 (Appolinaire et al., 2012; Lemoalle, 2015). The lake, which is a shallow freshwater layer that has experienced significant fluctuations in size over time due to climate variability and human factors (see Chapter 2), is shared by Chad, Niger, Nigeria and Cameroon. Lake Chad receives its water supply primarily from Chari River, which encompasses Logone River and other tributaries in Chad, the Central African Republic and Cameroon (Lemoalle, 2015, pp. 20–22). As for the Nile River, various tributaries contribute to its overall flow in the part of its basin that falls in the Sahel. The White Nile, which descends from the equatorial highlands, meets at Khartoum with the Blue Nile, which arrives from the Ethiopian highlands. It thus forms the main Nile, which receives the Atbara River downstream as its last tributary (Sutcliffe, 2009). Notably, only a section of the aforementioned river basins lies within the pluviometric Sahel, as defined in Chapter 5 (see Table 7.1).

The Sahel has a single rainy season, crucial for agriculture and pastoralism, lasting from June to September, reaching its peak between late July and August (Elagib

Table 7.1 Total area and area within the pluviometric Sahel for each major basin, expressed in square kilometres and relative percentage.

Basins	*Total area (km^2)*	*Area in the Sahel (km^2)*	*Area in the Sahel (%)*
Senegal	458,028	394,618	86
Volta	412,649	220,378	53
Niger	2,123,002	1,046,647	49
Chad	2,471,050	1,079,272	44
Nile	3,353,211	1,336,450	40

et al., 2021) and progressively shortening towards the north (de Planhol & Rognon, 1970; Fensholt et al., 2013). As a remarkably unique region in hydrological terms, the Sahel is characterised by its distinctive dynamics, encompassing thousands of square kilometres with seasonal floods due to abundant rainfall in areas outside the region, carried into it by some of the above-mentioned major rivers. These could be called "displaced waters" compared with where the precipitation occurs. These waters reach the Sahel through river floods, fuelled by rains to the south or, in the case of the Blue Nile and Atbara River, from the Ethiopian highlands, with weeks or even months of delay due to the predominantly flat morphology of the area. The flood waves of these rivers slowly traverse the smooth lands of the Sahel, causing extensive inundations from September to January and creating valuable wetlands in the semi-arid or arid context. Considering the above pattern, for the purposes of the next analysis, here we introduce the concept of the "hydrological Sahel" as comprising the areas that experience flooding beyond the rainy season. Borrowing from Francophone literature, we refer to this period as the "counter-season", when moisture is thus available even beyond the rainy season due to the major river floods. However, a distinction must be made for Volta River. In fact, only one of its branches, the Black Volta (Mouhoun), follows the hydrological dynamics of the other major rivers of the Sahel. Its floodwaters are fed by the southern rains, which then follow the river's northward course before turning resolutely southwards. The "boucle" is the location of the junction with Sourou River, a minor tributary that, however, acts as a flood reservoir (as explained in Chapter 12). Apart from this situation, Volta River's hydrological pattern differs from those of the other major Sahelian rivers and is therefore not analysed in this chapter as far as seasonal flooding is concerned, while it remains part of the overall framework.

Using the Global Surface Water Explorer (GSWE) dataset (Pekel et al., 2016), a collaboration between the European Commission's Joint Research Centre (JRC) and Google Earth Engine, a series of global maps with 30-m resolution was utilised to track changes in surface waters over time globally, regionally or in specific areas between 1984 and 2021. The GSWE dataset defines "surface waters" as open water bodies visible in satellite imagery, including rivers, lakes, reservoirs and other inland water bodies. Open water refers to any stretch of water larger than 30 m × 30 m, exposed to the sky, encompassing both freshwater and saltwater environments. The dataset provides monthly, seasonal and annual composites of surface water extent.

Despite its extensive coverage, the GSWE dataset is subject to limitations that may affect its precision and comprehensiveness. These shortcomings encompass gaps (no data) in the observational data due to factors, such as cloud cover, persistent shadow conditions, sensor-related issues and the absence of Landsat imagery.

Utilising the above-mentioned dataset for the entire pluviometric Sahel made it possible to gain specific insights about surface waters in the region of interest. The area of the "maximum water extent" (MWE), defined as the cumulative surface area where water was detected in any month of any year over the entire observation period (1984–2021), is approximately 51,550 km², representing around 12.52% of the area of the entire Africa (411,500 km²).

However, the dynamics of surface waters for the Sahel show different patterns with respect to continental Africa. Using the "transition layer", which illustrates changes in surface water states over the observation period by comparing the water conditions of the first year with those of the last year, enables us to compute the extent of the changes in surface water resources over time across the Sahel and its major relevant basins. The GSWE transition map specifically identifies the following classes of changes, as described in Fig. 7.1:

- permanent water surface (unchanging, defined as an area that remains underwater throughout the year)
- new permanent water surface (conversion of a no-water/land place into a permanent water place)
- lost permanent water surface (conversion of a permanent water place into a dry place)
- unchanging seasonal water surface (characterised by being underwater for less than 12 months of the year)
- new seasonal water surfaces (conversion of a no-water/land place into a seasonal water place)
- lost seasonal water surface (conversion of a seasonal water place into a no-water place)
- conversion of permanent water into seasonal water
- conversion of seasonal water into permanent water
- ephemeral permanent water (permanent water detected during the observation period although not present at the beginning or the end of that period)
- ephemeral seasonal water (seasonal water detected during the observation period although not present at the beginning or the end of that period)

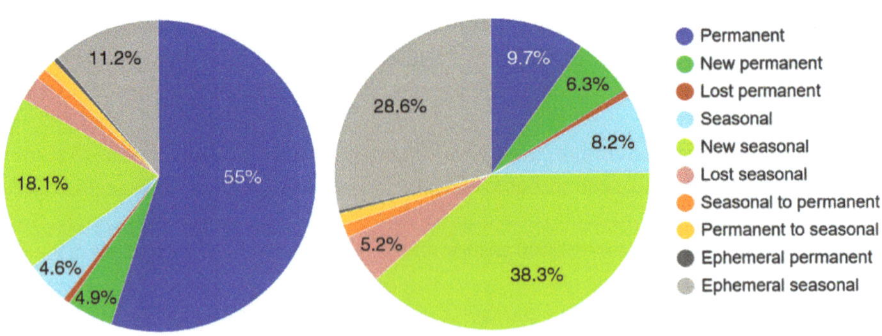

Figure 7.1 Summary of transition class areas for the entire African continent (left) and for the pluviometric Sahel (right).

In the Sahel region, the availability of permanent water is significantly low, constituting less than 10%, compared with 55% in continental Africa. This region particularly exhibits a high predominance of seasonal waters, with 24,494 km^2 of seasonal waters identified in 2021 and 42,590 km^2 that exhibited seasonal characteristics at any point between 1984 and 2021. A significant portion of these seasonal waters is ephemeral, covering 14,717 km^2, which represents 32% of the total ephemeral seasonal waters across Africa. The behaviour of these seasonal waters reflects the distinctive pattern of counter-season dynamics, as previously described. This pattern is also evident across all major basins of the Sahel, as illustrated by the charts in Fig. 7.2. The data reveal very low values of the total permanent water fraction (permanent and new permanent classes), never beyond

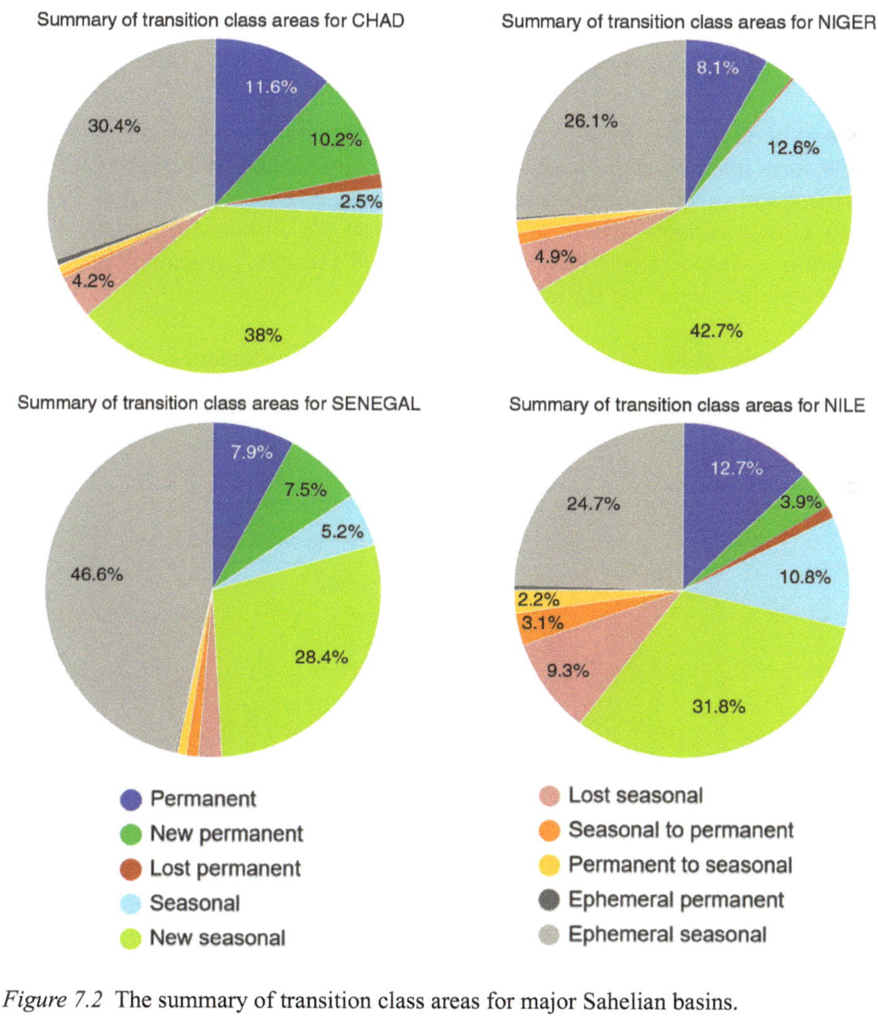

Figure 7.2 The summary of transition class areas for major Sahelian basins.

22%, and a predominance of seasonal waters across all four major basins. Notably, the new seasonal and ephemeral seasonal water classes collectively exceed 60% of the extents in each basin (Senegal reaches 75%). In contrast, the permanent and seasonal water classes (the only unchanging water classes) remain very low in all instances, highlighting the significant variability and pronounced seasonal behaviour of the water bodies in the Sahel region.

The Yearly Water History (v1.4) dataset, an alternative tool in the GSWE dataset, enables an annual examination of surface water trends. This dataset provides a classification of water seasonality based on detected occurrence values throughout the year, offering a comprehensive temporal breakdown of surface water distribution from 1984 to 2021. Each pixel is categorised into "no-water/land", "seasonal", "permanent" and "no-data" classes. A "permanent" water surface remains submerged throughout the year, whereas a "seasonal" water surface is underwater for fewer than 12 months annually. The dataset comprises 38 global images, one for each year spanning the period 1984–2021.

It is important to note the variations in the GSWE dataset's geographic coverage due to early imaging disparities, notably before 1999. Landsat 5 lacked on-board recorders and relied on limited receiving stations, causing gaps (Goward et al., 2006). Coverage improved in 1999 with Landsat 7's on-board recorders. These limitations are especially evident before Landsat 8's launch, particularly in parts of Africa with few receiving stations.

The time-series in Fig. 7.3 depicts yearly trends in surface water extents across the Sahel region, categorised into three classes: permanent water, seasonal water and

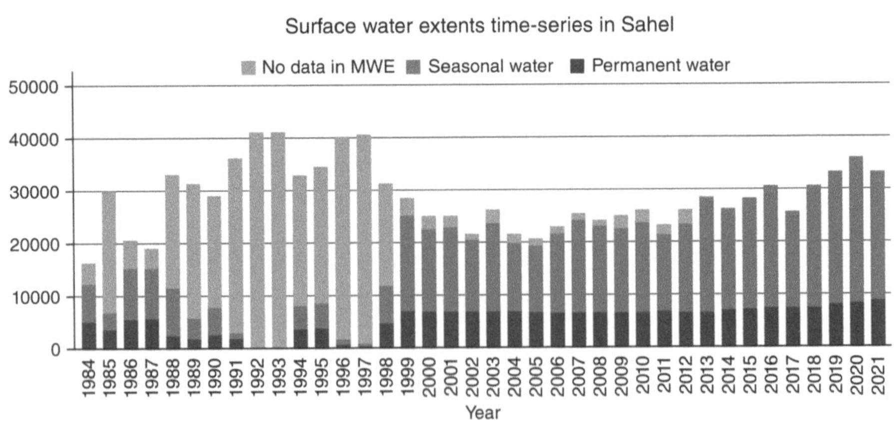

Figure 7.3 Yearly surface water extent (in square kilometres) time-series in the Sahel: permanent water (dark gray), seasonal (mid-gray) and no-data (light gray) extents within the maximum water extent (MWE).

no-data (considering the extent of missing data within the MWE, which is used as a proxy for uncertainty). The analysis reveals several key findings:

- From 1999 onwards, the notable increase in reliable data results in fewer instances of missing data.
- The proportion of permanent surface water remains relatively stable over time.
- The seasonal fraction of surface water indicates a discernible upward trend.

These trends reflect the distinctive seasonal patterns observed in water bodies in the Sahel region, as previously demonstrated. The trends are consistent when examining time-series at the basin level (Fig. 7.4), with the exception of Senegal, which exhibits even greater variability in its seasonal water fraction. This variability is likely influenced by the presence of regulated reservoirs and dams in the region, mainly the Manantali Dam (started operating in 1988).

The increase in seasonal surface water in the Sahel is fully consistent with the rise in surface runoff documented over recent decades. Some authors describe a trend called the "Sahelian hydrological paradox" (Descroix et al., 2013, 2018; Gal et al., 2017). During the period of rainfall recovery from the 1990s onwards, the runoff coefficient increased despite the general re-greening of the Sahel. This scenario is counter-intuitive because runoff is expected to decrease with vegetation recovery.

Researchers have been puzzled by this phenomenon, which is not fully understood but has been linked to various factors, such as changes in land cover (Aich et al., 2015; Descroix et al., 2013), soil properties, vegetation degradation (Yonaba et al., 2021), intensification of rainfall (Chagnaud et al., 2022; Ragettli et al., 2022) and development of drainage networks (Gal et al., 2016, 2017).

The charts presented above underscore the distinctive nature of counter-season dynamics in the Sahel region. To further analyse this phenomenon, it is essential to return to the definition of the hydrological Sahel as comprising areas flooded from September to January, beyond the rainy season. We therefore indicate as hydrological Sahel all non-permanently submerged areas that experienced flooding beyond the rainy season from 1984 to 2021. Using Google Earth Engine (Gorelick et al., 2017) in our study area (see Chapter 5) to identify flood zones between September and January, we could extract valuable insights. Specifically, pixels showing seasonal water presence from September to January were extracted for the entire historical series to highlight flood areas of the major Sahelian river systems during the flood peak months.

Figs. 7.5 and 7.6 provide maps of the hydrological Sahel, focusing on two distinct areas: the Inner Niger Delta in Mali and Lake Chad, respectively.

Pase et al. (2022) determined that the maximum extent of the hydrological Sahel covered an area of 41,905 km^2 in the period 1984–2019, which increased to

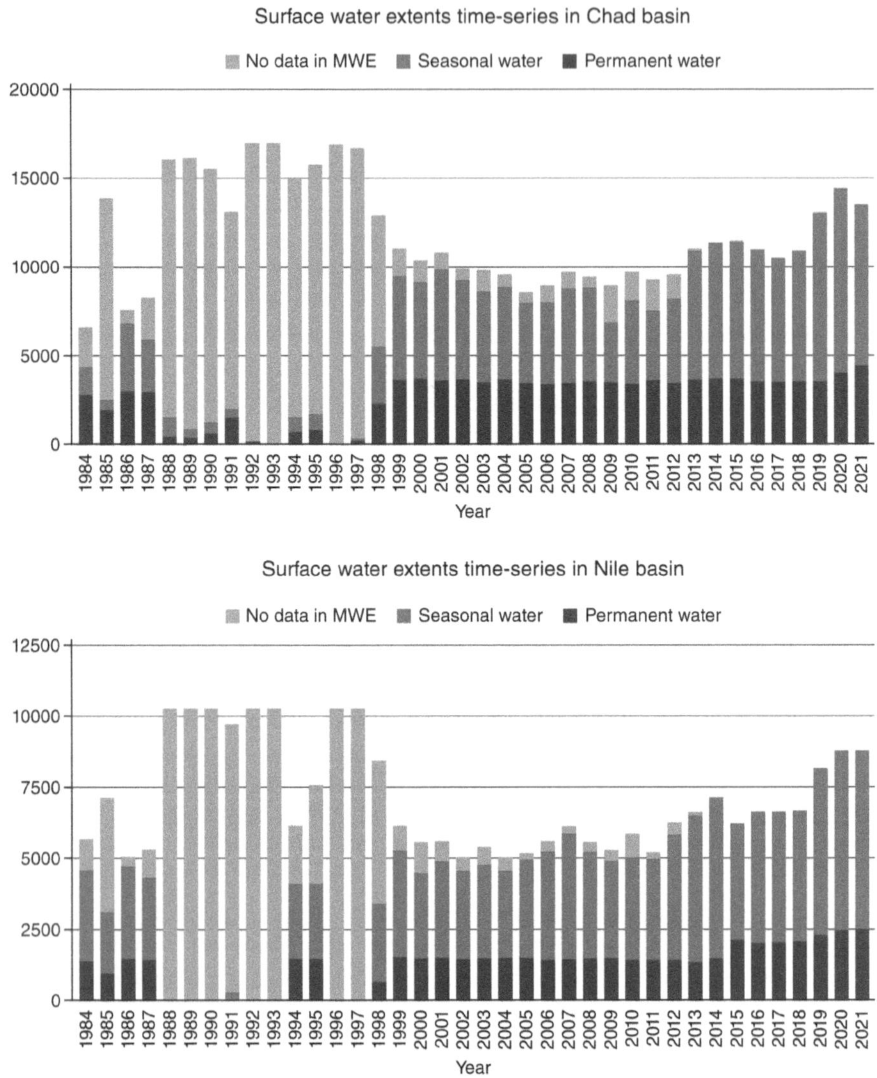

Figure 7.4 Yearly surface water extent (in square kilometres) time-series for Chad (a), Nile (b), Niger (c) and Senegal (d) basins: permanent water (dark gray), seasonal (mid-gray) and no-data (light gray) extents within the maximum water extent (MWE). *(Continued)*

46,450 km^2 upon re-evaluation in 2021, representing approximately 1.17% of the pluviometric Sahel (Table 7.2), confirming an upward trend consistent with the more general pattern of seasonal surface waters.

When disaggregating the MWE in areas inundated during the counter-seasonal period (1984–2021) at the basin level (Table 7.3), it is evident that the primary watershed in

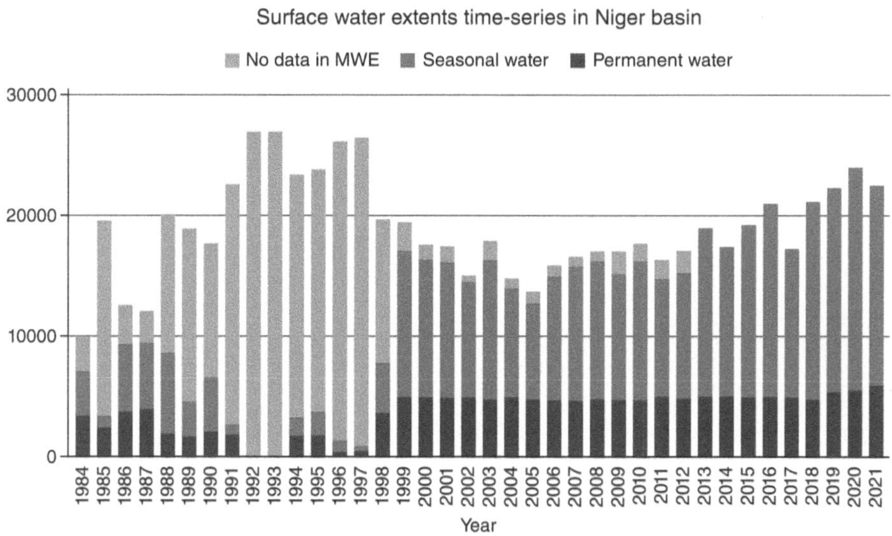

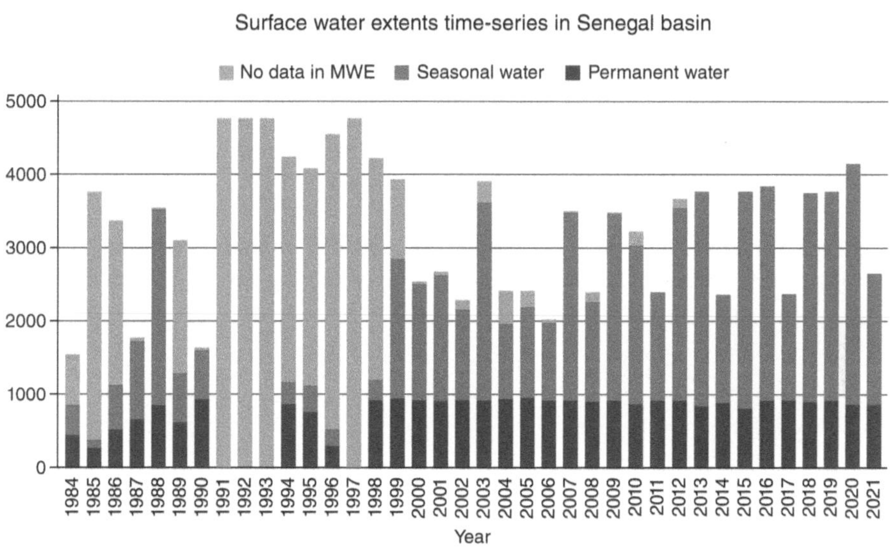

Figure 7.4 (*Continued*).

terms of impact is the Niger, followed by Chad and the Nile, collectively accounting for slightly over 80% of the total extent of the hydrological Sahel.

Examining the extent of flooded areas in the off-season—covering two biennial intervals with markedly different precipitation levels, the driest (1986–1987) and the wettest (2020–2021)—reveals a noticeable increase, exceeding twofold, over the wettest biennium (see the data presented in Table 7.4).

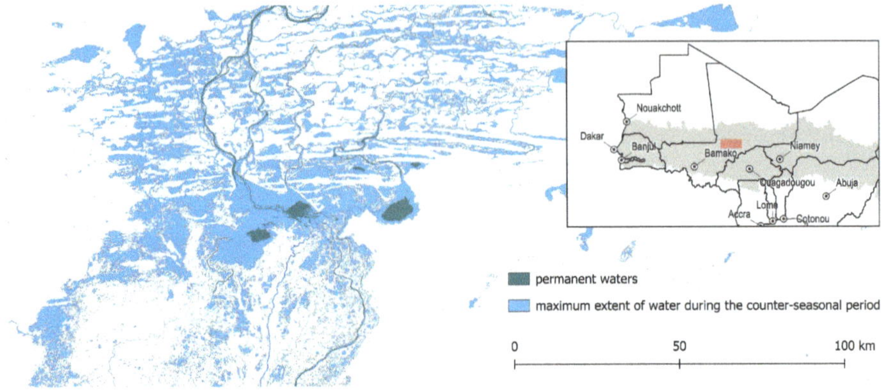

Figure 7.5 Details of the Inner Niger Delta (Mali) as part of the hydrological Sahel. The map illustrates the maximum water extent (MWE) in areas inundated during the counter-seasonal period (1984–2021) and the permanent waters.

It is likely, though not definitive, that the wettest years in the Sahel coincide with the wettest periods in the southern regions, where the flood wave originates. While we cannot explore this issue in detail at present, we plan to conduct further research on the relation between the wettest years in the Sahel and the patterns in the watershed basins of these southern regions to determine whether these phenomena are correlated.

Tables 7.3 and 7.4 illustrate an intriguing dynamic, differentiating between counter-season areas that are consistently flooded—even during drought years and thus

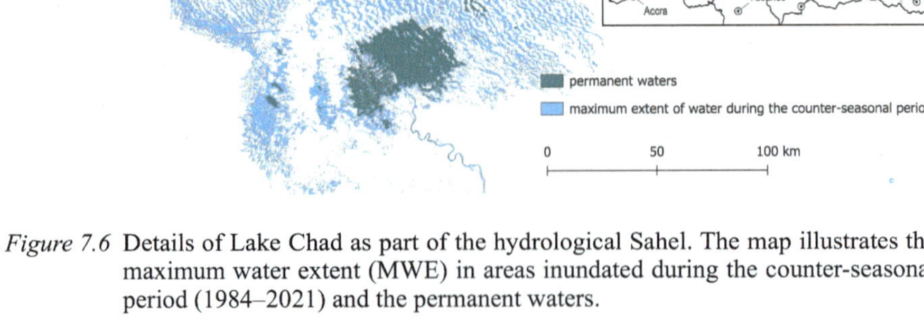

Figure 7.6 Details of Lake Chad as part of the hydrological Sahel. The map illustrates the maximum water extent (MWE) in areas inundated during the counter-seasonal period (1984–2021) and the permanent waters.

Table 7.2 Extents of the pluviometric and the hydrological Sahel, expressed in square kilometres and relative percentage.

Extent of pluviometric Sahel (km²)	*Extent of hydrological Sahel (km²)*	*Percentage of hydrological Sahel relative to pluviometric Sahel*
3,957,592	41,905 (1984–2019)	1.05
	46,450 (1984–2021)	1.17

Table 7.3 Maximum water extent (MWE) for each basin—considering only the areas in the region of interest (Sahel).

Major basins	*MWE (km²)*
Niger	~15,945
Chad	~11,976
Nile	~10,230
Senegal	~4446
Volta	~1629

Table 7.4 Extents of the hydrological Sahel for the driest biennium* (1986–1987) and the wettest biennium (2020–2021).

Period	*Extent of hydrological Sahel (km²)*	*Percentage of pluviometric Sahel*
1986–1987	~11,916	0.30
2020–2021	~31,743	0.80

* This analysis takes into account the periods during which the lack of data is not significantly critical; refer to Fig. 7.3.

particularly valuable due to their reliability for human activities—and areas that are only inundated during wetter years. As discussed in Chapter 5 regarding the pluviometric Sahel, we can observe a "hydrological core" region that is perpetually affected by floods and an expansion zone that extends only under specific conditions. We define this hydrological core as comprising areas with a high recurrence of counter-seasonal surface waters, reappearing annually with at least 90% frequency, which indicates that these areas consistently experience counter-seasonal surface waters year after year across the entire period (1984–2021).

The increase in counter-season surface waters is also evident when examining the time-series of monthly counter-season surface waters across the entire Sahel region (Fig. 7.7) and at the basin level (Fig. 7.8). The charts below illustrate these patterns, highlighting significant positive trends in recent years. However, in the Senegal basin, this trend is less pronounced, likely influenced by regulatory effects from reservoirs and dams (especially the Manantali dam).

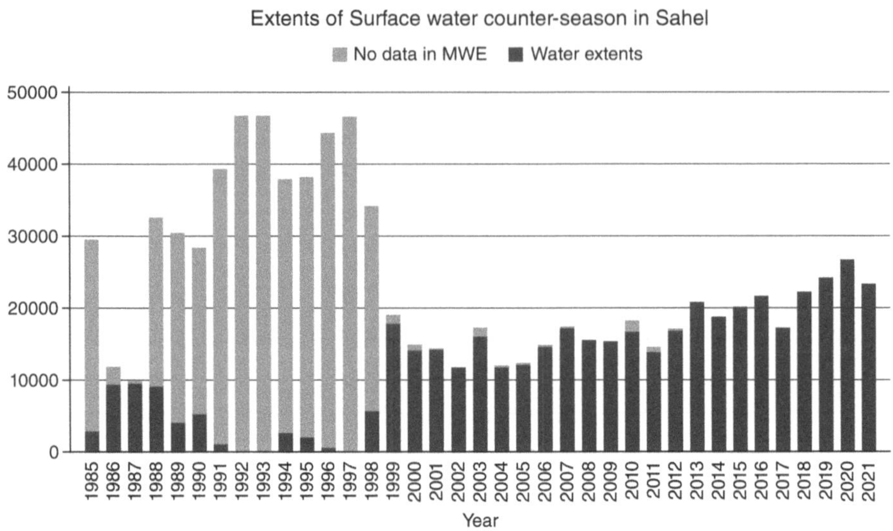

Figure 7.7 Annual extents of counter-seasonal surface waters in the Sahel, measured in square kilometres.

This analysis is significant as it underscores the vital role of these flooded areas in supporting local populations during the dry season, despite covering a very small portion of the entire Sahel area. However, these areas play a pivotal part in the livelihood of the populations by ensuring the availability of resources inaccessible during the dry season. This calculation does not cover all the low-lying areas flooded in the rainy season; rather, it includes only those that remain inundated in September, the month when the rainy season ends. These areas are accounted for and visible on our map—due to substantial water accumulations in ponds or semi-permanent wetlands.

However, various situations can also occur, differentiated in space and time, such as particularly abundant rains in a single section of one of the involved hydrographic basins. Indeed, empirical evidence reveals a higher level of complexity and variability when examining spatial and temporal fluctuations: the cumulative extent (MWE) encompassing the initial three decades (1984–2014) amounts to 35,493 km^2 (Table 7.5), whereas during the subsequent phase spanning eight years (2014–2021), a discernible augmentation is validated, resulting in a total coverage of 40,823 km^2 (Table 7.5). The proportional contributions originating from distinct basins exhibit a relatively consistent pattern across the various temporal segments under scrutiny. Niger consistently retains its status as the predominant source across the two delineated periods, with Chad and Nile collectively accounting for approximately 80% of the aggregate counter-seasonal water surface area.

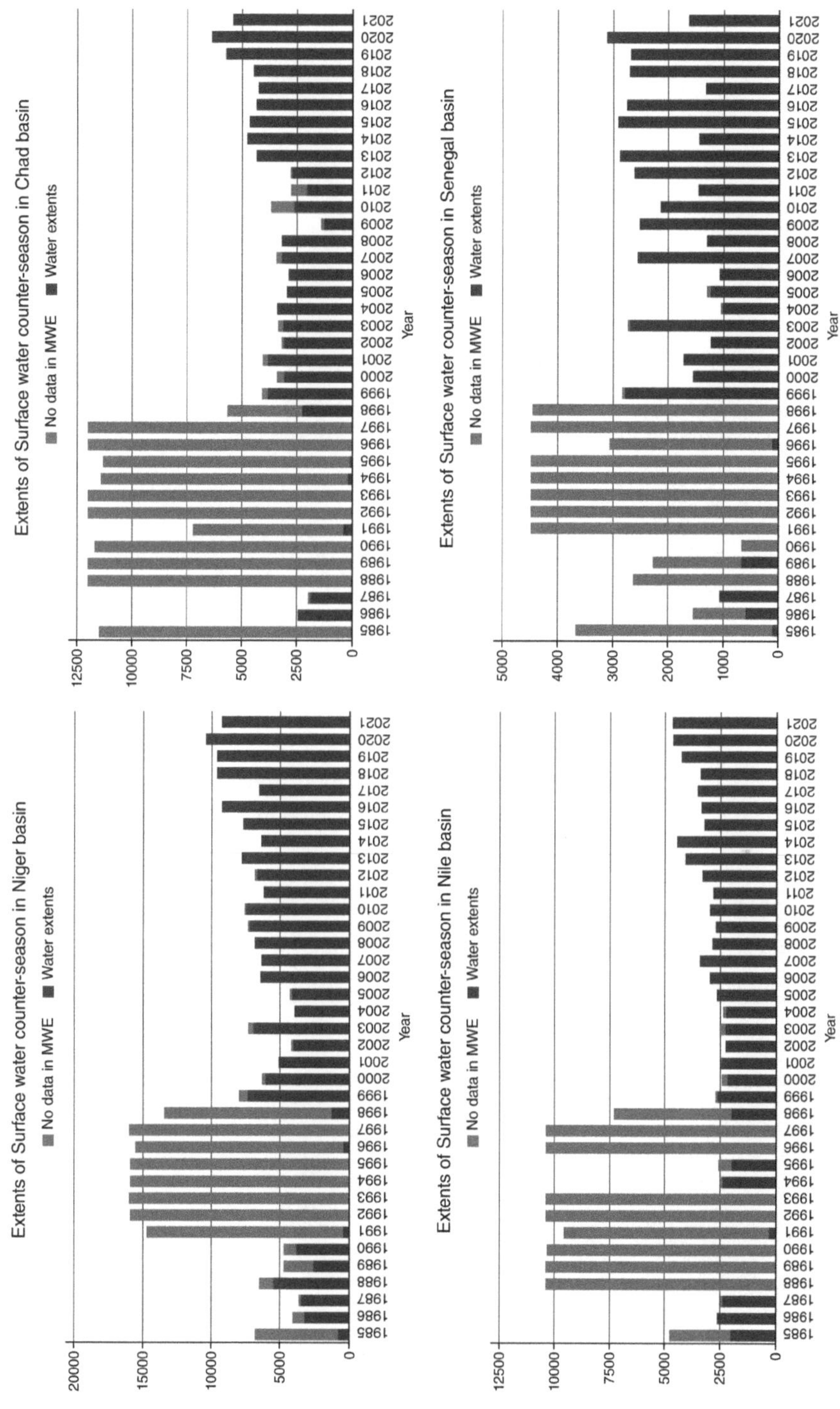

Figure 7.8 Annual extents of counter-seasonal surface waters feed major Sahelian, measured in square kilometres.

Table 7.5 Cumulative extent (maximum water extent [MWE]) of counter-seasonal surface waters: initial three decades (1984–2014) versus subsequent eight years (2014–2021) with proportional contributions from distinct basins.

Region	*MWE 1994–2014 (km²)*	*MWE 2014–2021 (km²)*
Niger basin	12,323	14,674
Chad basin	8876	10,635
Nile basin	7213	8648
Senegal basin	4065	3788
Volta basin	993	1488
Hydrological Sahel	35,493	40,823

The rising trend of counter-seasonal surface water presence over time indicates the emergence of water in regions where such occurrences were previously absent, referred to as "new water". When evaluating the contribution per basin of such "new water" presence, the distribution of these counter-seasonal waters displays a novel pattern compared with the consistent distribution exhibited by the MWE over time.

The analysis of the data provided in Table 7.6 reveals a discrepancy in the proportional contributions of new water with respect to the hydrological extent discussed earlier, focusing on the influx of counter-seasonal surface water per basin over the most recent five-year period (2017–2021). In this time span, the Chad basin experienced a relatively higher influx of new water compared with the Niger and the Nile basins.

These findings indicate the Chad basin's notably more dynamic behaviour over the past five years compared with other basins. The statistical evidence about the

Table 7.6 Surface of counter-season new water occurrence per basin in the last five years (2017–2021), where "new water" refers to the emergence of surface water during the counter-season in areas where it was previously never detected across the time window (1984–2021).

Region	*New surface water in last five years (2017–2021) (km²)*	*Proportional contribution of new water with respect to total* new water in last five years (2017–2021) (%)*
Niger basin	2272	28
Chad basin	2530	32
Nile basin	2355	29
Senegal basin	205	2.5
Volta basin	502	6.2
Hydrological Sahel	***7997**	–

* The final figure represents the total extent of the hydrological Sahel within the pluviometric Sahel and does not directly sum up the individual basin extents shown because it includes smaller basins that are not individually accounted for. Similarly, the percentages do not add up to 100% due to these unaccounted basins.

Table 7.7 Extents of the "hydrological core" in square kilometres for major basins and for the whole hydrological Sahel with their respective proportional contributions.

Region	*Hydrological core extents*	*Proportional contribution of core with respect to total** *(1984–2021) (%)*
Niger basin	683	47
Chad basin	103	7.1
Nile basin	534	37
Senegal basin	80	5.5
Volta basin	36	2.5
Hydrological Sahel	***1443**	–

* The final figure represents the total extent of the hydrological Sahel within the pluviometric Sahel and does not directly sum up the individual basin extents shown because it includes smaller basins that are not individually accounted for. Similarly, the percentages do not add up to 100% due to these unaccounted basins.

hydrological core, high recurrent counter-seasonal surface water areas, further underscores the dynamic nature of counter-seasonal surface waters. Over the 38-year observation period, the Niger basin accounted for nearly half of the high recurrent counter-seasonal surface waters in the Sahel region. In contrast, the Chad basin contributed only 103 km^2, representing just 7.1% of the hydrological core observed from 1984 to 2021 (Table 7.7).

In conclusion, we can observe a distinct pattern with the *hydrological core* consistently impacted by counter-seasonal floods and an expanding area that enlarges only under specific conditions. This phenomenon is illustrated in the maps in Figs. 7.9 and 7.10, depicting the Inner Niger Delta (Mali) and Lake Chad (Chad), respectively.

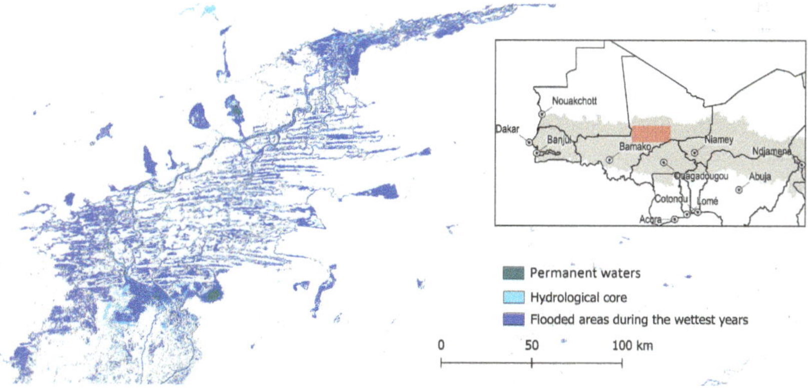

Figure 7.9 The hydrological Sahel—details of the Inner Niger Delta (Mali). The map represents the difference between the hydrological core that remains flooded even during the driest years and the areas flooded only during the wettest two-year period and the permanent waters.

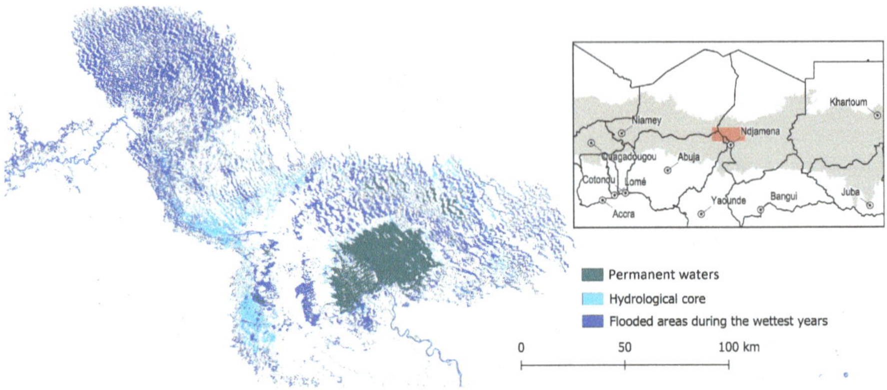

Figure 7.10 The hydrological Sahel—details of Lake Chad (Chad). The map represents the difference between the hydrological core that remains flooded even during the driest years and the areas flooded only during the wettest two-year period and the permanent waters.

Taming Sahelian rivers: Dams and reservoirs

Dams and reservoirs in the Sahel region play a crucial role in water storage and management, affecting both water availability and quality. For the purposes of this chapter, their impacts on floods (duration and peaks) and thus on the extent of the hydrological Sahel are of particular relevance. While these structures provide clear benefits, the environmental and social consequences of these dams are sources of substantial controversy. Dams have led to changes in land cover, such as the expansion of agriculture, loss of wetlands and deforestation (Barnieh et al., 2020; Fu et al., 2023). These changes have influenced river connectivity, flow regimes and water abstractions, impacting rivers and wetlands in the Sahel and surrounding semi-arid regions (McClain, 2012; Wetlands International, 2017), which in turn affect local agriculture, fishing and pastoralism. Major dams and water infrastructure constructions could lead to mass migration due to drying up of wetlands and changing hydrological conditions, impacting people's lives along the river (Xu & Famiglietti, 2023). Studies have shown that sediment incursion is a significant issue, reducing reservoir capacities in Sahelian regions such as Kano State, with reservoirs losing millions of litres of storage annually (Yakubu et al., 2015). Additionally, the construction of hydraulic installations on rivers such as Senegal, although it has led to stable water levels, has also altered the hydrological regime and water quality, which in turn has increased vegetation growth and contributed to the spread of diseases (Cogels et al., 1997). The environmental and human impacts of dam construction in the Sahel can be classified into various categories, such as climatic, hydraulic, biological, social and cultural, emphasising the need for comprehensive measures to minimise adverse effects on nature and humans (Cogels et al., 1997).

Despite these challenges, dams and reservoirs remain essential in the Sahel region, addressing water scarcity challenges and supporting various aspects of life in this semi-arid zone (Guiraud, 1990; Kibret et al., 2018; Owusu et al., 2022). These structures are relevant for managing water resources, offering benefits, such as irrigation for agriculture, hydroelectric power generation, flood control and enhanced water security.

The controversy surrounding dams underscores the need to carefully weigh their benefits against their significant environmental and social costs. Despite these concerns, the strategic construction of dams and reservoirs remains a key component of development narratives and regional plans (see Chapter 2; FAO, 2024). For example, Burkina Faso has significantly increased the number of its small reservoirs from 200 in 1974 to approximately 1650 in 2008 to address food security issues (de Fleury et al., 2023).

However, the remaining deficiency in the comprehensive assessment of these infrastructures in the Sahel region highlights the need for more thorough evaluations. To begin addressing this gap, we utilised two recent global datasets to conduct a thorough analysis of dams and reservoirs in the region.

The first dataset is from the Global Dam Watch (GDW) consortium, which can be accessed at www.globaldamwatch.org (Lehner et al., 2024). This dataset was initiated through a collaboration among several academic institutions and NGOs. The consortium aims to fill critical gaps in global dam and reservoir information. The current version of the GDW database (version 1.0 delta) includes 41,145 barrier and dam locations with 35,295 associated reservoir polygons and a wide range of attribute information about both dams and reservoirs.

The second dataset is provided by the Global Water Watch (GWW) platform and partnership, available at www.globalwaterwatch.earth (Donchyts et al., 2022). This dataset offers reservoir locations and statistics based on surface extent time-series and integrates the results of multiple studies to identify a total of 71,208 small- and medium-sized reservoirs.

After scrutinising the reservoirs located in the Sahel region based on the aforementioned datasets, we conducted a quality assessment to eliminate redundancies and erroneously included reservoirs. The GWW dataset is particularly vulnerable to such errors; hence, we prioritised the reservoirs present in the GDW dataset in cases of duplication due to numerous attributes that would subsequently be utilised. Afterwards, these reservoirs were consolidated into a single set totalling 523 (Fig. 7.11), with 336 originating from GWW and the remaining 187 from GDW.

In this dataset, it is necessary to include the Grand Ethiopian Renaissance Dam (GERD), which Ethiopia began building in 2011 on the Blue Nile close to the Sudanese border for the purpose of generating electricity, with plans to impound a lake with a capacity of 74 billion cubic metres of water. The initial filling of the dam's reservoir commenced in July 2020.

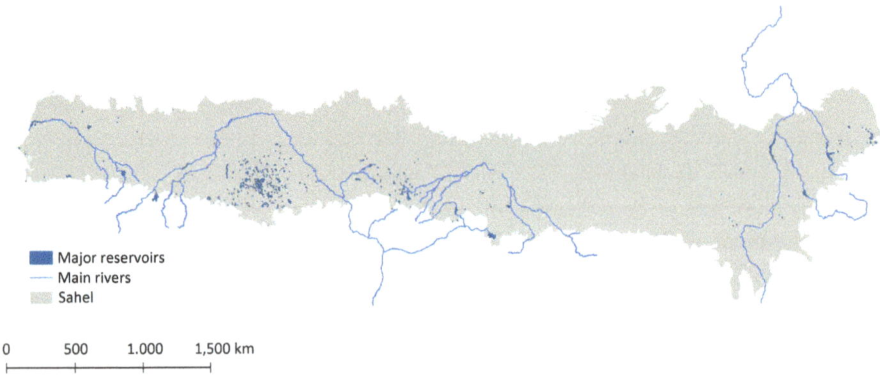

Figure 7.11 The major reservoirs (in blue) in the pluviometric Sahel.

Furthermore, it is crucial to consider significant reservoirs located beyond the boundaries of the Sahel region, as defined in Chapter 5, which are closely related to the control of floods. These reservoirs play an essential role in the overall hydrological system of the region, highlighting the interconnected nature of water resources management in this area.

In the tabular representation (Table 7.8), the major reservoirs are delineated along with their corresponding maximum storage capacity, as reported in the GDW dataset attribute (CAP_MCM), along with an indication of whether the reservoirs are outside the Sahel, as defined in Chapter 5.

It is important to note that not all reservoirs have reported storage capacities; only those from the GDW dataset include this attribute, totalling 195 reservoirs. Considering both the reservoirs in the region of interest and the most significant ones (even if beyond the limits of the pluviometric Sahel), we have a total of 531 reservoirs with a combined storage capacity of 308,932 million cubic metres (mcm) (including the GERD). In the pluviometric Sahel region, the total storage capacity is 125,627 mcm. In contrast, the 24 major reservoirs listed in Table 7.8 alone have a combined storage capacity of 296,817 mcm.

This substantial storage capacity holds significance due to its mitigating effect on drought and flood hazards at a local level, particularly in arid regions. Reservoirs play a crucial role in regulating both local and regional drought and flood patterns, as highlighted in Brunner's (2021) research. Their ability to mitigate the intensity of these natural occurrences comes with the trade-off of prolonging their duration (Brunner, 2021), with various factors such as reservoir function and climatic conditions contributing to their overall influence, as discussed in two studies (Brunner, 2021; Figueiredo & Becker, 2018). Table 7.9 shows that 99 reservoirs have storage capacities under 10 mcm, a large portion of 195 reservoirs with reported storage capacities (GDW dataset). Furthermore, 29 reservoirs fall

Table 7.8 Representative maximum storage capacity (CAP) in million cubic metres (from GDW dataset attributes) of the major reservoirs located in the Sahel's main basins.

Reservoirs	*CAP_MCM*
Maga	50
Shiroro**	59.7
Sennar	100
Markala	175
Diama	300
Bakalori	450
Kiri Dam**	615
Goronyo	974
Challawa Gorge	1000
Khashm el Girba	1300
Merowe**	1930
Tiga	1968
Selingue	2170
Upper Atbara e Setit	2700
Dadin Kowa	2855
Roseires	3000
Jebel Aulia	3500
Jebba**	3600
Lagdo	7800
Manantali	11,270
Kainji**	15,000
Aswan**	162,000
GERD*	74,000
Total capacity	**296,816.7**

* Grand Ethiopian Renaissance Dam (GERD) total capacity (Melesse et al., 2021).

** Reservoirs outside the pluviometric Sahel, as defined in Chapter 5, but relevant to the overall functioning of hydrographic systems.

Table 7.9 Number of reservoirs grouped by significant storage capacity classes (in million cubic metres), based on the CAP_MCM attribute from the GDW dataset, in the pluviometric Sahel, with added major dams selected (see Table 7.8).

Storage capacity range (CAP_MCM) (mcm)	*Number of reservoirs*
<10	99
10–20	29
<20	128
>20	67
>500	23

Note: The total capacity is derived from 195 reservoirs in the GDW dataset.

between 10 and 20 mcm, 67 reservoirs exceed 20 mcm and 23 reservoirs surpass 500 mcm, indicating a smaller but significant number of larger reservoirs.

Given our interest in examining the impacts of substantial infrastructures, we solely consider reservoirs with total capacities surpassing 20 mcm as a reasonable threshold when analysing the total capacity.

Among the attributes present in the GSW dataset are the estimated year of construction of each dam and the country where the reservoir is located. By utilising these attributes, the distribution of reservoirs based on both the country and the year of construction can be analysed. Considering the above threshold of 20 mcm, it is evident that Nigeria has the highest number of major reservoirs, in contrast to Egypt, which possesses only one, among those we considered (Fig. 7.12).

However, information on the total capacity of reservoirs per country (total capacity derived from 195 reservoirs in the GDW dataset) is useful in clarifying the big picture (Fig. 7.13).

When considering hydrological basins rather than national boundaries, the Niger basin contains the highest number of significant reservoirs (above the threshold of 20 mcm), totalling 23. Additionally, the Chad, Volta and Nile basins stand out with 14, 13 and 10 reservoirs, respectively (Fig. 7.14). However, when assessing total installed capacity, the dynamics shift significantly. The Nile basin emerges as dominant, boasting an impressive capacity of nearly 250,000 mcm, whereas the Niger basin's capacity is substantially lower, at less than 35,000 mcm (Fig. 7.15).

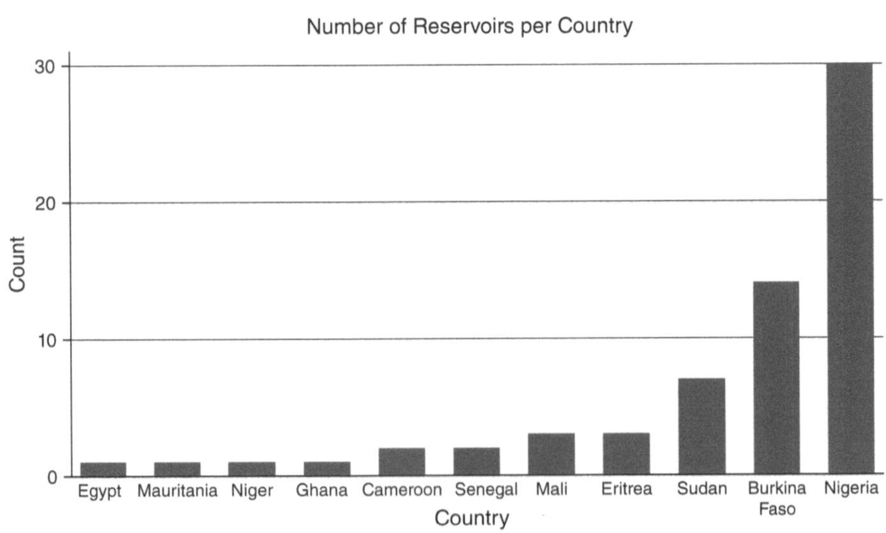

Figure 7.12 Number of reservoirs per country.

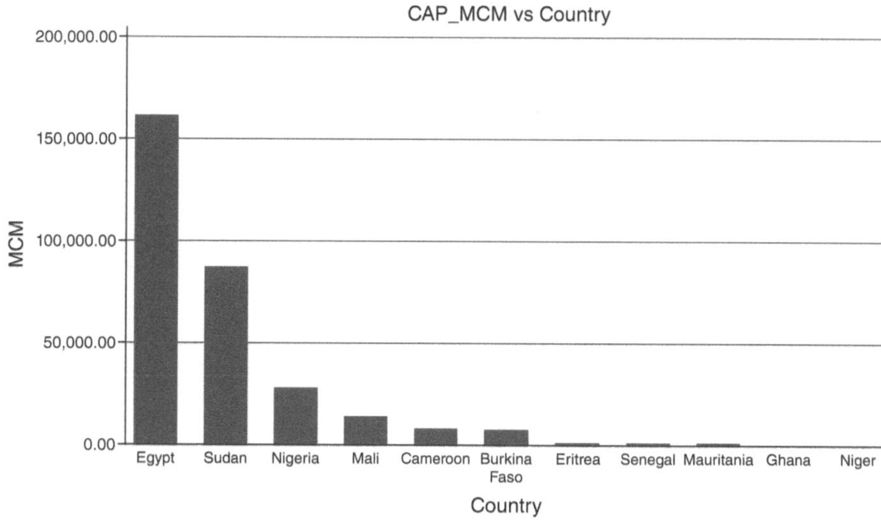

Figure 7.13 Cumulative total capacity (CAP_MCM) of reservoirs per country.

When examining the attribute Year_dam (according to the GDW database—technical documentation—version 1.0 delta, it could be the year of construction, year of completion or year of commissioning, either reported or estimated), a notable peak is observed in the 1970s, extending up to the year 2000 (Fig. 7.16). It should also be noted that the construction of dams was already significant in

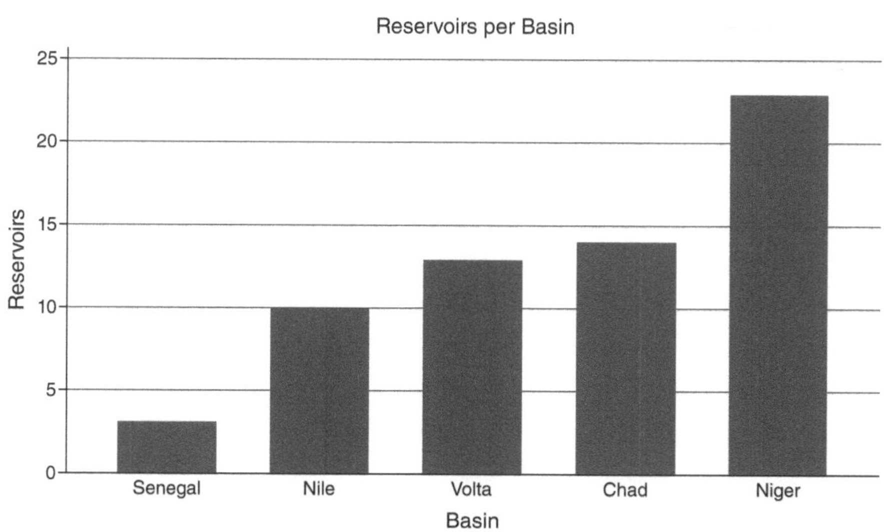

Figure 7.14 Number of reservoirs per basin.

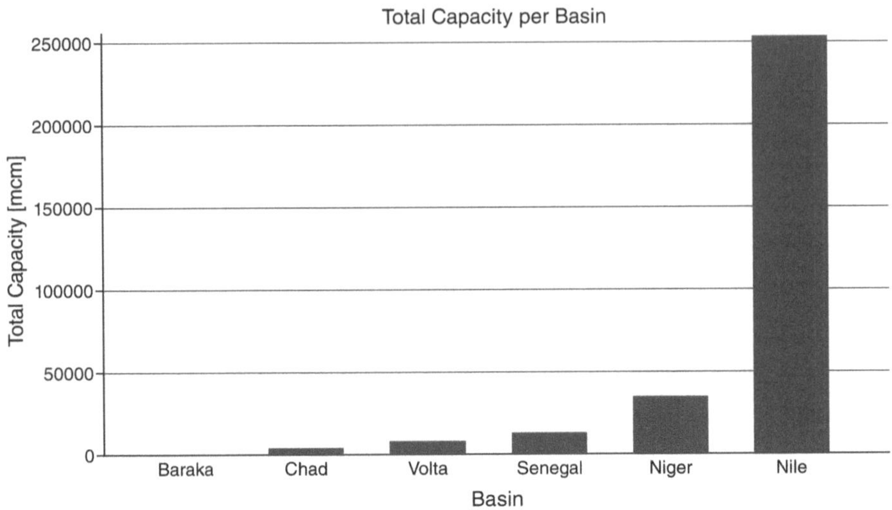

Figure 7.15 Total capacity of reservoirs per basin.

colonial times and that the data pertaining to the year originate from a total of 135 reservoirs in the GDW database.

People and rivers: Footprint on water

In this chapter on the hydrological dynamics of the Sahel, we have attempted to analyse the human impact on watercourses by calculating the human footprint on water resources. If there is one thing on which a person cannot leave one's

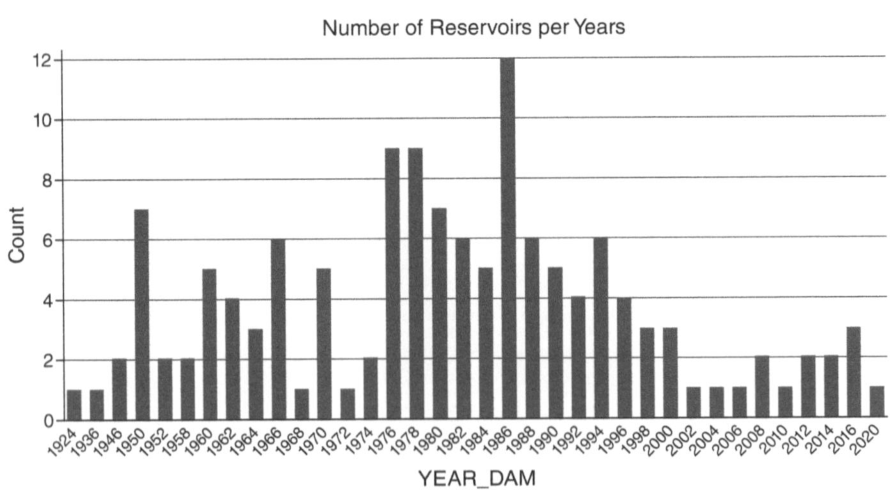

Figure 7.16 Annual distribution of reservoir construction in the Sahel considering the attribute Year_dam (GDW database).

footprint, it is water. The metaphor we propose is therefore paradoxical. However, it seems significant to us in identifying the "weight" that human communities exert on water resources and, in particular, on the great Sahelian rivers. This aspect includes the regulation of rivers for irrigation and energy production purposes, as outlined in the previous section, which, in turn, affects water resource management and supply for local populations. In this last section, we intend to provide a concise depiction of how anthropogenic pressure increases from upstream to downstream, following the major rivers flowing into the Sahel region. We consider these rivers: Senegal, Niger, certain stretches of the Nile (both the White Nile and the Blue Nile and the Atbara) and finally—regarding the Lake Chad basin—Chari-Logone and Komadugu.

In the map shown in Fig. 7.17, the focus shifts to representing the increase in the water flow of the rivers, segmented at the confluence points of the major tributaries. The map indicates major dams that regulate and manage the flow. While the map shows the positions of the dams, it does not account for the changes in water flow caused by them.

This visual representation (Fig. 7.17) provides insights into how the flow of watercourses evolves as they receive additional water input from tributaries along their course. At each junction of a tributary to one of the major rivers, and taking into account roughly the size of the contribution, the blue line outlines the river's flow becoming proportionally thicker.

At this point, we have tried to represent how anthropogenic pressure increases as the rivers progress downstream. To do so, we first segmented each river into stretches between the main "breaks" that characterise them: the confluence of the main tributaries, on one hand, and the major dams, on the other hand. For each river segment (as defined, between artificial dams or confluences), a buffer zone of 10 km from each bank was calculated to define the area more directly affected by the proximity of the river. The resident population in each of these buffer zones was then determined using data from the JRC Global Human Settlement Layer (GHSL; Schiavina et al., 2023). The width of the black line broadly indicates the size of the

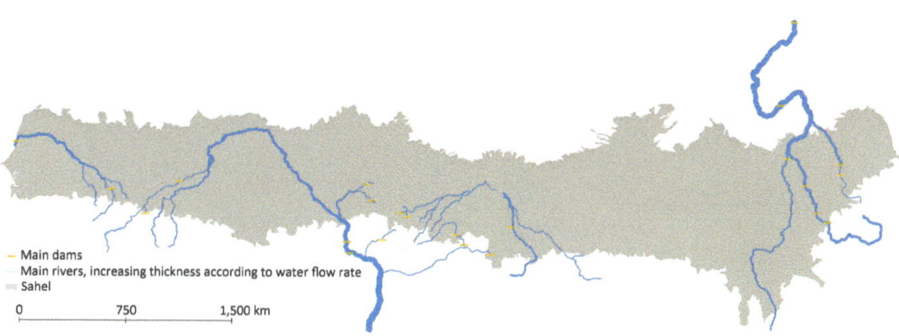

Figure 7.17 Main dams and river flow.

Table 7.10 Population sizes on the river segments considered (10-km buffer zone).

River	*Population in the first considered section*	*Population in the last section*
Senegal	86,628	2,400,384
Chari-Logone	8968	4,213,526
Niger	1,953,534	27,982,905
Nile	3,035,465	17,532,370
Komadugu	1,063,328	12,204,703

population that gravitates to those waters. Subsequently, the population data for each river segment were aggregated by adding the subsequent segments following the river's course (Table 7.10). With this aggregation, we aim to illustrate how the population that is dependent on the river increases along its length. The resulting data were used to create a map, presented in Fig. 7.18, where each river segment thickens in proportion to the total population considered (buffer zone) within its upstream area.

By comparing both maps—one depicting the increase in water flow and the other showing the population gravitating to these waters—we can obtain a comprehensive, though certainly rough, view of the interactions between human activities and water systems. This is obviously just a graphic proposal to show at a glance another reason why the waters of the Sahel are so troubled.

Conclusion

In this chapter, we have provided the first examination of the hydrological dynamics in the Sahel, focusing on the seasonal fluctuations and the behaviour of major river systems, such as Senegal, Niger, Chad and Nile. Through the use of satellite data (30 m/px), we have observed an increase in seasonal water bodies,

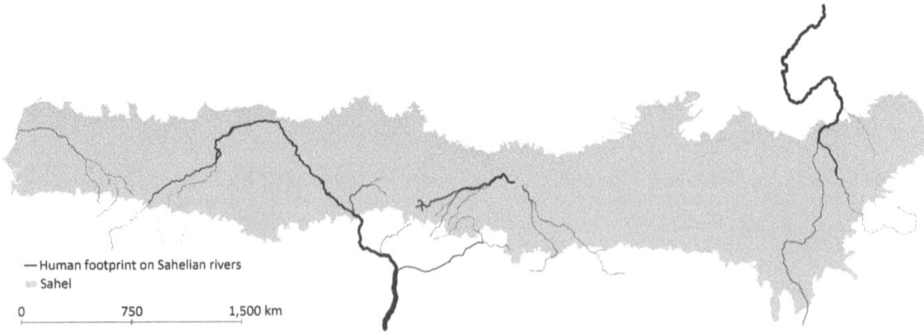

Figure 7.18 Human footprint on Sahelian rivers.

particularly beyond the rainy season, contributing to what we have termed the "hydrological Sahel". This phenomenon highlights the region's specific water dynamics, where significant flooding occurs due to rainfall in areas outside the Sahel, subsequently affecting the region with a delay of the flood waves lasting for weeks or even months.

However, the findings reveal a complex interplay of factors leading to greater variability and unpredictability in water availability. While an increase in seasonal water resources is noted, it emphasises the growing challenges posed by the increasing variability and the altered distribution, influenced by both natural factors and human interventions.

The expansion of human activity, particularly through large-scale irrigation and the development of infrastructures such as dams, has significantly impacted these water systems. While these interventions have allowed the regulation and utilisation of water resources beyond the rainy season, they have also introduced new challenges, such as the disruption of natural hydrological processes and the degradation of vital ecosystems. From this perspective, the troubled outcomes of large-scale irrigation schemes may appear paradoxical (see Chapter 1).

We have also highlighted the growing human footprint along the Sahel's rivers, with an increasing population dependent on these waters for its livelihood. This intensification of anthropogenic pressure poses a significant challenge to the sustainability of the region's water resources, as the demands of development often clash with the need to preserve these ecosystems for future generations.

In conclusion, the hydrological dynamics of the Sahel are becoming increasingly troubled due to both natural variability and human intervention. While the data do not indicate a decrease in the total amount of available water, they bring attention to the growing challenges in managing these resources sustainably. In this unique and sensitive Sahelian context, policymakers and stakeholders must focus on balancing the technicist approach, oriented primarily to economic factors, with the long-term sustainability of water resources to ensure that the region can adapt to its changing hydrological landscape without compromising its future.

References

Aich, V., Liersch, S., Vetter, T., Andersson, J., Müller, E., & Hattermann, F. (2015). Climate or land use?—Attribution of changes in river flooding in the Sahel zone. *Water*, *7*(12), 2796–2820. https://doi.org/10.3390/w7062796

Appolinaire, R., Soumaïne, A. M., Borgoto, A., & Simon, A. D. (2012). Lake Chad: Meanwhile waiting for the safeguard, stopping on the management of the potentialities of the area. *OIDA International Journal of Sustainable Development*, *4*(11), 63–94.

Ardagh, J. C. (1889). Nilometers. *Proceedings of the Royal Geographical Society and Monthly Record of Geography*, *11*(1), 28–38.

Barnieh, B. A., Jia, L., Menenti, M., Zhou, J., & Zeng, Y. (2020). Mapping land use land cover transitions at different spatiotemporal scales in West Africa. *Sustainability*, *12*(20), Article 8565. https://doi.org/10.3390/su12208565

Brunner, M. I. (2021). Reservoir regulation affects droughts and floods at local and regional scales. *Environmental Research Letters*, *16*(12), Article 124016. https://doi.org/10.1088/1748-9326/ac36f6

Chagnaud, G., Panthou, G., Vischel, T., & Lebel, T. (2022). A synthetic view of rainfall intensification in the West African Sahel. *Environmental Research Letters*, *17*(4), Article 044005. https://doi.org/10.1088/1748-9326/ac4a9c

Cogels, F. X., Coly, A., & Niang, A. (1997). Impact of dam construction on the hydrological regime and quality of a Sahelian lake in the River Senegal basin. *Regulated Rivers Research & Management*, *13*(1), 27–41. https://doi.org/10.1002/(SICI)1099-1646(199701)13:1<27::AID-RRR421>3.0.CO;2-G

de Fleury, M., Kergoat, L., & Grippa, M. (2023). Hydrological regime of Sahelian small waterbodies from combined Sentinel-2 MSI and Sentinel-3 Synthetic Aperture Radar Altimeter data. *Hydrology and Earth System Sciences*, *27*(11), 2189–2204. https://doi.org/10.5194/hess-27-2189-2023

de Planhol, X., & Rognon, P. (1970). *Les zones tropicales arides et subtropicales*. Colin.

Descroix, L., Bouzou, I., Genthon, P., Sighomnou, D., Mahe, G., Mamadou, I., Vandervaere, J. P., Gautier, E., Faran, O., Rajot, J. L., Malam, M., Dessay, N., Ingatan, A., Noma, I., Souley, K., Karambiri, H., Fensholt, R., Albergel, J., & Olivry, J. C. (2013). Impact of drought and land – Use changes on surface – Water quality and quantity: The Sahelian paradox. In *InTech eBooks*. https://doi.org/10.5772/54536

Descroix, L., Guichard, F., Grippa, M., Lambert, L., Panthou, G., Mahe, G., Gal, L., Dardel, C., Quantin, G., Kergoat, L., Bouaita, Y., Hiernaux, P., Vischel, T., Pellarin, T., Faty, B., Diongue-Niang, A., Wilcox, C., Abdou, M. M., Mamadou, I., … Paturel, J. E.. (2018). Evolution of surface hydrology in the Sahelo-Sudanian strip: An updated review. *Water*, *10*(6), Article 748. https://doi.org/10.3390/w10060748

Descroix, L., Mahé, G., Lebel, T., Favreau, G., Galle, S., Gautier, E., Olivry, J. C., Albergel, J., Amogu, O., Cappelaere, B., Dessouassi, R., Diedhiou, A., Breton, E. L., Mamadou, I., & Sighomnou, D. (2009). Spatio-temporal variability of hydrological regimes around the boundaries between Sahelian and Sudanian areas of West Africa: A synthesis. *Journal of Hydrology*, *375*(1–2), 90–102. https://doi.org/10.1016/j.jhydrol.2008.12.012

Donchyts, G., Winsemius, H., Baart, F., Dahm, R., Schellekens, J., Gorelick, N., Iceland, C., & Schmeier, S. (2022). High-resolution surface water dynamics in Earth's small and medium-sized reservoirs. *Scientific Reports*, *12*(1). https://doi.org/10.1038/s41598-022-17074-6

D. S. R. (1955). The Cairo Nilometer, studies in Ibn Taghrî Birdî's chronicles of Egypt. *Bulletin of the School of Oriental and African Studies*, *17*(1), 185. https://doi.org/10.1017/s0041977x00106627

Elagib, N. A., Zayed, I. S. A., Saad, S. A., Mahmood, M. I., Basheer, M., & Fink, A. H. (2021). Debilitating floods in the Sahel are becoming frequent. *Journal of Hydrology*, *599*, Article 126362. https://doi.org/10.1016/j.jhydrol.2021.126362

FAO. (2024). *Regional initiative for the Sahel*. Retrieved May 3, 2024, from https://www.fao.org/hand-in-hand/hih-IF-2023/regional-initiative-for-the-sahel

Fensholt, R., Rasmussen, K., Kaspersen, P., Huber, S., Horion, S., & Swinnen, E. (2013). Assessing land degradation/recovery in the African Sahel from long-term Earth observation-based primary productivity and precipitation relationships. *Remote Sensing*, *5*(2), 664–686. https://doi.org/10.3390/rs5020664

Figueiredo, A. D. V., & Becker, V. (2018). Influence of extreme hydrological events in the quality of water reservoirs in the semi-arid tropical region. *RBRH*, *23*. https://doi.org/10.1590/2318-0331.231820180088

Fu, S., Zhou, Y., Lei, J., & Zhou, N. (2023). Changes in the spatiotemporal of net primary productivity in the conventional Lake Chad basin between 2001 and 2020 based on CASA model. *Atmosphere*, *14*(2), Article 232. https://doi.org/10.3390/atmos14020232

Gal, L., Grippa, M., Hiernaux, P., Pons, L., & Kergoat, L. (2016). *Modeling the paradoxical evolution of runoff in pastoral Sahel. The case of the Agoufou watershed, Mali*. https://doi.org/10.5194/hess-2016-623

Gal, L., Grippa, M., Hiernaux, P., Pons, L., & Kergoat, L. (2017). The paradoxical evolution of runoff in the pastoral Sahel: Analysis of the hydrological changes over the Agoufou watershed (Mali) using the KINEROS-2 model. *Hydrology and Earth System Sciences*, *21*(9), 4591–4613. https://doi.org/10.5194/hess-21-4591-2017

Gorelick, N., Hancher, M., Dixon, M., Ilyushchenko, S., Thau, D., & Moore, R. (2017). Google Earth Engine: Planetary-scale geospatial analysis for everyone. *Remote Sensing of Environment*, *202*, 18–27. https://doi.org/10.1016/j.rse.2017.06.031

Goward, S., Arvidson, T., Williams, D., Faundeen, J., Irons, J., & Franks, S. (2006). Historical record of Landsat global coverage. *Photogrammetric Engineering & Remote Sensing*, *72*(10), 1155–1169. https://doi.org/10.14358/pers.72.10.1155

Guiraud, R. (1990). Small dams and subsurface dams in arid and semi-arid areas: The example of Saharan and Sahelian Africa. In R. Paepe, R. W. Fairbridge, & S. Jelgersma (Eds.), *Greenhouse effect, sea level and drought*. NATO ASI Series, vol 325 (pp. 513–522). Springer. https://doi.org/10.1007/978-94-009-0701-0_31

Kibret, S., Wilson, G. G., Ryder, D., Tekie, H., & Petros, B. (2018). Can water-level management reduce malaria mosquito abundance around large dams in sub-Saharan Africa? *PLoS ONE*, *13*(4), e0196064. https://doi.org/10.1371/journal.pone.0196064

Lehner, B., Beames, P., Mulligan, M., Zarfl, C., De Felice, L., van Soesbergen, A., Thieme, M., Garcia de Leaniz, C., Anand, M., Belletti, B., Brauman, K. A., Januchowski-Hartley, S. R., Mandle, L., Mazany-Wright, N., Messager, M. L., Pavelsky, T., Pekel, J.-F., Wang, J., Wen, … Higgins, J. (2024). The Global Dam Watch database of river barrier and reservoir information for large-scale applications. *Nature Scientific Data*, *11*(1), 1069.

Lemoalle, J. (2015). Le bassin du lac Tchad: Plusieurs définitions possibles. *Passages*, 20–22. https://horizon.documentation.ird.fr/exl-doc/pleins_textes/divers21-02/010064582.pdf

McClain, M. E. (2012). Balancing water resources development and environmental sustainability in Africa: A review of recent research findings and applications. *AMBIO*, *42*(5), 549–565. https://doi.org/10.1007/s13280-012-0359-1

Melesse, A. M., Abtew, W., & Moges, S. A. (2021). *Nile and Grand Ethiopian Renaissance Dam*. Springer Nature. http://books.google.ie/books?id=1mxAEAAAQBAJ&printsec=frontcover&dq=Nile+and+Grand+Ethiopian+Renaissance+Dam:+Past,+Present+and+Future&hl=&cd=1&source=gbs_api

Ndehedehe, C. E., Agutu, N. O., Okwuashi, O., & Ferreira, V. G. (2016). Spatio-temporal variability of droughts and terrestrial water storage over Lake Chad basin using independent component analysis. *Journal of Hydrology*, *540*, 106–128. https://doi.org/10.1016/j.jhydrol.2016.05.068

Owusu, S., Cofie, O., Mul, M., & Barron, J. (2022). The significance of small reservoirs in sustaining agricultural landscapes in dry areas of West Africa: A review. *Water*, *14*(9), Article 1440. https://doi.org/10.3390/w14091440

Pase, A., Gianoli, F., De Felice, L., Bertoncin, M., Cherlet, M., & Kronenburg Garcìa, A. (2022). Il respiro del Sahel. Rappresentazioni di uno spazio in movimento. *Rivista Geografica Italiana*, *CXXIX*(1), 32–59.

Pekel, J. F., Cottam, A., Gorelick, N., & Belward, A. S. (2016). High-resolution mapping of global surface water and its long-term changes. *Nature*, *540*(7633), 418–422. https://doi.org/10.1038/nature20584

Ragettli, S., Donauer, T., Molnar, P., Delnoije, R., & Siegfried, T. (2022). Unraveling the hydrology and sediment balance of an ungauged lake in the Sudano-Sahelian region of West Africa using remote sensing. *Earth Surface Dynamics*, *10*(4), 797–815. https://doi.org/10.5194/esurf-10-797-2022

Rameshwaran, P., Bell, V. A., Davies, H. N., & Kay, A. L. (2021). How might climate change affect river flows across West Africa? *Climatic Change*, *169*(3–4). https://doi.org/10.1007/s10584-021-03256-0

Schiavina, M., Freire, S., Carioli, A., & MacManus, K. (2023). *GHS-POP R2023A – GHS population grid multitemporal (1975–2030)*. European Commission, Joint Research Centre (JRC). http://data.europa.eu/89h/2ff68a52-5b5b-4a22-8f40-c41da8332cfe, https://doi.org/10.2905/2FF68A52-5B5B-4A22-8F40-C41DA8332CFE

Sutcliffe, J. V. (2009). The hydrology of the Nile basin. In *Monographiae biologicae* (pp. 335–364). https://doi.org/10.1007/978-1-4020-9726-3_17

Torrey, C. C. (1952). Review of *The Cairo Nilometer: Studies in Ibn Taghrī Birdī's Chronicles of Egypt*, by W. Popper. *Speculum*, *27*(3), 417–419.

Wetlands International. (2017). *Water shocks: Wetlands and human migration in the Sahel*. Wetlands International, The Netherlands.

Xu, L., & Famiglietti, J. S. (2023). Global patterns of water-driven human migration. *Wiley Interdisciplinary Reviews Water*, *10*(4). https://doi.org/10.1002/wat2.1647

Yakubu, M. L., Abdurrahman, U. T., Ahmed, M. T., Aliyu, A. S., Zango, M. U., & Maihula, A. S. (2015). The progressive response of Sahelian dams to sediment intrusion: Case of Kano State between 1976 and 2009. *Applied Mechanics and Materials*, *773–774*, 1301–1306.https://doi.org/10.4028/www.scientific.net/amm.773-774.1301

Yonaba, R., Biaou, A. C., Koïta, M., Tazen, F., Mounirou, L. A., Zouré, C. O., Queloz, P., Karambiri, H., & Yacouba, H. (2021). A dynamic land use/land cover input helps in picturing the Sahelian paradox: Assessing variability and attribution of changes in surface runoff in a Sahelian watershed. *The Science of the Total Environment*, *757*, Article 143792. https://doi.org/10.1016/j.scitotenv.2020.143792

8 Shapes of the water: the rectangle and the circle

A tentative survey on irrigation in the Sahel

Ludovica Crocitto, Federico Gianoli and Andrea Pase

Introduction

Drawn on maps or observed from satellite photos, the large-scale irrigation schemes (the history of which is presented in Chapter 1 of this volume) offer a well-differentiated appearance. The prevailing shapes are rectangular meshes formed by the different levels of canals in surface irrigation systems: from the major to the minor ones and then to the field channels. For example, in the Gezira irrigation system, we have many levels from the top to the fields: the hydraulic network has as its axes the two main canals that bring water from Sennar and Roseires dams. The Main Canals are the sources of the Major Canals, which, in turn, lead to the Minor Canals:

> A Minor Canal feeds to a set of 90 feddan fields (known as "numbers"). The water is taken off from the Minor Canal through a channel called an *abu 'ishreen*, which feeds a whole "number". And the tenancies [*howasha*] within a "number" are fed by smaller channels, called *abu sitta*.
>
> *(Barnett, 1977, pp. 10–11)*

The supply of these hydraulic networks, which grow in complexity as the irrigated area increases, comes from water reservoirs (e.g., in the case of the Gezira), thresholds on rivers (e.g., in the case of the Office du Niger, with the barrage of Markala) or pumping systems (e.g., the four stations of *Secteur expérimental de modernisation de la riziculture de Yagoua* [SEMRY] I). The pumping stations can be placed in series to overcome major differences in height, as is the case with the Kenana Sugar Company's six pumping stations along the 37-km span of the Main Canal, which allows the water of the White Nile to irrigate sugarcane fields up to 46 m above the river (Bertoncin et al., 2019; Ganawa & Awad Allah, 2013).

Water can be distributed via open channels (made of clay or concrete) or pipes, as is the case with the hydro-flume system, with PVC pipes used particularly for sugarcane cultivation to avoid water dispersion and to facilitate distribution. Either way, the result is a well-characterised hierarchical grid of irrigated spaces, the

DOI: 10.4324/9781003429111-11

minimum size of which is based on the irrigated parcel assigned to the individual allottee (usually 1 or 0.5 ha or even less) or on the requirement for processing specific crops, as is the case with sugarcane.

The areas irrigated with the system that has a central pivot, where water arrives from a well or an underground pipe, appear different. The sprinkler pipes are positioned above mobile towers on rotating wheels, driven by electricity in a coordinated manner around the central pivot. These systems are also suitable for undulating terrain and allow the cultivation of large areas (in the Sahel, the usual sizes are 34–63 ha per central pivot). Viewed from above, the results of this type of irrigation (developed in the USA in the 1940s and the 1950s and then widely spread worldwide) appear as green circles, particularly noticeable when placed in arid areas or even in the middle of the desert. In fact, a characteristic of these systems is their flexibility—that is, they do not necessarily depend on the proximity of a river but can also use groundwater, for example, in the case of the Nubian aquifer in Sudan. By enabling precise and uniform control over water distribution, central pivot irrigation significantly boosts crop yields, proving especially valuable in areas with limited or erratic rainfall patterns (Gaznayee et al., 2023).

Surface irrigation, with its rectangular networks, is the oldest and most widespread in the Sahel (see Chapter 1), while the central pivot system is more recent. The first large-scale experimentation with the latter began in the 1970s in Chad, south of Sarh, actually just outside the pluviometric Sahel as we define it (see Chapter 5). Located in Banda, this area encompasses around 3500 ha cultivated with sugarcane by what is now the Compagnie Sucrière du Tchad (CST). Apart from this first example, this irrigation system has mostly been spread since 2000 in Sudan, with the expansion of domestic and foreign investments, especially from Gulf countries, in fodder cultivation (see Chapter 10). The central pivots, the green circles in the desert, are therefore the "guiding fossils", the evident traces of a precise process on the ground, linked to land and water grabbing (see Chapters 1 and 10). The distinction between "rectangles" and "circles" is thus important in the irrigated landscape of the Sahel.

In this chapter, we aim to use remote sensing and mapping techniques to understand how these different "irrigation geometries" evolve over time. Satellite data allow us to cover large spaces and thus potentially the entire Sahel. As we shall show, it is actually crucial to combine the aerial view with the field research, the latter being decisive for a sound interpretation of the information drawn from the former.

The rest of the chapter is organised into three sections. In the first section, we offer a concise overview of remote sensing and explain the calculation of the Normalized Difference Vegetation Index (NDVI). We discuss how NDVI has been applied to the Sahel region using Google Earth Engine (GEE), specifying precise areas of interest (AOIs) that were utilised as separate masks for the temporal evolution calculations of selected individual or combined irrigation systems. In the second section, we examine the chronological development of irrigation systems, focusing on the expansion and subsequent reduction of circular and rectangular patterns, occasionally leading to their complete disappearance over the years under investigation. Our analysis provides data that could be useful for a comprehensive explanation from historical, environmental and political perspectives. In the third and final section, we present maps of selected areas to visually illustrate their temporal evolution.

Mapping Sahelian vegetation: Leveraging NDVI and remote sensing with GEE

Remote sensing is a technology that enables the acquisition of detailed information about the Earth's surface through sensors on satellites, aircraft or drones, without the need for humans' physical presence, making it indispensable in fields, such as agriculture, meteorology and natural resource management (Lillesand et al., 2007).

The NDVI is one of the indicators used to quantify vegetation greenness and is useful in understanding vegetation density and assessing changes in plant health (U.S. Department of the Interior, 2024). The NDVI is calculated as a ratio between the red (R) and near infrared (NIR) bands:

$$(NIR - R)/(NIR + R)$$

The NDVI values, ranging from −1 to +1, indicate different levels of vegetation cover and vigour, from bare soil to dense, healthy vegetation. Widely employed in various contexts, this indicator is instrumental in detecting and monitoring green areas, including urban green spaces and natural or human-modified vegetative regions, providing valuable insights into vegetation density and health (Tamiminia et al., 2020).

In this study, the NDVI is specifically utilised to identify and analyse irrigated areas within two different AOIs. As explained in the next subsection, AOIp refers to the AOI associated with pivot irrigation systems, while AOIsu denotes the AOI for surface irrigation. These AOIs are defined starting from an existing mask of the Sahel region, as discussed in Chapter 5. The NDVI calculations were performed using the GEE platform, ensuring precise and efficient analysis of the targeted regions.

GEE's vast public data repository contains a wide range of geospatial information. The GEE catalogue has a collection of over 40 years of spatial data, enabling global-scale temporal analysis of environmental changes. GEE provides powerful computing resources to process large geospatial datasets, continuously updated with satellite data (Gorelick et al., 2017). Key sources include Landsat (since 1982) and Sentinel (since 2015), with Sentinel-2 images updated every 5 days at a 10-m resolution and Landsat images updated every 16 days at a 30-m resolution. Remote sensing through GEE facilitates the detection and analysis of irrigated areas in the

Sahel region, which can then be further analysed using open-source geographic information system (GIS) tools (Amani et al., 2020).

The GEE platform was initially used to calculate the NDVI at the maximum resolution available across the entire Sahel region. However, it was observed that applying the Sahel mask (see Chapter 5) was too computationally intensive for GEE, resulting in unsuccessful calculations. Consequently, the decision was made to subdivide the Sahel mask into numerous hexagons with a side length of 30 km and to select only those containing irrigated areas. To further reduce computational load, hexagons containing only circles and those containing only rectangles were selected, and separate masks were created for each. The procedure is further explained in the next subsection.

Spatial analysis using hexagonal grids: Defining AOIp and AOIsu in the Sahel

Using a hexagonal grid for spatial analysis and geographic data management offers numerous benefits. The hexagonal shape ensures a regular and uniform distribution of points across the grid, minimising distortion and ensuring an even representation of elements within the analysed area. Compared to other grid shapes, hexagons are geometrically efficient and cover space more evenly, reducing distortion in representations of geographic data. The hexagonal layout also facilitates effective management of spatial relationships among points, which is crucial for spatial analyses and GISs. This arrangement enhances the precision of spatial analyses, leading to more accurate evaluations of geographic and environmental phenomena. For the reasons outlined above, the entire Sahel region has been subdivided into hexagons with a side length of 30 km, each covering an area of approximately 75,000 ha.

The hexagonal mask and all subsequent processing steps, including the delineation of the AOIs, were carried out using the open-source QGIS software. The decision to individuate a 30-km-side hexagon was based on the goal of creating a straightforward mask for clipping our AOIs during remote sensing analyses. Defining a precise AOI is essential for obtaining clean and accurate boundaries to better delineate the irrigated areas, specifically those related to the two different geometric shapes of irrigation—rectangles and circles.

To analyse the irrigation systems separately, two distinct AOIs were created: one for the pivot systems and the other for the surface systems. Central pivot systems were detected throughout the Sahel primarily using a visual method: aerial photographs were analysed starting from 2024 and going back in time to the 1980s. The results were compared with the information from the fieldworks carried out by the research team (see Chapters 1 and 10). Thus, only the hexagonal grid cells containing the pivot systems were selected, resulting in the creation of the pivot AOI, referred to as AOIp.

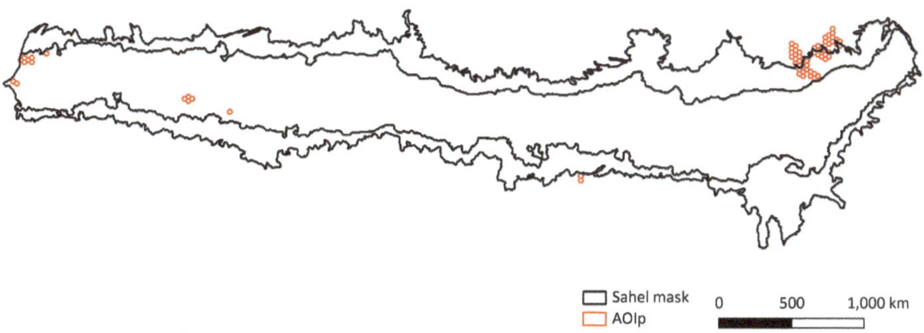

Figure 8.1 Identification of area of interest for pivot irrigation systems (AOIp).

As shown in Fig. 8.1, in some cases, the analysis extended beyond the boundaries of the pluviometric Sahel, following the principle of proximity. In other words, we have found it useful to keep irrigated areas within the limits we have assigned to the Sahel together with those immediately adjacent, which, in fact, constitute a whole in terms of organisational systems of production. This extended setting is visible in Sudan, especially in the Nile State. By focusing on proximity, we capture the whole spatial context of the selected sample areas, ensuring that each analysis considers not only the part within the Sahel boundary but also its adjacent areas.

While it was relatively easy to identify an appropriate AOI for the circles, primarily because the pivots were clearly identifiable due to their size and immediately recognisable shape, and secondarily because they were significantly fewer than the rectangles, it was a significant challenge for the surface irrigation systems. The study started from an initial hypothesis, called the "main area of interest", which included, on one hand, the large-scale irrigation schemes known to the research team and the most studied in the literature, and, on the other hand, a buffer zone around the main rivers running through the entire Sahel region to verify the possible presence of other minor irrigation schemes. The hexagons representing the main irrigated areas and all hexagons located near the Sahelian rivers were therefore selected, as illustrated in Fig. 8.2.

It was observed that many of these hexagons included extensive wetlands, as expected given the importance of seasonal flood zones in the Sahel (see Chapter 7). Calculating the areas whose NDVI was above the threshold within these hexagons would also have included wetlands, thus distancing us from our goal of monitoring irrigated areas. Only extremely detailed work would allow us to correctly distinguish wetlands from smaller irrigated schemes, which was beyond the scope of our work; however, it would be interesting research in the future. To address this issue, hexagons containing predominantly wetland areas were visually identified and excluded. This new AOI was defined as the "selected area of interest" (Fig. 8.3).

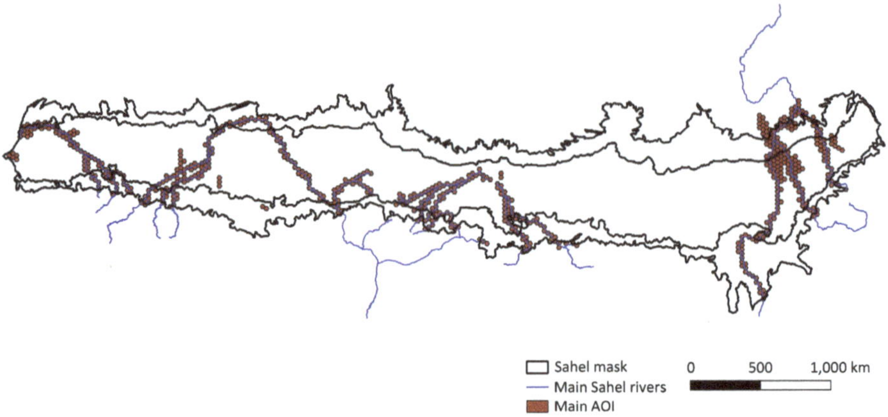

Figure 8.2 Main area of interest (AOI).

In the selected AOI, which is smaller than the one initially identified and therefore easier to deal with, we attempted to distinguish irrigated areas from wetlands, bearing in mind the particularities of the Sahelian climatic and hydrological conditions. Using satellite images, the NDVI was calculated for the selected AOI, filtered for different periods. Therefore, we first excluded the rainy season, a period in which the NDVI largely exceeds the threshold. We then excluded the months in which the floods of the major Sahelian rivers inundate neighbouring areas (see Chapter 7). Thus, we focused on the dry season at the end of the floods, that is, the period between February and April. Despite these efforts, it became apparent that it was impossible to completely separate wetlands from irrigated areas within individual hexagons without resorting to the kind of detailed analysis that is beyond our capacity.

Therefore, we decided to reduce the AOI again, limiting it to the main irrigation systems in the Sahel region, known to the research team from direct investigation or from existing literature.

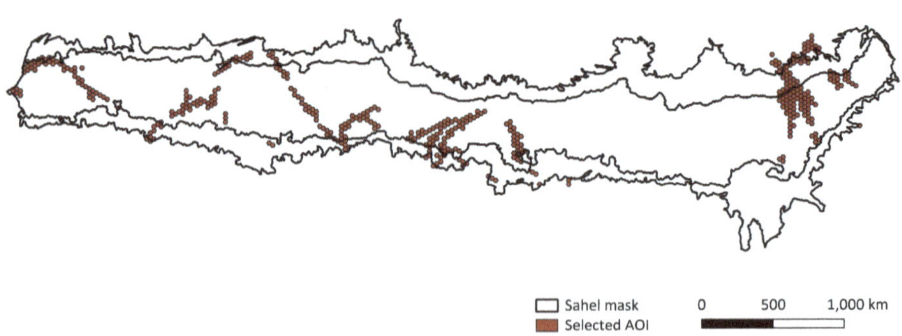

Figure 8.3 Selected area of interest (AOI).

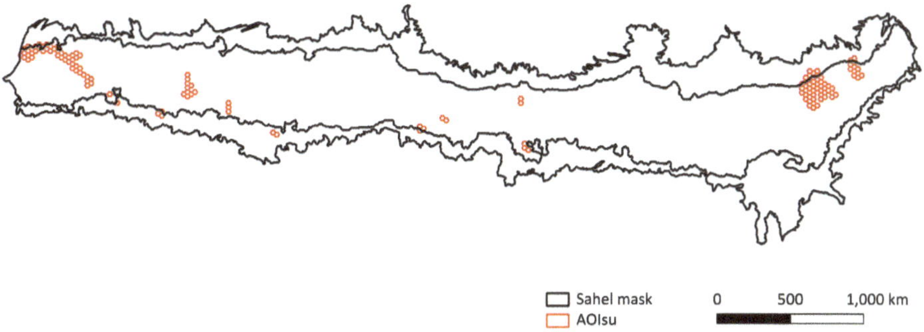

Figure 8.4 Identification of area of interest for surface irrigation (AOIsu).

Hence, we chose to identify the following irrigated areas: Senegal Valley (Senegal and Mauritania), Sourou Valley and the Bagré dam irrigated zone (Burkina Faso), the Office du Niger and the Banguinéda perimeter (Mali), the schemes downstream of the Tiga dam and of the barrage threshold of the Hadejia (Nigeria), the Bol polders (Chad) and SEMRY I and II (Cameroon), the Gezira, Rahad and Khashm El Girba schemes, and the sugarcane perimeters (Sudan). This list is limited to those areas where we feel confident, based on our experience in the field, that we can reliably identify the extent of irrigated areas. It is therefore far from an exhaustive list. However, we believe that the identified zones represent the most significant irrigated areas in the Sahel. In the future, it will be possible to add other areas to the analysis. The result is the map shown in Fig. 8.4, defined as the AOI for surface irrigation (AOIsu).

After generating a mosaic of hexagons, resulting in two AOIs, one for the pivot systems (AOIp) and the other for the rectangular areas (AOIsu), both AOIs were subsequently modified to separate irrigated areas from wetland areas. This step is crucial and of fundamental importance in data processing since the inclusion of wetland areas would have clearly altered the final data.

The chosen approach involved manually delineating the main irrigated areas, including both circular and rectangular plots, with careful attention to accurate clipping to exclude wetland areas, except the Senegalese rectangles, where detailed clipping was applied using an existing mask related to the irrigation infrastructure along Senegal River (Zwart, 2017). The result is shown in Fig. 8.5, following the delineation of AOIp and AOIsu.

Irrigation over time: Why circles and rectangles need different timeframes and thresholds

The objective is to propose a tentative survey of changes over time for both pivot and surface irrigation systems, focusing specifically within these two AOIs. This allows an assessment, albeit in broad terms, of the fluctuations of irrigated areas

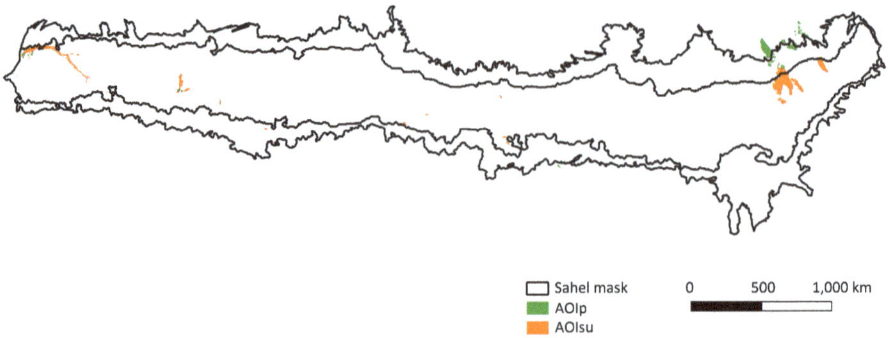

Figure 8.5 AOIp and AOIsu after contour cut.

over time and their hypothetical correlations with various phenomena, including anthropogenic events, such as mismanagement or conflicts, and natural occurrences, for example, water scarcity or exceptional floods.

A temporal evolution was calculated starting from 2000 for pivot irrigation and from 2016 for surface irrigation. The analysed years and the satellite series that were used are listed in Table 8.1.

The decision to start from 2000 for pivot irrigation and from 2016 for surface irrigation, as well as to use the Landsat collection rather than the Sentinel collection for each, was based on several observations, primarily related to the spatial resolution of the satellite used and the knowledge of the ages of the different irrigation systems.

Pivot systems became more widespread in the early 2000s, with the exception of the area in southern Chad for CST's sugarcane scheme, the oldest pivot systems in the Sahel still in operation. Thus, we decided to start our historical series from 2000.

While a single pivot, approximately 400–500 m in diameter, can be detected by a satellite with a 30-m spatial resolution (Landsat), the rectangular areas have a more detailed partition, showing both irrigated and less irrigated zones within the same rectangle. The reason is that the sizes of the plots, as mentioned above, are highly

Table 8.1 Years analysed, satellite collection (Landsat/Sentinel) and spatial resolution for pivot and surface irrigation. Periods: March–May (pivot) and February–April (surface).

Satellite series	*Years investigated*	*Agriculture*	*Spatial resolution (m/pixel)*
Landsat 7	2000–2023	Pivot	30
Landsat 8	2024	Pivot	30
Sentinel-2	2016–2024	Surface	10

variable and can even be less than 0.5 ha. Furthermore, within the same hydraulic unit, defined as the rectangular shape formed by the field channels, some plots may be cultivated, while others may not be tilled in a given season. Finally, in surface irrigation, the parts of the hydraulic unit closest to the field channels are favoured if levelling has not been done well or if it is not renewed periodically.

It is therefore clear that investigating only objects larger than 30 m—and thus using the Landsat series for surface irrigation—is unreliable given the required level of detail. In contrast, Sentinel-2 can capture finer details, potentially leading to a different perception of land features, especially in areas with mixed or fragmented vegetation. Consequently, the temporal evolution of gravity-fed areas was calculated using the Sentinel-2 collection, which has a 10-m resolution. Nevertheless, the script was run in GEE with both Landsat and Sentinel, using known irrigation systems (e.g., the Gezira in Sudan) as case studies. It was observed that when using Landsat collections, the calculated area was significantly lower than the expected data, in terms of both the area and the resulting NDVI map. Conversely, with Sentinel-2, the data were more reliable. Sentinel-2 has been operational since 2016; hence, the series for surface irrigation systems was calculated from 2016 onwards.

Using GEE, the NDVI was calculated for March–May for pivot systems, while for gravity-fed crops, it was calculated for February–April. This difference in calculation periods reflects the varying irrigation needs and seasonal conditions specific to each type of agricultural system. We always used the time of the year furthest from the rainy season to avoid distortions on the NDVI. During this period, cultivation is only possible through irrigation. Additionally, residual vegetation from the rainy season has largely dissipated, allowing more precise detection of actually irrigated areas. The warm counter-season (March–May) is always potentially productive for crops, such as fodder, that use the pivot system. Higher temperatures and increased solar radiation promote vigorous plant growth, making the NDVI a particularly sensitive indicator of vegetative activity. Moreover, pivot systems, being highly controllable, can be managed to provide optimal irrigation during this crucial period, maximising water use efficiency. For surface irrigation, bringing forward the observation period by one month (February–April) offers a better way to capture irrigated areas for different types of crops with varying periods of vegetative growth and harvesting. The multiplicity of crops is usual in surface irrigation schemes, which often (as in Gezira) implement crop rotation. In this way, the periods selected for NDVI calculation reflect the optimal seasonal conditions for detecting and distinguishing the vegetative activity associated with each irrigation system. This method enables a more accurate and representative assessment of agricultural practices in the Sahel.

A threshold NDVI value was selected to distinguish between the two irrigation systems. A threshold is a specific value that serves as a cutoff point, determining whether a pixel in a satellite image is classified as having significant vegetation (irrigated area) or otherwise. For pivot irrigation systems, an NDVI threshold of

0.5 was chosen, while for gravity-fed irrigation systems, a lower threshold of 0.4 was utilised.

The decision to use the threshold of 0.4 is consistent with the value applied to various agricultural types, as reported in the established literature. The maximum NDVI values for different crop classes, such as banana, maize, rice and others, range from 0.44 to 0.55, suggesting that a threshold of 0.4 effectively captures the lower limit of vegetation activity within these classes (Liepa et al., 2024).

In fact, there are several reasons for the different NDVI values obtained in the two types of irrigation, including irrigation efficiency, vegetation cover and uniformity and water stress. Pivot irrigation systems tend to distribute water more uniformly and in a controlled manner compared with gravity-fed systems. This efficient distribution results in more consistent and vigorous vegetation growth, often producing higher NDVI values. In contrast, surface-irrigated areas may have less dense and more irregular vegetation cover; as observed, they are more susceptible to water stress due to less efficient water distribution. Given this difference, a lower NDVI threshold ensures that areas with slightly lower vegetation density are still identified as irrigated.

Dual approach to NDVI: Sentinel for surface and Landsat for pivot irrigation

The codes we developed were designed to analyse and monitor irrigated areas in the Sahel region using satellite data. For NDVI calculations related to pivots, the Landsat collection was used, while for the rectangular areas, the Sentinel collection was employed.

Regarding the rectangles, our objective was to calculate the irrigated area for each year from 2016 to 2024. The script began by defining a specific period for analysis, from February 1 to April 30 of each year. It then loaded the Sentinel-2 image collection, filtering to include only images with less than 20% cloud cover and those that captured the AOIsu. For each image in the collection, the NDVI was calculated using Sentinel-2 bands 8 and 4, specifically chosen due to their relevance in assessing vegetation health. Band 8 (NIR) captures wavelengths strongly reflected by healthy vegetation, while Band 4 (R) captures wavelengths absorbed during photosynthesis (European Space Agency [ESA], 2021). The difference between these two bands forms the basis of the NDVI calculation:

$$\text{NDVI} = (\text{Band}\,8 - \text{Band}\,4)/(\text{Band}\,8 + \text{Band}\,4)$$

allowing a clear distinction between vegetated and non-vegetated areas. The resulting NDVI images were then filtered to include only values greater than 0.4. The mean NDVI was computed from these filtered images, producing a single representative image for the period under consideration. Finally, using the mean NDVI image, the script calculated the total irrigated area in hectares.

With respect to the circles, the script was developed to calculate the NDVI using Landsat 7 imagery. In contrast to Sentinel-2, Landsat 7 utilises different bands for calculating the vegetation index (Band 4 on Landsat 7 corresponds to the NIR spectrum and Band 3 corresponds to the R spectrum); therefore, the formula is expressed as follows:

$$\text{NDVI} = (\text{Band}\,4 - \text{Band}\,3) / (\text{Band}\,4 + \text{Band}\,3)$$

The NDVI images were filtered for values above 0.5, clipped to the AOIp and averaged to create a mean NDVI image. The script calculated the total area of pixels with an NDVI value greater than 0.5, converting it into hectares.

In conclusion, the scripts ran in a loop, processing images from March to May each year from 2000 to 2023 using Landsat 7 and, for 2024, using Landsat 8, specifically for the circles. Meanwhile, images from February to April, covering the period from 2016 to 2024, were processed using Sentinel-2.

Tracing the evolution of irrigation patterns in the Sahel

We now present the results derived from the codes implemented in GEE. This section includes tables representing the areas covered by central pivot and surface irrigation systems. These areas are subsequently plotted on graphs to illustrate the temporal evolution of both irrigation systems.

Specifically, our analysis began with a comprehensive examination of all central pivot systems, focusing on the overall evolutionary trend of their aggregated areas. Next, we conducted a comparative analysis between the "total pivots" and the "Sudanese pivots", emphasising the significance of the latter in the context of a critical historical period.

A similar approach was applied to rectangular surface irrigation systems. Initially, we selected all the rectangles contained within a specific basin. Noting the Nile River basin's significant weight compared to the total because it contains the Gezira irrigation scheme, the largest in the entire Sahel, we subsequently studied the weight of the Gezira in relation to all other Sahelian surface irrigation systems by summing those of all basins. Focusing on the Sahel mask, it was also observed that the vast majority of surface-irrigated areas were located within the core of the Sahelian pluviometric zone, as defined in Chapter 5.

Pivot systems over time and space: Analysing the circles

As previously stated, pivot irrigation systems were analysed starting from the year 2000. For this analysis, Landsat 7 was used for the period from 2000 to 2023. Due to the historical significance of 2024 (war in Sudan), Landsat 8, which has been operational since 2013, was chosen to capture the data for that year. To assess the reliability of using data from both satellites, we calculated the area from each for the overlapping years between Landsat 7 and Landsat 8 and compared the results. The findings are presented in Table 8.2 (area in hectares):

The discrepancies in area calculations between Landsat 7 and Landsat 8 can be attributed to some factors. First, the sensors and bands differ between the two satellites: Landsat 7 and Landsat 8 use the ETM+ sensor and the OLI sensor, respectively, which have different spectral responses and band specifications, leading to variations in NDVI values (Roy et al., 2014). Second, the calibration processes for Landsat 7 and Landsat 8 differ, resulting in variations in the captured data (Li et al., 2013). Third, differences in atmospheric correction methods can influence the NDVI values since variability in atmospheric conditions, such as aerosols and cloud cover, can affect the data (Li et al., 2024). Additionally, the preprocessing algorithms, including radiometric, geometric and atmospheric corrections, differ between the two satellites, impacting the final NDVI calculations (Vermote et al., 2016). Seasonal and annual changes also play a role because variability in vegetation

Table 8.2 Area per year (in hectares), showing the total pivot systems identified across the Sahel using the AOIp mask, calculated with Landsat 7 (L7) and Landsat 8 (L8) data (period: March–May).

Year	*Area L7 (Ha)*	*Area L8 (Ha)*
2000	*1028*	
2001	*796*	
2002	*1222*	
2003	*1138*	
2004	*906*	
2005	*884*	
2006	*581*	
2007	*923*	
2008	*781*	
2009	*1121*	
2010	*2223*	
2011	*3425*	
2012	*4605*	
2013	*7264*	
2014	*9041*	
2015	*11,066*	*13,665*
2016	*14,556*	*17,999*
2017	*15,807*	*21,817*
2018	*18,103*	*22,610*
2019	*16,031*	*21,520*
2020	*20,149*	*22,073*
2021	*17,329*	*20,536*
2022	*11,788*	*15,068*
2023	*14,591*	*20,367*
2024		*11,946*

dynamics and meteorological conditions between years can lead to differences in NDVI values, even if the same period is used in the comparison (Zhu et al., 2015). Instrumental degradation is another factor; for instance, Landsat 7 has experienced issues with the Scan Line Corrector (SLC) since 2003, introducing data gaps that may affect NDVI calculations despite correction methods, and sensor degradation over time can also affect data quality (Storey et al., 2016). Temporal and acquisition angle differences between the satellites can influence NDVI values due to varying sun angles and illumination conditions (Flood, 2014). Lastly, small variations in the implementation of NDVI calculation algorithms or the parameters used can produce different outcomes (Weier & Herring, 2000).

Despite the substantial volume of the data involved, spanning thousands of hectares, the calculations derived from Landsat 7 and Landsat 8 proved remarkably consistent. Between 2013 and 2023, the difference in the areas measured by the two satellites ranged from approximately 3000 to 6000 ha. It is noteworthy that by comparing data from both satellites, a similar trend emerges. The evolution of pivot areas over time follows the same trajectory for both Landsat 7 and Landsat 8 (Fig. 8.6), with the only distinction being a consistent shift in the number of hectares recorded each year.

It was thus possible to create a graph illustrating the areal evolution of pivot irrigation systems throughout the Sahel. Note that the area refers to the NDVI calculated within the delineated AOIp mask (Fig. 8.6).

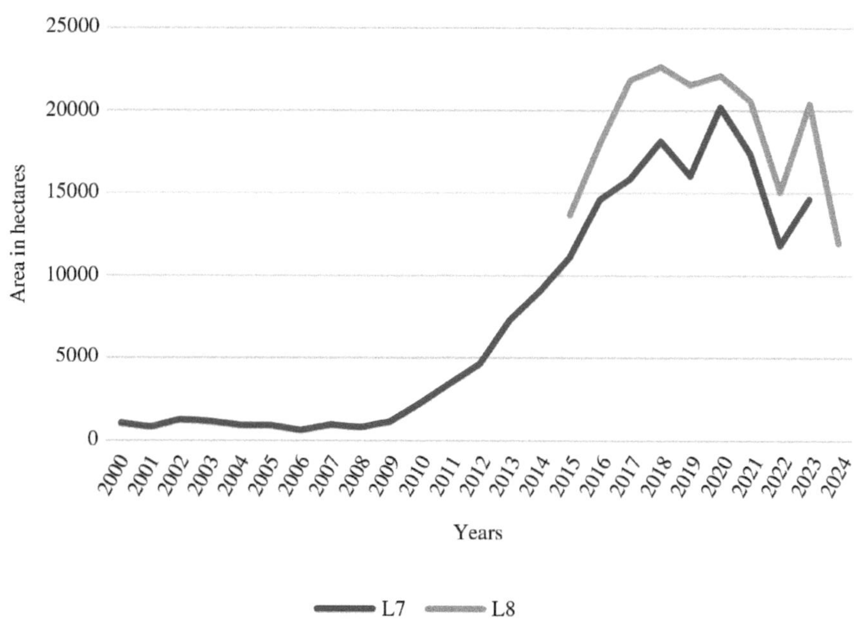

Figure 8.6 Temporal evolution of the pivot series from 2000 to 2024 using Landsat 7 (L7) and Landsat 8 (L8) satellites (period: March–May).

As depicted in both curves, a steep rise is observed until 2018, followed by a slight dip in 2019, and then another increase in 2020. From 2020 onwards, a significant drop is evident until 2022, followed by a subsequent increase until 2023 and a further rapid decline in 2024, with the latter available only for Landsat 8. This last collapse raised concerns about its possible connection to the conflict in Sudan, that the impact might be confined to the AOIs in Sudan. Therefore, specific calculations for the Sudanese areas were undertaken.

For the calculation in Sudan alone, Landsat 7 and Landsat 8 were both used. The evolutionary trends from 2013 to 2023 were compared. It would not matter if Landsat 8 provided higher values than those of Landsat 7 (as thoroughly explained above). We focused on Sudan solely to study the trends (Table 8.3).

The evolutionary trends remain consistent for both Landsat 7 and Landsat 8, as shown in Fig. 8.7.

For a broad comparison up to 2024, we decided to compare Sudanese pivot systems with the rest of the Sahelian ones using Landsat 8; the results are shown in Fig. 8.8.

In the graph (Fig. 8.8), the line representing the area of Sudan pivot systems shows a significant expansion between 2013 and 2020, followed by a decline in 2022, a recovery in 2023 and a major reduction in 2024. In contrast, the line for Sahelian pivot systems, excluding Sudan, remains relatively stable throughout the same period, with only minor fluctuations in area. This finding indicates that the decrease observed in Fig. 8.6, from 2023 to 2024, is almost entirely due to the Sudanese pivot systems. This result is consistent with, and can be explained by, the ongoing war in the region.

Table 8.3 Total area of the Sudanese pivot systems, divided by Landsat 7 (L7) and Landsat 8 (L8) (period: March–May).

Sudanese pivot systems		
Year	*Area L8 (Ha)*	*Area L7 (Ha)*
2013	*4434*	*5321*
2014	*7701*	*7483*
2015	*11,327*	*10,229*
2016	*12,821*	*11,492*
2017	*16,943*	*13036*
2018	*19,021*	*16,059*
2019	*18,135*	*17,796*
2020	*19,223*	*17,506*
2021	*16,327*	*14,796*
2022	*10,884*	*10,637*
2023	*15,795*	*12,154*
2024	*8228*	

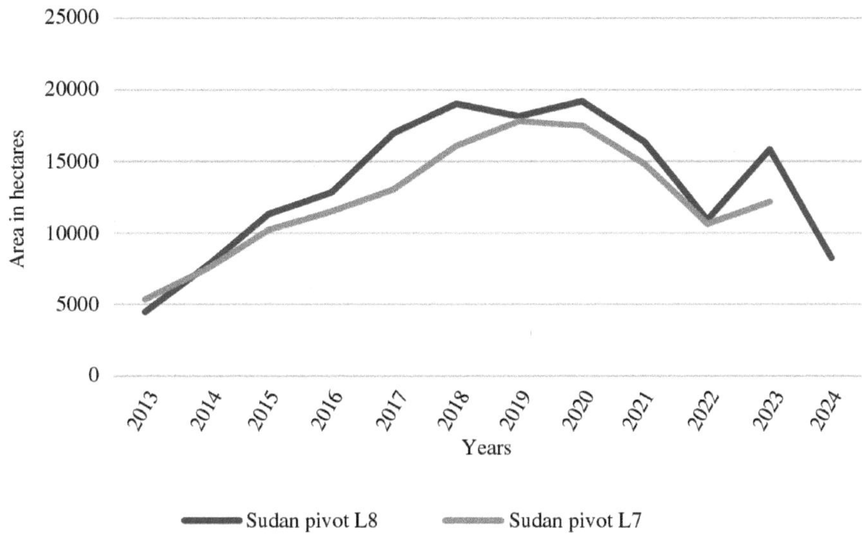

Figure 8.7 Temporal evolution of Sudanese pivot systems from 2015 to 2024 comparing Landsat 7 (L7) and Landsat 8 (L8) collections (period: March–May).

Figure 8.8 Comparative analysis of the evolution of Sudanese pivot systems versus other Sahelian pivot systems from 2013 to 2024 using Landsat 8 (L8) collection (period: March–May).

Evolution of surface-irrigated areas: A spatial and temporal analysis of rectangles

The analysis of surface irrigation systems began in 2016. In this study, we utilised Sentinel-2 satellite imagery, renowned for its high resolution and frequent revisit times, providing comprehensive coverage and detailed insights into changes in agricultural practices over time. The choice to start from 2016 was directly linked to the availability of 10 m resolution imagery provided by Sentinel-2, as previously discussed in detail. Consequently, the areas of the major surface irrigation systems were calculated from 2016 to 2024. We found it interesting to analyse the major irrigation systems within each river basin and to study their "breathing", their expansion and contraction, from 2016 to 2024. Specifically, the basins of Lake Chad and Senegal, Niger, Volta and Nile Rivers were analysed, as shown in Fig. 8.9.

The major irrigation systems under study (as listed above) are therefore grouped according to the aforementioned basins (Table 8.4).

From the graph in Fig. 8.10, it is evident that the majority of the irrigated areas fall within the Nile River basin because it contains the Gezira irrigation system, the largest one in the Sahel. The question was then raised about the significance of the Gezira irrigation system relative to the total, as explained in Table 8.5.

Figure 8.11 represents the graphical "response". The graph effectively illustrates the dominant role of the Gezira irrigation system in the Sahel. It highlights the Gezira system as the major contributor to the overall irrigated area in the region, and any analysis of irrigation in the Sahel must account for this significant system.

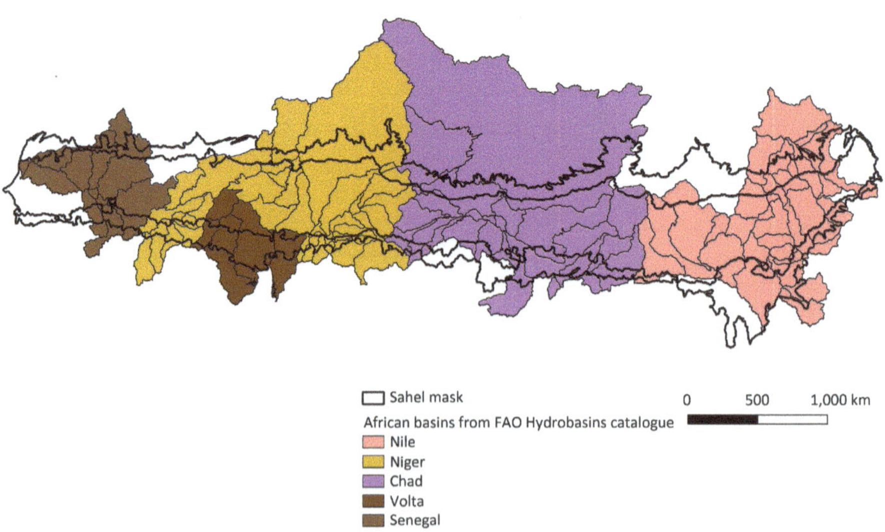

Figure 8.9 Main river basins in the Sahel.

Table 8.4 Areas (in hectares) of surface irrigation systems within each river basin in the pluviometric Sahel, analysed over time (period: February–April).

Years	*Senegal*	*Niger*	*Volta*	*Chad*	*Nile*
	Senegal Valley	*Office du Niger and the Banguinedà perimeters*	*Sourou Valley and Bagrè dam*	*Tiga dam, Hadejia, Bol polders, SEMRY*	*Gezira, Rahad, Kashm El Girba schemes and the sugarcane perimeters*
2016	*16,262*	*8230*	*1802*	*4760*	*174,469*
2017	*18,728*	*6507*	*588*	*7522*	*224,108*
2018	*23,813*	*11,698*	*2546*	*6740*	*247,462*
2019	*27,116*	*16,658*	*2057*	*6588*	*226,110*
2020	*22,729*	*20,029*	*2081*	*4664*	*278,322*
2021	*23,474*	*23,484*	*3073*	*4318*	*278,176*
2022	*29,641*	*19,787*	*1230*	*1975*	*236,655*
2023	*32,787*	*29,313*	*857*	*5474*	*225,028*
2024	*25,233*	*30,665*	*2751*	*3869*	*206,750*

The trends suggest that while other irrigation systems contribute, they do not match the scale of the Gezira system, which remains a critical component of irrigated agriculture.

Additionally, if we focus on the Gezira system from 2021 to 2024, a noticeable decline is evident, although not as pronounced as in the pivot systems, which, being geared towards export, have experienced an even greater impact caused by

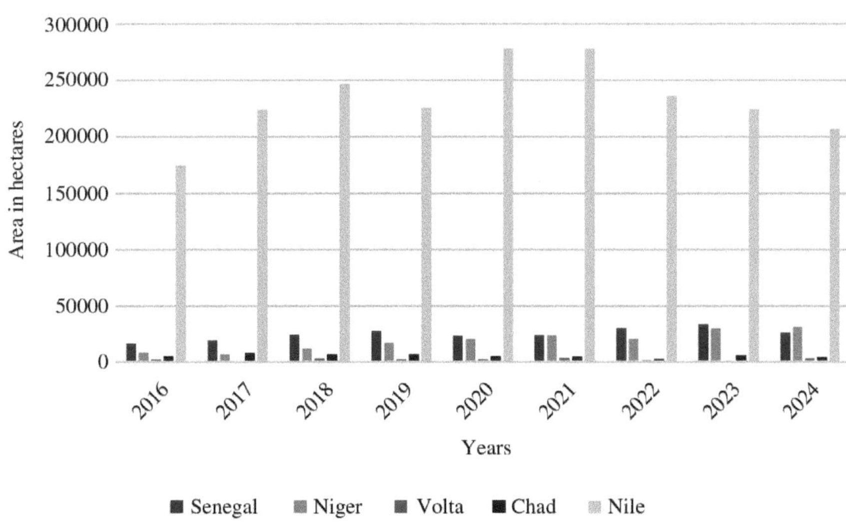

Figure 8.10 Areas of surface irrigation systems across five river basins (Senegal, Niger, Volta, Chad and Nile) from 2016 to 2024 (period: February–April).

Table 8.5 Area of the Gezira irrigation system compared with the sum of the areas (Ha) of all irrigation systems (Gezira included), divided by year (period: February–April).

Year	*Gezira*	*Total*
2016	*145,629*	*205,523*
2017	*169,452*	*257,453*
2018	*185,595*	*292,259*
2019	*174,050*	*278,529*
2020	*223,168*	*327,825*
2021	*216,863*	*332,524*
2022	*185,958*	*289,289*
2023	*174,240*	*293,459*
2024	*153,806*	*269,268*

the socio-political turmoil in the region. While there is still production in Gezira, since mid-December 2023, it has also become a conflict zone (see Chapters 1 and 10). Consequently, we anticipate a significant decline in next year's data. It is our intention to monitor developments in Gezira and, more broadly, in all Sudanese irrigated areas, which are crucial for the country's economy and, more importantly, for the subsistence of the population.

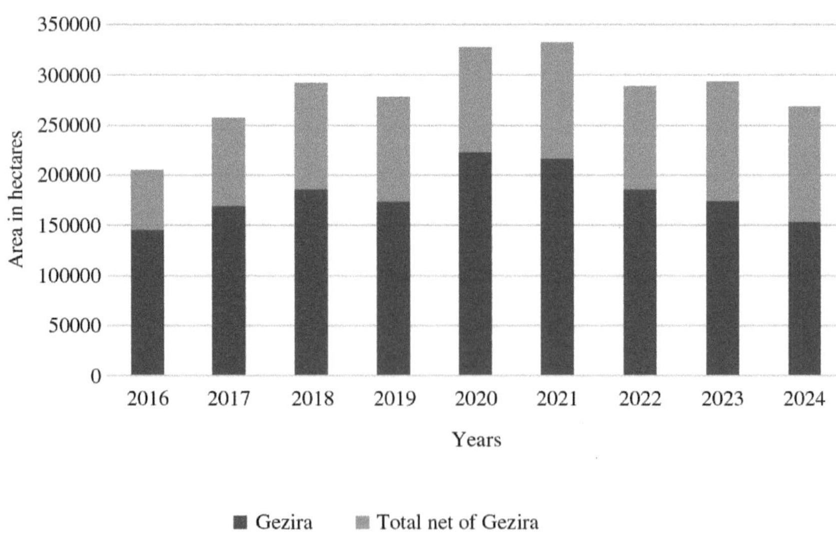

Figure 8.11 Evolution of the Gezira irrigation system versus the total of all other irrigation systems in the Sahel over the years 2016–2024 (period: February–April).

Evolving landscapes of pivot and rectangular irrigation systems

In this section, we analyse selected case studies of both circular and rectangular irrigation systems. Two cases were chosen for the central pivot systems and two for the surface irrigated systems, specifically the sites of GLB and Waha in Sudan (discussed in Chapter 10) and the irrigation systems of Sourou Valley and SEMRY (discussed in Chapters 11, 12 and 15). Considering Figs. 8.6 and 8.10 from the previous sections, it was deemed reliable to represent the years 2009, 2012, 2015, 2018, 2021 and 2024 for the pivot systems (this choice is linked to the fact that the circles began to appear in 2009 in the selected areas) and 2016, 2020 and 2024 for the surface irrigation systems (using the Sentinel-2 series, we have no data prior to 2016) to capture the changes in each selected system.

The circles: Two case studies

The graph in Fig. 8.12 illustrates the analysis of the expansion and contraction dynamics of the selected pivot systems.

The Waha system shows significant growth beginning around 2010. From 2010 to 2016, there is a sharp increase in the irrigated area, peaking at approximately 3500 ha in 2018. However, after reaching this peak, a gradual decline in the irrigated area begins, with notable fluctuations between 2019 and 2023. The most significant change occurs between 2023 and 2024, where the irrigated area drops sharply from around 3000 ha to 0. This indicates the complete cessation of the Waha system's operation in 2024, rather than a mere reduction in activity, marking a critical point in its evolution where no land is irrigated at all.

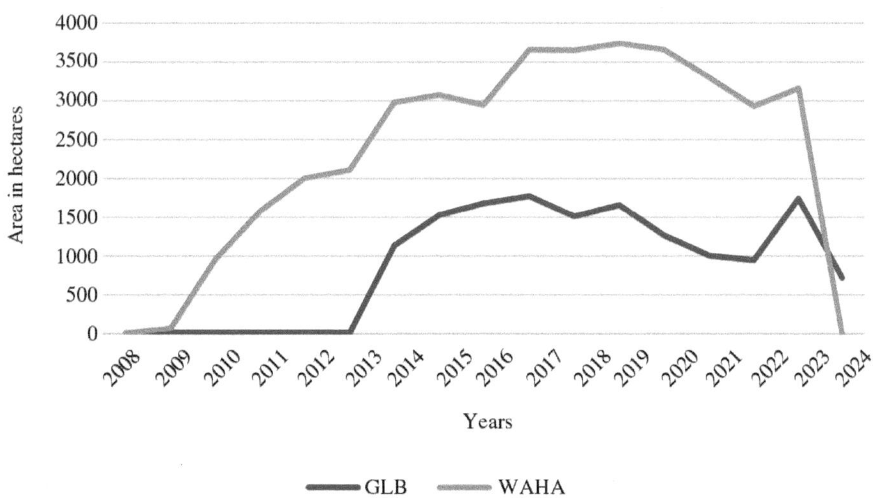

Figure 8.12 Evolution of GLB and Waha pivot irrigation systems (period: March–May).

The GLB system shows a much slower initial growth compared to Waha, beginning its expansion around 2012. The GLB irrigation area increases steadily until around 2016, where it stabilises just below 1700 ha. In contrast to Waha, the GLB system remains relatively stable over the years, with small fluctuations until a sharp rise occurs between 2022 and 2023, again reaching a peak of approximately 1700 ha. However, the GLB system experiences a decline in 2024, similar to the Waha system, though not as drastic. It still maintains some operational capacity, irrigating around 700 ha by the end of the period.

To represent the data in Fig. 8.12, we deemed it appropriate to begin the mapping of Waha in 2009, that is, from the start of its expansion. For GLB, it only made sense to display the map starting from 2015 since no relevant features were observed before this date.

Only for the GLB system, the NDVI was entirely extracted using Landsat 8 until 2024, unlike other systems where the NDVI was calculated using Landsat 7 until 2021 and Landsat 8 in 2024. The reason was that Landsat 7 likely passed over the GLB site less frequently, and the extracted image contained some "data gaps" or "blank stripes", most likely due to the problems that L7's sensor had with the SLC.[1]

Figure 8.13 illustrates the spatial and temporal evolution of the Waha irrigation system from 2009 to 2024 in three-year intervals. In 2009, the project covered a limited area, with only a few scattered irrigated zones visible. By 2012, the system had expanded significantly, showing a marked increase in both the number and density of irrigated zones. This expansion continued into 2015, with the irrigation network becoming more widespread and denser, covering a larger area. By 2018, the system had reached its peak, as evidenced by a high concentration of irrigated plots across the region, reflecting the project's maximum development. However, by 2021, the map reveals a modest decrease in the density of irrigated areas, indicating the onset of a reduction in activity or contraction of the system. Finally, by 2024, the Waha system experienced a dramatic decline, signalling the collapse of irrigation activity in the area.

Similarly, Fig. 8.14 presents the development of the GLB irrigation system over the period from 2015 to 2024, in three-year intervals. In 2015, the GLB project was well underway, following its establishment in 2012, and was characterised by approximately 1500 ha of cultivated land. Over the next three years, by 2018, the system remained relatively stable. However, by 2021, the map shows a reduction in the number of irrigated plots, suggesting a phase of contraction in the system's growth. Interestingly, despite this contraction, there was a noticeable peak in irrigation activity between 2022 and 2023, as shown in Fig. 8.12, where the irrigated area temporarily increased again before experiencing a decline in 2024.

In both cases, the most striking pattern is the initial expansion of the irrigation systems, followed by a period of relative stability and eventually a sharp decline, which can be

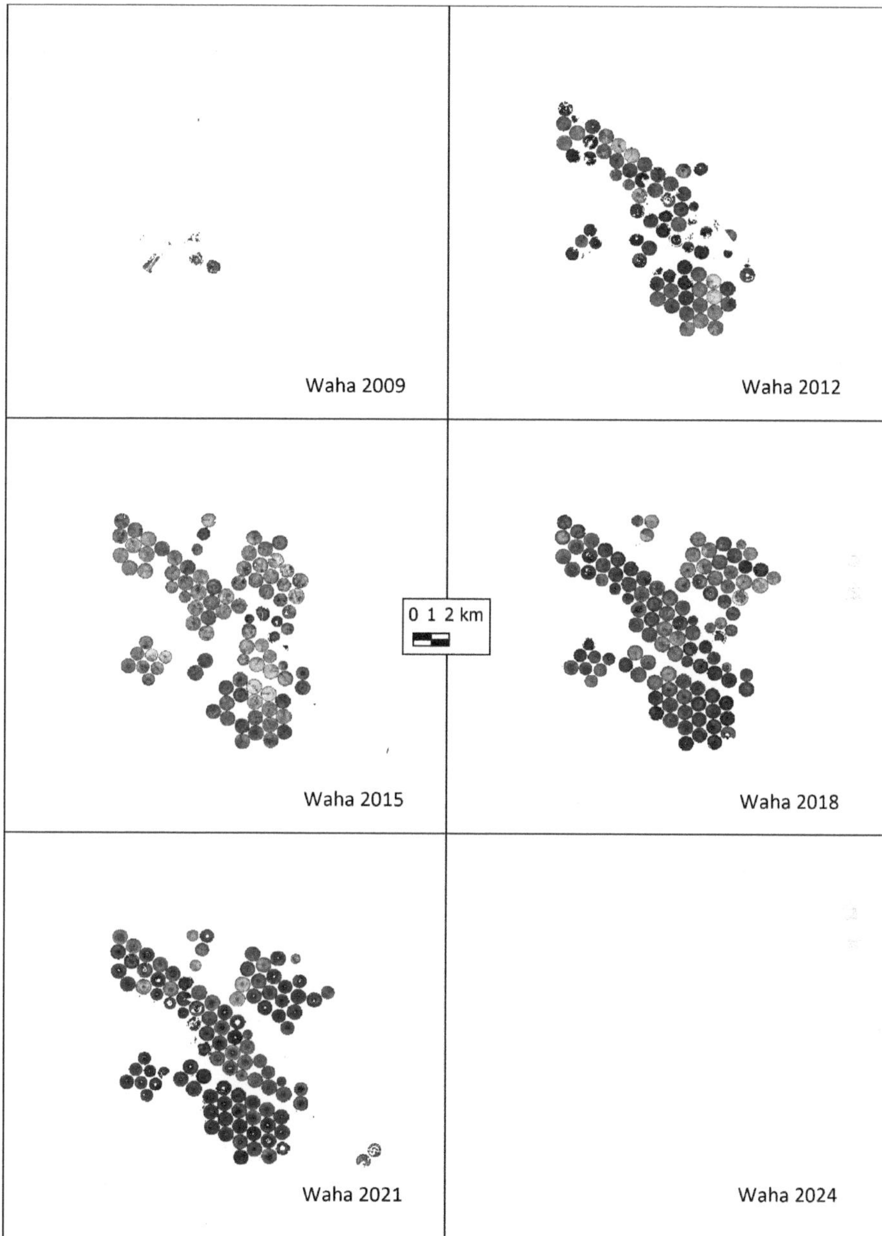

Figure 8.13 Evolution of Waha project from 2009 to 2024 (period: March–May).

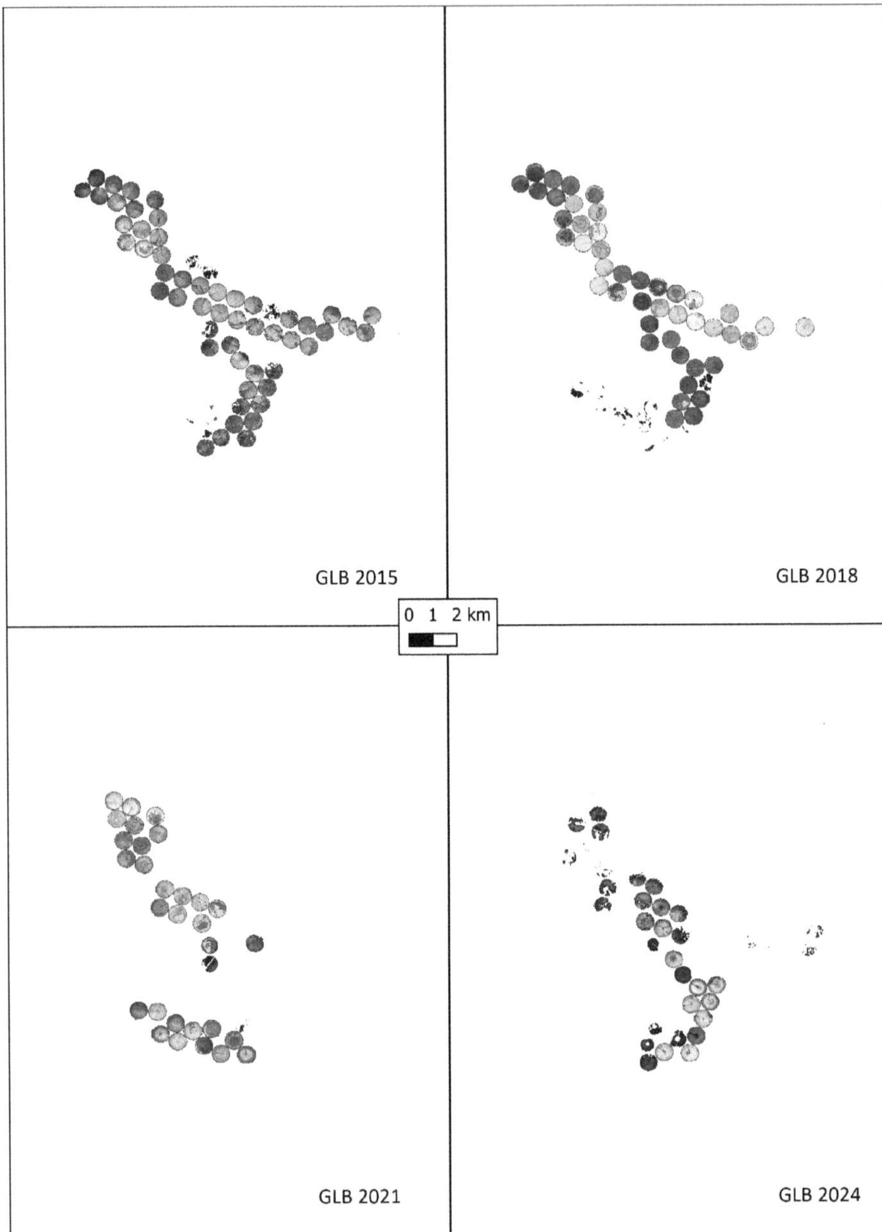

Figure 8.14 Evolution of GLB project from 2009 to 2024 (period: March–May).

attributed to the impacts of war. This is clearly reflected in the 2024 maps, where GBL shows a significant reduction in the number of irrigated plots, and the Waha project is completely disrupted, with the looting and destruction of every infrastructure. The collapse of these irrigation systems underscores the vulnerability of agricultural projects to socio-political instability and highlights the devastating impact that conflict can have on food security and rural livelihoods in the Sahel region.

The rectangles: Two case studies

The graphs in Figs. 8.15 and 8.16 illustrate the evolution of the irrigated areas for the SEMRY and the Sourou Valley projects from 2016 to 2024, focusing on the period between February and April.

In the case of the SEMRY project (Fig. 8.15), the irrigated areas show an increase from around 1500 ha in 2016 to nearly 2500 ha in 2017 until 2020. However, the project begins to experience a sharp decline by 2021, with a significant drop in the irrigated areas to just 500 ha. There is a brief recovery in 2022 and 2023, but by 2024, the project undergoes another decline, indicating persistent challenges that have affected its capacity.

In contrast, the Sourou Valley project (Fig. 8.16) shows a more gradual increase in irrigated areas, starting from a lower base in 2016 and 2017, growing to around 1400 ha in 2018. After reaching its peak in 2021, there is a sharp drop in 2022 and 2023, but the project shows resilience, recovering significantly by 2024, when the irrigated areas climb back to approximately 1500 ha.

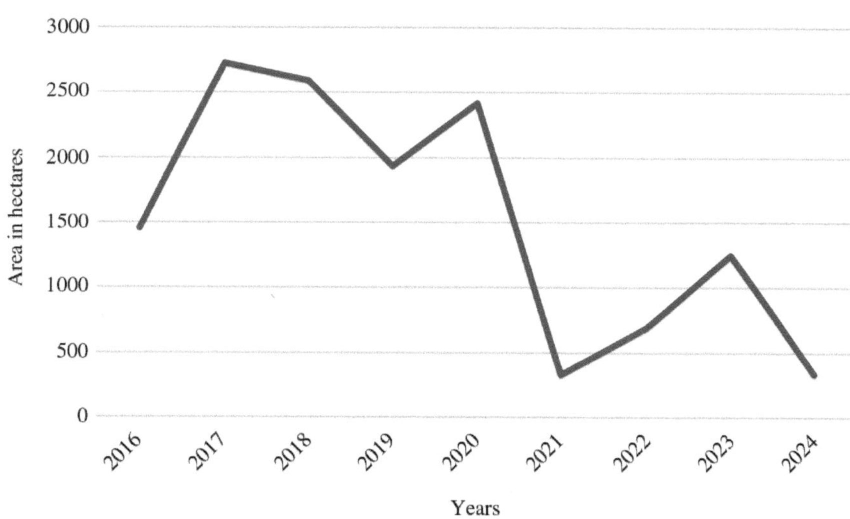

Figure 8.15 Evolution of the SEMRY project (period: February–April).

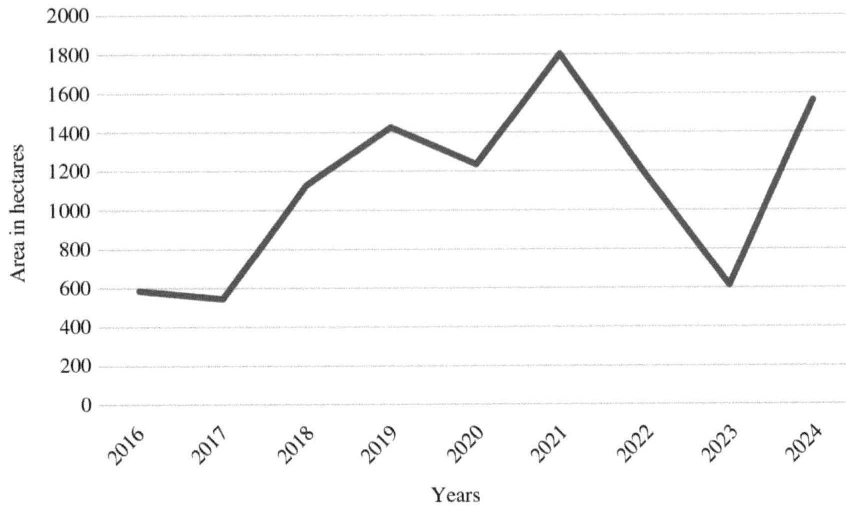

Figure 8.16 Evolution of the Sourou Valley project (period: February–April).

These trends suggest that while external factors—such as water availability or socio-political conditions—likely affected both systems, the Sourou Valley project might have been better able to adapt or recover from disruptions, whereas the SEMRY project struggled to regain a minimum of results.

The systems are mapped in Figs. 8.17 and 8.18, covering the period from 2016 to 2024.

For better visual clarity, only the central section of the irrigation system is shown, while the southernmost part of the project is not displayed (investigated period: February–April).

In Fig. 8.17, the SEMRY project demonstrates a clear and organised irrigation network in 2016 and 2020, with a structured layout, particularly in SEMRY II (right side of the map). However, in 2024, a significant reduction in the overall irrigated area becomes apparent. The project appears to have contracted, with fewer irrigated zones and a noticeable decline in density, especially in SEMRY I (left side of the map).

Figure 8.18 illustrates the central region of the Sourou Valley project. A significant improvement is evident beginning in 2020, particularly in the northern area. When comparing the maps from 2020 and 2024, it is clear that the irrigated area remains stable, with no reduction in coverage, indicating the system's resilience over time.

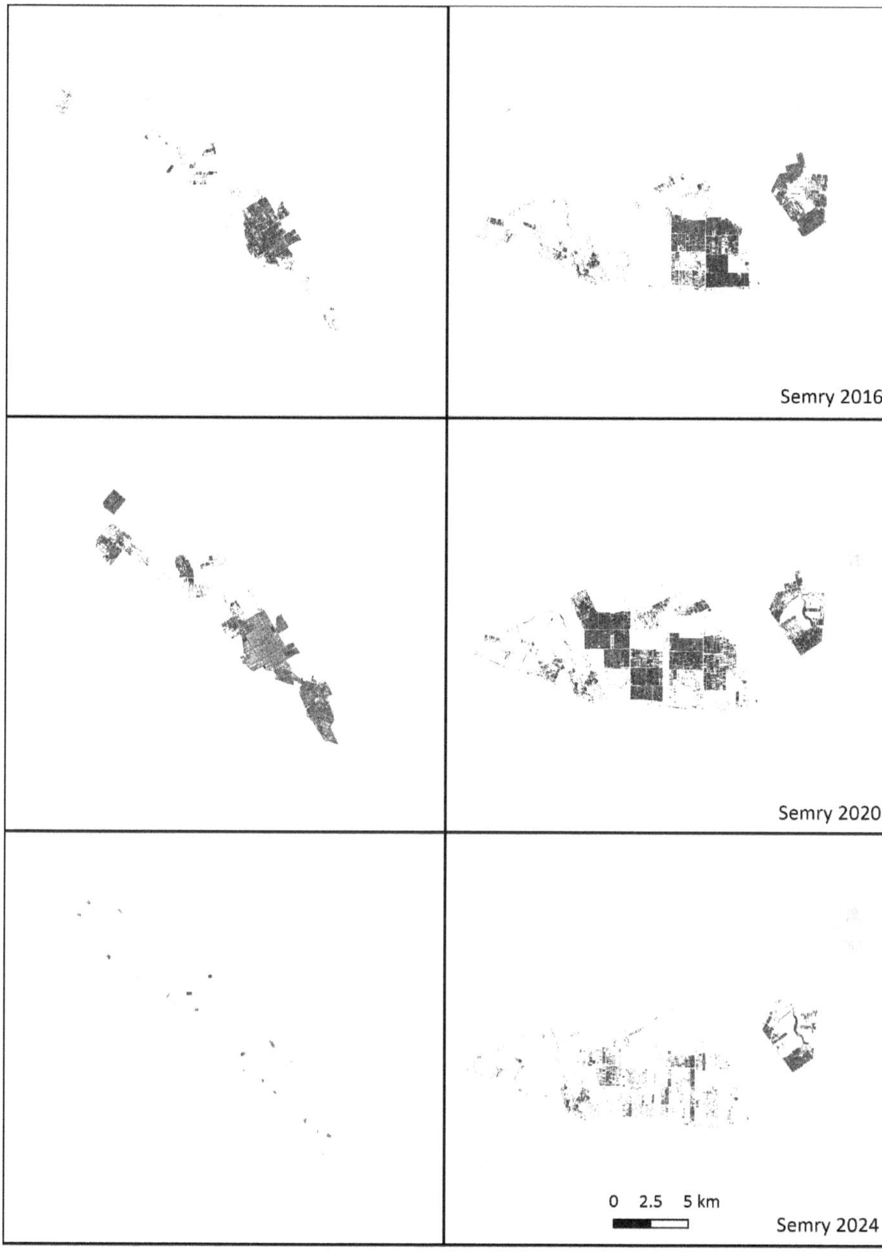

Figure 8.17 Mapping of the SEMRY project. SEMRY 1 (on the left) and SEMRY 2 (on the right) have been split on the map for better visualisation (period: February–April).

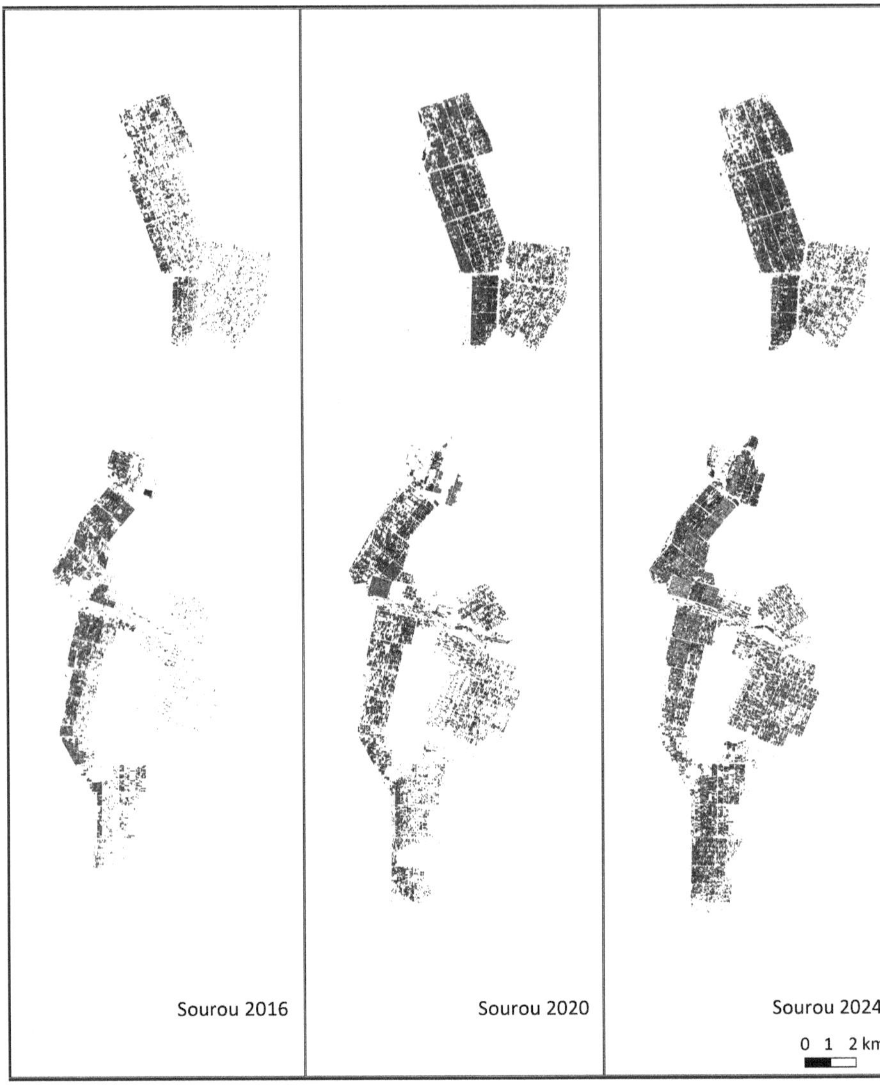

Figure 8.18 Mapping of Sourou Valley (period: April–May).

Conclusion

In this chapter, we have provided an analysis of the spatial and temporal dynamics of the two key irrigation systems in the Sahel—central pivot (circular) and surface (rectangular) systems—using remote sensing technologies, particularly the GEE platform. The study spans the period from 2000 to 2024 for pivot systems and from 2016 to 2024 for surface irrigation systems, using satellite data from Landsat and Sentinel. Our analysis reveals that pivot irrigation systems had a steady increase in coverage until 2018, followed by a sharp decline, particularly in Sudan, where the ongoing war has severely impacted agricultural infrastructure. The findings highlight this decline's close link to political instability and conflict, emphasising the vulnerability of such irrigation systems in conflict zones.

For surface irrigation, the Gezira scheme in Sudan emerges as the dominant system, significantly contributing to the overall irrigated area in the Sahel. However, even in this scheme, the effects of the political instability and then the conflict are becoming apparent, with a gradual reduction in irrigated areas since 2021.

In terms of prospects, we emphasise the need for continued monitoring of both pivot and rectangular irrigation systems, focusing on conflict-affected regions such as Sudan, where the war poses an ongoing threat to agricultural sustainability. Additionally, the implementation and expansion of the AtlaSahel web GIS (atlasahel.it) are crucial for improving data accessibility.

Note

1 https://www.usgs.gov/faqs/what-landsat-7-etm-slc-data (accessed September 26, 2024).

References

Amani, M., Ghorbanian, A., Ahmadi, S. A., Kakooei, M., Moghimi, A., Mahdavi, S., Ghahremanloo, M., Parsian, S., & Mirmazloumi, S. M. (2020). Google Earth Engine Cloud computing platform for remote sensing Big Data applications: A comprehensive review. *IEEE Journal of Selected Topics in Applied Earth Observations and Remote Sensing*, *13*, 5326–5350. https://doi.org/10.1109/JSTARS.2020.3021052

Barnett, T. (1977). *The Gezira scheme. An illusion of development.* Frank Cass. https://doi.org/10.4324/9781315827957

Bertoncin, M., Pase, A., Quatrida, D., & Turrini, F. (2019). At the junction between state, nature and capital: Irrigation mega-projects in Sudan. *Geoforum*, *106*, 24–37. https://doi.org/10.1016/j.geoforum.2019.07.015

European Space Agency (ESA). (2021). *Sentinel-2 user handbook.* https://sentinels.copernicus.eu

Flood, N. (2014). Continuity of reflectance data between Landsat-7 ETM+ and Landsat-8 OLI, for both top-of-atmosphere and surface reflectance: A study in the Australian landscape. *Remote Sensing*, *6*(9), 7952–7970. https://doi.org/10.3390/rs6097952

Ganawa, E., & Awad Allah, F. (2013). *Adoption of new engineering systems for base maps production and crop monitoring, case study: Kenana Sugar Cane Corp.* University of Khartoum. https://core.ac.uk/reader/71673257

Gaznayee, H. A. A., Zaki, S. H., Al-Quraishi, A. M. F., Aliehsan, P. H., Hakzi, K. K., Razvanchy, H. A. S., Riksen, M., & Mahdi, K. (2023). Integrating remote sensing techniques and meteorological data to assess the ideal irrigation system performance scenarios for improving crop productivity. *Water (Switzerland)*, *15*(8), Article 1605. https://doi.org/10.3390/w15081605

Gorelick, N., Hancher, M., Dixon, M., Ilyushchenko, S., Thau, D., & Moore, R. (2017). Google Earth Engine: Planetary-scale geospatial analysis for everyone. *Remote Sensing of Environment*, *202*, 18–27. https://doi.org/10.1016/j.rse.2017.06.031

Li, Y., Liu, Q., Chen, S., & Zhang, X. (2024) An improved gap-filling method for reconstructing dense time-series images from LANDSAT 7 SLC-Off data. *Remote Sensing*, *16*, 2064. https://doi.org/10.3390/rs16122064

Li, F., Masek, J. G., Roy, D. P., & Shuai, Y. (2013). Impact of Landsat 7 ETM+ SLC-off on image classification: Assessing the tradeoff between reducing scan line noise and information loss. *Remote Sensing of Environment*, *133*, 121–127. https://doi.org/10.1016/j.rse.2013.02.001

Liepa, A., Thiel, M., Taubenböck, H., Steffan-Dewenter, I., Abu, I., Singh Dhillon, M., Otte, I., Otim, M. H., Lutaakome, M., Meinhof, D., Martin, E. A., & Ullmann, T. (2024). Harmonized NDVI time-series from Landsat and Sentinel-2 reveal phenological patterns of diverse, small-scale cropping systems in East Africa. *Remote Sensing Applications: Society and Environment*, *35*, Article 101230. https://doi.org/10.1016/j.rsase.2024.101230

Lillesand, T. M., Kiefer, R. W., & Chipman, J. W. (2007). *Remote sensing and image interpretation* (6th ed.). Wiley.

Roy, D. P., Wulder, M. A., Loveland, T. R., Woodcock, C. E., Allen, R. G., Anderson, M. C., Helder, D. L., Irons, J. R., Johnson, D. M., Kennedy, R. E., Scambos, T. A., Schaaf, C. B., Schott, J. R., Sheng, Y., Vermote, E. F., Belward, A. S., Bindschadler, R. A., Cohen, W. B., Gao, F., & Hipple, J. D. (2014). Landsat-8: Science and product vision for terrestrial global change research. Remote Sensing of Environment, *145*, 154–172. https://doi.org/10.1016/j.rse.2014.02.001

Storey, J., Choate, M., & Lee, K. (2016). Landsat 8 Operational Land Imager on-orbit geometric calibration and performance. *Remote Sensing*, *8*(6), Article 540. https://doi.org/10.3390/rs61111127

Tamiminia, H., Homayouni, S., Safari, A., & McCarty, J. (2020). Google Earth Engine for Geo-Big Data applications: A meta-analysis and systematic review. *ISPRS Journal of Photogrammetry and Remote Sensing*, *164*, 152–170. https://doi.org/10.1016/j.isprsjprs.2020.04.001

U.S. Department of the Interior. (n.d.). *Landsat Normalized Difference Vegetation Index (NDVI)*. U.S. Geological Survey. Retrieved August 24, 2024, from https://www.usgs.gov/landsat-missions/landsat-normalized-difference-vegetation-index

Vermote, E., Justice, C., Claverie, M., & Franch, B. (2016). Preliminary analysis of the performance of the Landsat 8/OLI land surface reflectance product. *Remote Sensing of Environment*, *185*, 46–56. https://doi.org/10.1016/j.rse.2016.04.008

Weier, J., & Herring, D. (2000). *Measuring Vegetation (NDVI & EVI)*. NASA Earth Observatory. https://earthobservatory.nasa.gov/features/MeasuringVegetation

Zhu, Z., Wang, S., & Woodcock, C. E. (2015). Improvement and expansion of the Fmask algorithm: Cloud, cloud shadow, and snow detection for Landsats 4–7, 8, and Sentinel 2 images. *Remote Sensing of Environment*, *159*, 269–277. https://doi.org/10.1016/j.rse.2014.12.014

Zwart, S. (2017). *A detailed map of irrigation infrastructure along the Senegal River by 2015*. 10.13140/RG.2.2.28009.49761.

9 The breath and the cage

Re-imagining the Sahel through mapping

Andrea Pase and Federico Gianoli

Introduction

In Part 2, its many authors attempt to produce "fresh" maps of the Sahel. This endeavour starts with the delimitation of the area study, a crucial and sensitive procedure for processing and discussing data from remote sensing and georeferenced databases (Chapter 5). In Chapter 6, we then proceed to address many different issues, somehow always related to the focus of this volume—water and land. This is accomplished by using the convergence of evidence methodology to identify critical hotspots where multiple environmental crisis factors, related to land degradation, proliferate. In Chapter 7, we continue concentrating on the major Sahelian rivers and the rhythmic movements of these water bodies and, in particular, by analysing the areas of seasonal floods. The data on dams and reservoirs provide an initial overview of the main anthropogenic modifications that alter hydrological dynamics. We end that chapter with an attempt to visualise the "human footprint" of water. In Chapter 8, we then focus on the Sahelian large-scale irrigation schemes, the true heart of our research. By identifying the different geometric shapes (rectangles and circles) which irrigation schemes render on the ground, we are able to follow their recent evolution over time. The aim of this concluding chapter of Part 2 is to present a synthesis of the evidence observed and its implications for the Sahel, as well as some methodological observations—both imaginative and rigorous—derived from our collective experience.

A metaphorical approach to Sahel mapping

Many authors, particularly Retaillé and Walther (2019), have questioned the presumption of objectivity or even the usefulness of mapping an ever-changing reality such as the Sahelian one. Well aware of this scepticism, in Chapter 5, we have nonetheless chosen to continue constructing representations—even cartographic ones of the regions—yet with an attitude that we describe as metaphorical. Our references are the French historian F. Braudel (1995) and his propensity to multiply the points of view from which to observe a given historical-geographical space, and the Italian geographer G. Dematteis (1985), who reflected on metaphors as key tools for describing the world. Following their insights, we have approached

DOI: 10.4324/9781003429111-12

mapping the Sahel in a way that would lead to drawing many different Sahels, overlapping and intertwined, and always conscious of their relativity. Instead of proposing what is considered as the most correct, right or adequate representation, we set the goal of proposing many, even infinite, possibilities. The metaphorical approach to mapping the Sahel consists of two essential aspects: multiplication (many Sahels, many different maps to represent them) and the use of imagination. The capacity for using the imagination to produce a gap between the established clichés, the "given-for-granted" in the interpretation of reality, is inherent to geography (Dematteis, 2021) as well as to other social science disciplines. This is made possible by offering new unusual words suggestive of different perceptions and horizons, that is to say, metaphors. Our intention has therefore been to try to produce "fresh maps", along the lines of Davidson's (1978) "fresh metaphors" as they are reread by Mari (2001) and Rorty (1989), capable of producing a novelty, of disarticulating the established. It seems to us that in Part 2, there are several passages in which this metaphorical, multiplicative and imaginative propensity is visible: the three pluviometric Sahels, the unpacking of political boundaries, the pulsating of the hydrological Sahel, the graphical visualisation of rivers flowing from upstream to downstream, and—in the same framework—the identification of human-made fractures, the very idea of a human footprint on rivers and its visualisation, the rendering through geometric forms (rectangles and circles) of large-scale irrigation schemes.

The metaphorical approach certainly means proposing suggestions rather than assertions, hypotheses rather than "self-proclaimed" truths, but it does not suggest acting in an approximate manner. Instead, there is a precise explanation of the methods used, rigorous use of data, transparency in the code we have written (with an available open source, therefore replicable) and an intention of honesty in highlighting the problems encountered (those that can be solved and those that are unsolvable).

The inspiration for this metaphorical approach to mapping the Sahel actually originates from joining together the potentialities offered by satellite remote sensing—the vast harvest of georeferenced data and the power of geographic information systems (GISs)—with the insights gained from the research group's fieldwork experiences, extended in time and space. Perhaps the most innovative aspects of Part 2 are derived from the surprise that arises from the comparison between the top-down and bottom-up views. On one hand, there is the possibility of mutual control: the interpretation of remote sensing data is guided by a certain awareness of what is moving on the ground. On the other hand, the results of the remote analysis also raised questions that required on-site verification, thanks to the involvement of the group's co-researchers and the corroboration by external colleagues and experts. The disconnections between the two types of sources (from above and below) were spontaneously creative because they were unexpected and therefore effectively fresh. Furthermore, the chapters in this volume prioritised a dialogue between the text and the maps, thus illuminating aspects that might otherwise have remained in the shadows. Although it has taken us a long time to become progressively closer to obtaining somewhat useful results, we believe that it is important not to leave all the influence that would come from the ability to construct representations of the territory in the hands of the few (states, armies, big companies). It may perhaps be critical of the pervasive power of "cartographic reason" (Farinelli, 2009) without giving up producing images of the world because, without imagination, it would be difficult even to think of different futures.

Two maps

At the (provisional, obviously) end of this journey, we have drawn two maps that allow us to highlight the two opposing polarities that identify the present-day Sahel. The first map (Fig. 9.1) brings together what contributes to the fluidity of the Sahel and the "flow" within it, clearly only for the aspects we have investigated. We intend to examine other aspects of mobility (e.g., mobility of humans and other living things) in the future or at least indicate them as objectives for those who wish to pursue this direction of investigation. As far as the proposals in Part 2 are concerned, we have selected the three pluviometric Sahels and the major Sahelian rivers. The liminal Sahel, the core and the buffer zone indicate the continuous swaying of the shore between desert and humid climates. It is the breath of the Sahel, its expansion and contraction, advancing towards the desert or retreating into wetter areas, depending on the alternation of rainy or dry periods. Among the representations of the rivers, we have chosen the one related to the human footprint to account for (albeit only with an image), on one hand, the hydrological dimension, the course of the major rivers and their role in connecting the different parts of the Sahel and, on the other hand, for the presence of human communities, somehow also symbolising their flow as well as the "weight" they exert on the water resource. Rivers bring with them another aspect of the Sahel's breath, that of the flood variability and thus the changing in extensions of the seasonally flooded areas. However, the choice of proposing the image of the footprint, to somehow

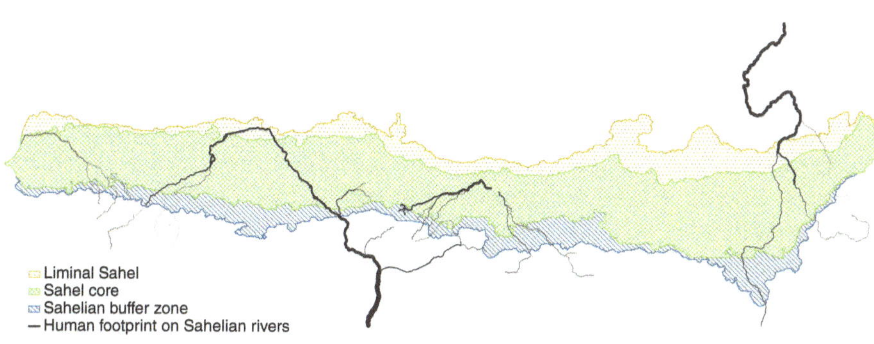

Figure 9.1 A cartographic synthesis: the breath of the Sahel.

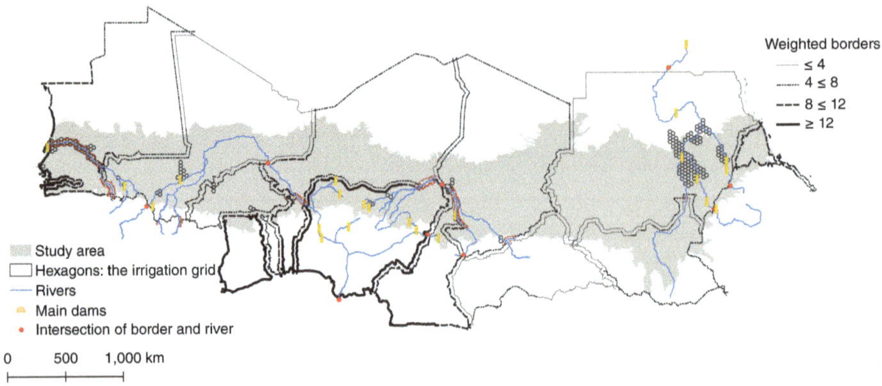

Figure 9.2 A cartographic synthesis: the cages.

highlight the role of human communities, indicates how an essential part of the Sahel's breath lies in the movement of its people, movement in space and flexibility in the use of resources, to adapt to the pluviometric and hydrological variability.

The second map (Fig. 9.2) includes all the aspects that stifle the Sahel, forcing its breath to be short, constrained within rigid cages. Among these, we have first selected all of the state boundaries, which carve the Sahel into territorial traps (Agnew, 1994, 2015). Then, we have chosen other traps that capture water, the dams on the major rivers, also indicating all of the points where the rivers intercept state borders. Finally, we have selected the large-scale irrigation schemes; to make them more visible on the entire regional scale, we have used the hexagons indicating the areas of interest (AOIs) for both pivot and surface irrigation schemes (see Chapter 8). The large-scale agricultural projects (as shown in Chapter 1) are traps that capture human and "other-than-human beings and entities" (Ghosh, 2021, pp. 57–58). All these traps constitute the cages that are fatal to the fluidity and flows in the Sahel, impeding the movements of people and water, communities and individuals, goods and trade. Although operating on different scales and spaces, the different traps interact with one another. The division of rivers among several countries creates hydropolitical frictions; flood capture by dams is essential for large-scale irrigation schemes but drastically alters the possibility of other multiple uses of wetlands. However, it must be noted that all these traps face many problems. Borders are often permeable and porous (see Chapter 5). Dams fill with sediment; large-scale irrigation schemes are subject to frequent failures (see Chapter 1). Cages rust quickly in the Sahel. The immensity of the Sahelian spaces and the difficulties in movement pierce the cages; no fence ultimately seems able to permanently block the flows. The political dimension itself often takes the form of movement rather than fixity (Deleuze & Guattari, 1980, pp. 434–527; Mbembe, 2012, pp. 180–190), and those who "do not want to be governed" find a crucial ally in the friction of the terrain (Scott, 2009, pp. 40–50, 129).

A dynamic that involves the whole region, at another level, is the intersection between Sahel's breath and cages, between the will of one polarity to move freely and that of the other to constrict or to block. More than a little of the present Sahelian vicissitudes can be found in these profound movements.

On the edge of endless change: Slender spiderwebs swaying in the wind

This metaphorical approach to mapping seems particularly apt for describing the Sahel whose very term is coined on the metaphor of the "shore" (in Arabic) of that immense sea that is the desert. The same idea of a map can be illuminated through a metaphor, that of a trap, as already used in the previous section. However, this time, we summon a particular kind of trap: the spider's web.[1] The anthropologist A. Corsín Jiménez (2018) suggests that "the spiderweb … offers an apposite metaphor for a world that holds itself in precarious balance, that tenses itself with violence and catastrophe but also grace and beauty" (p. 53), a metaphor certainly appropriate for the Sahel in this context as well. Nonetheless, for Corsín Jiménez, the spiderweb serves above all as a "method of description" (p. 54), as "a productive anthropological technology" (p. 75). "Traps have spatial, temporal, and ontological effects" (p. 75). Someone weaves a web to catch the prey and then waits until it is entangled. "The spider and the fly … mirror each other in the ecological interface of the web" (Corsín Jiménez, 2018, p. 63). For some, that setting could seem deadly, yet it is also important to take into account the fly's situation (Pase, 2020). Similar to the anthropologists, the geographers weave their webs, which tend to take the form of a map, with its web of meridians and parallels (Pase, 2020, pp. 42–43). From this perspective, the metaphor of the spiderweb fits well with our attempt to draw maps that capture different aspects of the Sahel without taking on the rigidity of modern cartography; they are not rigid, blocky grids. Rather, they resemble spiderwebs in the wind, continually undone and redone the next day to try to capture some meaning and to describe the trace of some movement in this changing world that is the Sahel. These light, fragile spiderwebs neither block nor claim any prey but only record the direction of a trajectory.

Our mapping is characterised by impermanence and relativity; that is why we have entitled Part 2 "Cartographic Sketches". Quick sketches are roughly drawn at the moment, always to be redone, updated and rethought. Our maps always intend to identify the in-between spaces, which are normally hidden in maps. For this reason, we have separated the border into two different lines, which create precisely an in-between space of relationships, which changes in meaning with the shift in the balance of power between states (Chapter 5). In this regard, we have investigated the seasonal and ephemeral amphibious spaces provided by the changing floods of the major Sahelian rivers (Chapter 7). In short, it is in the empty space between the threads of the spiderwebs that more appropriate views of the Sahel can be found. We have not let ourselves be caught in our own webs either. When it seems appropriate, as in the case of the irrigated areas with central pivots (Chapter 8), we have stepped outside the territorial limits within which we have defined our area study. What is of interest is not the web we have spun but rather the meanings we perceive; it has made more sense to go beyond the limits of the pluviometric Sahel. This is perhaps the lesson that the Sahel can teach in the domain of area studies and, at the same time, in map production: limits are needed, but the ability to criticise and go beyond them, following the phenomenon and yet conforming to the object, is always necessary. The spiderweb is useful in intercepting the meanings hidden in the spaces of the world. Nonetheless, the focus is always on what we are trying to intercept, not on the tool we have constructed to capture it, knowing that sometimes, like flies from spiderwebs, even the most intimate sense of reality will escape whatever traps we manage to weave.

GIS and cartographic challenges

One of the major challenges we have faced throughout this chapter has been the imagination of new maps, novel ways of drawing and representing boundaries, as well as analysing data from a critical standpoint. This approach stems from recognising the limitations of usual representations, allowing us to offer our readers an unconventional vision of the Sahel. We have aimed to highlight the complexity of vast and diverse spaces from climatological, geographical and political perspectives.

The software tools we utilised were quite varied, as we applied them in out-of-the-box ways. Only through the flexibility of each code were we able to accomplish, or at least we hope to have done so, our goals of clarity and effective communication. We also recognise that much of the analyses and insights featured in the different chapters of Part 2 would have been unattainable just a few years ago. The notion of performing large-scale analyses and simultaneously focussing on detailed aspects would have been impossible due to the limited availability of data, software and hardware resources, which, at the time, were accessible to only a few.

We used different types of software, from QGIS to Google Earth Engine (GEE), to find and analyse data from a wide variety of sources. It is likely due to GEE's data catalogue and resources that many of our ideas have been brought to fruition. The flexibility of GIS software allowed us to conduct historical analyses of time-series data, such as identifying the boundaries of the hydrological Sahel (see Chapter 7) or categorising different types of irrigated agriculture (see Chapter 8). Representing these results proved to be a complex challenge, especially since we are accustomed to the usual method of cartography, with its strict rules of representing territories, and most conventional tools are designed with those rules in mind. In this context, the flexibility of the instruments used allowed us to represent territorial dynamics that are not objective or static elements but are changing properties and perceptions of socio-political phenomena. We spent significant efforts on overcoming these representation issues, a clear example being the portrayal of boundaries—doubled, fragmented and depicted in unusual ways. Although the result may seem simple, this operation forced us to develop our own tools and procedures in QGIS to achieve an outcome that was further refined using other types of graphic software.

These processes have nonetheless produced a methodology, briefly described in the chapters of this book, as well as Python, SQL and JavaScript codes that are available in an open-source repository or from us upon request.

In Part 2, we have condensed years of research, reflections and knowledge of data and technologies that have enabled us to tackle this cartographic challenge. As Monmonier once wrote, we have had to "lie with maps" (1991). This does not mean we have made mistakes—at least not intentionally—rather, we have resorted to bending the rules of cartography to convey the message we want to communicate. In doing so, we have highlighted aspects that conventional cartography, constrained by topologically correct lines, would not have allowed us to show. In this sense, we have "lied" by doubling the boundary lines to give each reality its own dimension, weight and geopolitical significance. We have redrawn the waterways and accentuated their breakpoints and shifts in both flow and significance, while emphasising the impact of human populations on these waterways. Through these choices, we have aimed to not only represent the complexities of the Sahel but also push the boundaries of cartography, challenging conventions and offering readers a more nuanced and multifaceted way of understanding the socio-political and environmental dynamics of this region.

One last step remains in concluding this Part 2 devoted to cartographic representations of the Sahel. We look back at how the Western eye, over time, has tried to capture the Sahel in the grid of meridians and parallels. The chapter that follows is precisely dedicated to the historical cartography of the Sahel belt, investigated with the aid of the maps stored at the University of Padua's Museum and Library of Geography. From the 18th century onwards, in the centre of the African continent, enormous white spaces appeared in maps—empty spaces that colonisers would fill with their surveys (Pase, 2024), but not only their surveys. The once empty spaces will be filled in, among other things, with those large-scale irrigation schemes that precisely mark the history of when colonial times began.

Note

1 We thank Carla Braga for proposing this metaphor and suggesting that we read Corsín Jiménez's (2018) article.

References

Agnew, J. (1994). The territorial trap: The geographical assumptions of international relations theory. *Review of International Political Economy*, *1*, 53–80. https://doi.org/10.1080/09692299408434268

Agnew, J. (2015). Revisiting the territorial trap. *Nordia Geographical Publications*, *44*(4), 43–48. https://nordia.journal.fi/article/view/64824

Braudel, F. (1995). *The Mediterranean and the Mediterranean world in the age of Philip II*. Vol. I. University of California Press.

Corsín Jiménez, A. (2018). Spiderweb anthropologies. Ecologies, infrastructures, entanglements. In M. De La Cadena, & M. Blaser (Eds.), *A world of many worlds* (pp. 53–82). Duke University Press.

Davidson, D. (1978). What metaphors mean. *Critical Inquiry*, *5*, 31–47.
Deleuze, G., & Guattari, F. (1980). *Mille plateaux. Capitalisme et schizophrénie*. Les Éditions des Minuit.
Dematteis, G. (1985). *Le metafore della Terra. La geografia umana tra mito e scienza*. Feltrinelli.
Dematteis, G. (2021). *La geografia come immaginazione. Tra piacere della scoperta e ricerca di futuri possibili*. Donzelli.
Farinelli, F. (2009). *La crisi della ragione cartografica*. Einaudi.
Ghosh, A. (2021). *The nutmeg's curse. Parables for a planet in crisis*. Murray.
Mari, G. (2001). *I vocabolari di Braudel. Lo spazio come verità della storia*. Luciano.
Mbembe, A. (2012). *Out of the dark night*. Columbia University Press.
Monmonier, M. (1991). *How to lie with maps*. University of Chicago Press.
Pase, A. (2020). *Geografly. La mosca e la mappa*. Bette.
Pase, A. (2024). Emptying and filling. Maps of inland Africa. In T. Rossetto & L. Lo Presti (Eds.), *Routledge handbook of cartographic humanities* (pp. 235–245). Routledge. https://doi.org/10.4324/9781003327578
Retaillé, D., & Walther, O. (2019). *Mapping the Sahelian space*. https://arxiv.org/abs/1906.02223
Rorty, R. (1989). *Contingency, irony, and solidarity*. Cambridge University Press.
Scott, J. C. (2009). *The art of not being governed: An anarchist history of upland Southeast Asia*. Yale University Press.

Interlude

The Sahel in historical atlases (1493–1789)

Chiara Gallanti and Chiara Forner

Introduction

Thanks to the extensive collection of cartographic materials available at the University of Padua, the AtlaSahel project has the opportunity to include a Historical Cartography section. This collection represents the legacy of 150 years of geographic studies, preserved and enriched over time by the Library of Geography. Since its creation in 2019, the Museum of Geography has also been collaborating in this mission, receiving new donations as well, with the most notable contributions coming from the private collector Armando Morbiato (1933–2023).[1]

In the spring of 2021, a comprehensive census was conducted which included a photoshoot of the African maps found in *both* the 39 atlases or other map-equipped books donated by Morbiato *and* the library's original atlas collection. The objective of the census was to determine whether the identified maps, considering their number, chronological distribution and representativeness, could serve as a suitable foundation for a diachronic comparative study on the European perception of the African continent, particularly the Sahel region, from a cartographic perspective.

A total of 79 maps were surveyed, spanning the period from 1493 to 1990. Among these, we examined a chosen selection of 18 maps created before 1800. These represent the original core of the Historical Cartography section of the AtlaSahel website, which is currently being expanded to incorporate maps produced up to the early 20th century.

On the website, the entry page of the Historical Cartography section provides the first chronological overview of the selected maps, showing the evolution in the European portrayal of Africa, with reference to its shape, dimensions, place names, regional partitions and, of course, the number and precision of the included geographic features. However, these modifications did not necessarily correspond to an actual refinement of knowledge. It was only from the mid-17th century that a gradual omission of elements can be observed, rooted in long-held beliefs but not supported by direct knowledge (Betz, 2007), often resulting in a parallel enlargement of blank spaces in maps (Pase, 2024), as seen in Bonne's (1789) map.

DOI: 10.4324/9781003429111-13

From the entry page, each map can be explored by observing its high-resolution version through the Viewer tool or by reading its descriptive record.

Representations of Africa in the AtlaSahel historical maps: General considerations

The Historical Cartography section of the AtlaSahel website provides an insight into essential information about the evolution in the European representation of Africa in the Early Modern Era, spanning the period from 1493 (when the world map found in the *Liber Chronicarum* was likely created by Hieronymus Münzer) to 1789 (when the Paduan edition of Rigobert Bonne's *Atlas encyclopédique* was published).

Münzer's famous world map does not reflect the knowledge gained from the recent Portuguese explorations. Instead, it follows the 2nd-century Ptolemaic worldview, rediscovered during the 15th century (Woodward, 2007): Africa has no peninsular nature, appears too short and tilted to the East, and is connected to Asia southwards by a long strip of land. Furthermore, its northern coast continues to appear disproportionately wide, a misconception that would only be rectified by Guillaume Delisle in 1700, despite Arab cartographers having already corrected Ptolemy's (1477) inaccurate representation long before that time (Betz, 2007).

Important corrections in the shape of Africa are owed to Abraham Ortelius. The small map available on the AtlaSahel website, published in the 1724 Lovisa edition of the *Theatrum Orbis Terrarum*, faithfully reproduces the shape found in the original 1570 edition. Here, compared to previous depictions, the eastern extent of the continent is reduced, and the continent shows a more pointed shape (Betz, 2007).

While Portuguese discoveries provided new information regarding the African coastlines as early as the beginning of the 16th century, as reflected in the maps included in the AtlaSahel website, the reliance on Ptolemy continued for a considerable period when it came to representing the interior (Cusimano & D'Agostino, 1986). A notable case is that of the Mountains of the Moon. Originally placed by Ptolemy 7° south of the equator, these legendary mountains have been explained as an erroneous positioning of the Ruwenzori, which is actually located 0° 23′ north of the equator. Relying on Marinus of Tyre's account of the discoveries of the Greek merchant Diogenes, Ptolemy situated two lakes to the north of the Mountains of the Moon, fed by streams originating from the mountains themselves. These lakes were believed to be the sources of what is now known as the White Nile. In the maps included in the AtlaSahel website, the Mountains of the Moon are clearly depicted in Münzer's map, along with the twin lakes, as well as in the maps found in the works of Rosaccio and Ruscelli (1599), Clüver (1661) and Kircher (1682). In contrast, Ortelius omitted their representation while retaining the Ptolemaic lakes in the southern part of the continent, less symmetric in shape compared to those of previous models, with one larger lake situated in the west and a smaller one to the northeast (Mackraz, 2016). Various maps in the

Figure 9a.1 Nicolas Bion, *L'Afrique*, map included in N. Bion, *L'Usage des globes célestes et terrestres*, Paris, Jacques Guerin, 1751.

Source: Copyright: Università di Padova, Museo di Geografia

AtlaSahel website follow Ortelius's model, including those by Münster (1614), Botero (1671), Magini (1612) and Rosaccio (1598).

Ptolemy's legacy was definitely surpassed only after the European expeditions of the 1870s, although French cartographers had already begun to challenge it in the 17th century. Following Delisle's significant 1700 map, which removed the Mountains of the Moon and the Ptolemaic Lakes, these were largely absent from most 18th-century French maps, as evident in the AtlaSahel selection. Two exceptions are Bion's (1751) map (Figure 9a.1), which depicts the Mountains of the Moon, albeit without relating them to the Nile, and Janvier's (1776) map, which portrays both the mountains and the lakes as possible sources of the Nile. The Italian Zatta (1778–1785) also shows both features.

Zooming in on the Sahel and its prominent hydrographic features

Europe's understanding of the actual geography of the Sahel region was significantly delayed, needless to say, due to the formidable barrier imposed by the Sahara Desert. Moreover, an alternative access from the Atlantic coasts through the mouths of the rivers in the region was complicated by shallow waters and, in some cases, complex deltaic ramifications (Vallerani, 2022). Therefore, it is not surprising that many particularly significant aspects of the region remained misunderstood for a longer time, such as, in the field of hydrography: the number of rivers; the location of their sources; the location of the river mouths; the actual course; and even the direction of their flow.

The cases of Senegal River and Niger River are well known and emblematic. The first originates in Guinea's Fouta Djallon Highlands and spans approximately 1,790 km, passing through Mali before reaching its mouth at the Atlantic Ocean, along the border between Senegal and Mauritania. The Niger begins in the Guinea Highlands in southeastern Guinea and flows for 4,160 km in a crescent-shaped path through Mali, Niger and Nigeria before reaching the Gulf of Guinea. The Niger's peculiar course played a prominent role in causing past European misconceptions, which were gradually corrected, particularly between Mungo Park's first expedition in 1795 and Richard Lander's arrival at its mouth in 1830. The maps that we are analysing predate this phase of clarification and serve as evidence of the prevailing errors, frequently originating from classical sources. For example, Pliny the Elder (1938) believed the Niger to be an eastward-flowing river originating in current Mauritania, intermittently running underground and joining the Nile near Meroe (Betz, 2007). A fossilised remnant of this concept can explain the 60-mile-underground flow that characterises the Niger, as depicted in Münster's *Affricae tabula nova* (1614), originally published in 1588, although here the Nile is not involved, and the river flows northwest-west. However, the classical idea of a Nile tributary originating in central Sahel did not vanish from modern cartography and can be found in landmark maps, such as Mercator's (1569) and Blaeu's (1608). In the AtlaSahel selection, it is shown by Clüver (1661), Kircher (1682), Zahn (1696) and all 18th-century examples except Sepp's (1758). After 1772, this uncertain river began to overlap with the Bahr el Ghazal River, which was mapped that year by the French geographer Bourguignon d'Anville. This overlap can be seen in Zatta's map (1778–1785), which refers to the river as "Fiume Gazzella".

However, the dominant hydrographic feature of the region is the presence of a single river that flows westwards, combining elements from both the Senegal and the Niger, along with legendary information. Although the original cartographic source of this error can be traced back to Mohamed al-Idrisi's work (1154), a more direct and conscious influence is attributed to Giacomo Gastaldi (Betz, 2007), who, in his famous map of Africa (1564), depicted a single river in the region. Named "Niger f.", it originates from a lake north of the equator (Lac Niger) and flows northwards to a second lake, believed to be Lake Chad, before taking a turn westwards. It then passes through a third lake called Guarda or Guber, featuring a fortified island, before finally reaching the Atlantic Ocean. Ortelius (1724), Münster (1614), Magini (1612), Botero (1671) and Rosaccio (1598) follow this model, while Clüver (1661) aligns with it only in the portrayal of the lake with the fortified island.

This third lake, which is found in almost all European cartography from the 16th and the 17th centuries under different names and is included in the AtlaSahel historical maps, has been the subject of a specific study by Rizzo (2006). According to him, the fortified island can be identified as the city of Jenne (or Djenne), which was described in 15th-century accounts as built on marshy terrain with stone walls. The third lake thus corresponds to the "fluctuating area of lakes and branching streams in the middle course of Niger River, between Djenne and Timbuktu, known today as the Inland Niger Delta" (Rizzo, 2006, p. 80).

Niger River and Senegal River were finally depicted as separate entities by Delisle, who based his maps on French explorers' reports. After initially regarding them as two distinct rivers, one ending (the Niger) and the other originating (the Senegal) in Guarda Lake on his 1700 map (Betz, 2007), he separated their courses in his 1714 *Orbis veteribus noti tabula nova* and in his 1722 map of Africa, also subdividing Guarda Lake into two smaller lakes: the legendary Maberia, source of the westward-flowing Senegal River, and an unnamed lake, source of the eastward-flowing Guien (Niger) River (Rizzo, 2006). While Bion did not incorporate this model yet in 1751, the later 18th-century maps presented in the AtlaSahel did.

However, once again, the unusual shape of the Niger River generated different representation models. de Vaugondy and de Vaugondy (1762), Bonne (1780) and Desnos (1784) depicted Niger River as bending in an arc near Timbuktu and flowing into Lake Bournou, while Janvier (1776), Zatta (1778–1782) and Bonne (1789) showed it as bifurcating into the two lakes of Regebil (or Reghebil) and Semegonda (Forner, 2022). European uncertainties would only be resolved with the explorations of the 19th century.

Note

1 For additional information about Armando Morbiato, refer to Morbiato (2020).

List of cited maps

Al-Idrisi. (1154). *Tabula rogeriana*. Held at the National Library of France.
Bion, N. (1751). L'Afrique. In N. Bion (Ed.), *L'Usage des globes célestes et terrestres*. Jaques Guerin. https://atlasahel.it/en/historical_maps/lafrique-2/
Blaeu, W. J. (1608). *Africa*. Willem Janszoon Blaeu.
Bonne, R. (1780). Afrique. In G.T.F. Raynal (Ed.), *Atlas de toutes les Parties Connues du Globe*. Jean-Leonard Pellet. https://atlasahel.it/en/historical_maps/afrique-dressee-par-m-bonne-2/
Bonne, R. (1789). Afrique. In R. Bonne, & N. Desmarest (Eds.), *Atlas encyclopédique*. [s.n.]. (Map originally published in 1787). https://atlasahel.it/en/historical_maps/afrique-2/
Botero, G. (1671). Africa. In G. Botero (Ed.), *Relationi universali di Giovanni Botero benese*. Bertani. (Map originally published in 1595.) https://atlasahel.it/en/historical_maps/africa-4/
Clüver, P. (1661). Africa. In P. Clüver (Ed.), *Introductionis in universam geographiam tam veterem, quam novam libri sex*. Elsevier. (Map originally published in 1624.) https://atlasahel.it/en/historical_maps/africa-5/
Delisle, G. (1700). *L'Afrique dressée sur les observations de l'Academie Royale des Sciences, et quelques autres, et sur les memoires les plus recens*. Guillaume Delisle.

Delisle, G. (1714). *Orbis veteribus noti tabula nova*. Guillaume Delisle.
Delisle, G. (1722). *Carte d'Afrique*. Guillaume Delisle.
Desnos, L. C. (1784). Carte de l'Afrique. In L. C. Desnos (Ed.), *Atlas élémentaire de la géographie*. Bourgoin. https://atlasahel.it/en/historical_maps/carte-de-lafrique-2/
de Vaugondy, G. R., & de Vaugondy, D. R. (1762). *L'Afrique*. In G. R. de Vaugondy & D. R. de Vaugondy (Eds.), *Nouvel atlas portatif destiné principalement pour l'instruction de la jeunesse*. Robert de Vaugondy. (Map originally published in 1757.) https://atlasahel.it/en/historical_maps/barbarie-egypte-nigritie-guinee-nubie-abissinie-2/
Gastaldi, G. (1564). *Il disegno della geografia moderna de tutta la parte dell'Africa*. Fabio Licinio.
Janvier, J. (1776). L'Afrique. In F. Santini (Ed.), *Atlas universel dressé sur les meilleures cartes modernes*. Paolo Santini. https://atlasahel.it/en/historical_maps/lafrique-divisee-en-ses-principaux-etats-2/
Kircher, A. (1682). Tabula Geographico- Hydrographica Motus Oceani, Currentes, Abyssos, Montes Ignivomos. In A. Kircher (Ed.), *De Onderaardse Wereld*. Joannes Janssonius Van Waasberge. (Map originally published in 1665.) https://atlasahel.it/en/historical_maps/tabula-geographico-hydrographica-2/
Magini, G.A. (1612). Africa. In G. A. Magini (Ed.), *Geografia cioè Descrittione universale della Terra*. Paolo & Francesco Galignani. (Map originally published in 1596.) https://atlasahel.it/en/historical_maps/491/
Mercator, G. (1569). *Nova et Aucta Orbis Terrae Descriptio ad Usum Navigantium Emendate Accommodata* [s.n.].
Münster, S. (1614). Africae Tabvla Nova. In S. Münster (Ed.), *Cosmographey*. Sebastian Henricpetri. (Map originally published in 1588.) https://atlasahel.it/en/historical_maps/affricae-tabula-nova-2/
Münzer, H. (1493). Secunda etas mundi: Folium XIII. In H. Schedel (Ed.), *Liber Chronacarium*. Anton Koberger. https://atlasahel.it/en/historical_maps/secunda-etas-mundi-2/
Ortelius, A. (1724). Africae Tabvla Nova. In A. Ortelius (Ed.), *Teatro del mondo di Abraamo Ortelio: nel quale si dà notizia distinta di tutte le provincie, regni e paesi del mondo*. Lovisa. (Map originally published in 1570.) https://atlasahel.it/en/historical_maps/africae-tabula-nova-2/
Rosaccio, G. (1598). Figvra dell'Affrica e sve province. In G. Rosaccio (Ed.), *Teatro del cielo e della terra*. Domenico Farri. https://atlasahel.it/en/historical_maps/figura-dellaffrica-e-sue-provincie-figure-of-africa-and-its-provinces/
Rosaccio, G. & Ruscelli, G. (1599). Tabvla Africae IIII. In G. Rosaccio, & G. Ruscelli (Eds.), *Geografia di Claudio Tolomeo alessandrino*. Eredi Melchiorre Sessa. https://atlasahel.it/en/historical_maps/tabvla-africae-iiii-2/
Sepp, C. A. (1777). Africa XXI. In C. A. Sepp (Ed.), *Nuovo atlante portatile ovvero metodo facile per apprendere in breve la geografia*. Giammaria Bassaglia. https://atlasahel.it/en/historical_maps/africa-xxi-2/
Zahn, J. (1696). Facies Altera Hemispherii Terrestris. In J. Zahn (Ed.), *Specula Physico-Mathematico-Historica*. Johann Christoph Lochner. https://atlasahel.it/en/historical_maps/facies-altera-hemispherii-terrestris-2/
Zatta, A. (1778–1785). L'Africa. In A. Zatta (Ed.), *Atlante novissimo, illustrato ed accresciuto sulle osservazioni e scoperte fatte di più celebri e più recenti geografi*. Antonio Zatta. https://atlasahel.it/en/historical_maps/lafrica-divisa-ne-suoi-principali-stati-di-nuova-projezione-2/

References

Betz, R. L. (2007). *The mapping of Africa: A cartobibliography of printed maps of the African continent to 1700*. Hes & de Graaf.
Cusimano, G., & D'Agostino, G. (1986). *L'Africa ritrovata: Antiche carte geografiche dal XVI al XIX secolo*. Università di Palermo.

Forner, C. (2022). *L'Africa dal XV al XVIII secolo negli atlanti della Biblioteca di Geografia e del Fondo Morbiato (Museo di Geografia – Padova)* [Bachelor's thesis, Università degli Studi di Padova]. Padua Thesis and Dissertation Archive. https://thesis.unipd.it/retrieve/a8788456-e333-4b0f-aa81-e5b92ad88e8f/Forner_Chiara.pdf

Mackraz, B. (2016). *Early modern mapping of Africa*. Stanford Libraries. Retrieved July 27, 2023, from https://exhibits.stanford.edu/david-rumsey-map-collection/feature/early-modern-mapping-of-africa

Morbiato, A. (2020). *L'incanto del viaggiatore: Diari (1957–1967) e ricordi di un emigrante* (L. Morbiato, Ed.). Il Poligrafo.

Pase, A. (2024). Emptying and filling. Maps of inland Africa. In T. Rossetto & L. Lo Presti (Eds.), *The Routledge handbook of cartographic humanities* (pp. 235–245). Routledge. https://doi.org/10.4324/9781003327578

Pliny the Elder. (1938). *Natural history* (H. Rackham, Trans.). Harvard University Press. (Original work published AD 77–79.)

Ptolemy. (1477). *Geographia* (First illustrated ed.). Domenico de' Lapi.

Rizzo, G. J. (2006). The patterns and meaning of a great lake in West Africa. *Imago Mundi*, *58*(1), 80–89. https://doi.org/10.1080/03085690500362439

Vallerani, F. (2022). Acque continentali e misteri geografici: Gli esploratori fluviali del continente Africano. In M. Varotto (Ed.), *Giovanni Miani: Il leone bianco del Nilo* (pp. 18–25). Silvana Editoriale.

Woodward, D. (2007). Cartography and the renaissance: Continuity and change. In D. Woodward (Ed.), *The history of cartography: Vol. 3. Cartography in the European renaissance* (pp. 3–24). University of Chicago Press.

Part III

Complicated Muddles of Water and Land

10 Sudan

A 'picture' of irrigation spaces before April 15, 2023

Mariasole Pepa, Andrea Pase, Marina Bertoncin, Eltaib Saeed Ganawa, Abdelrahman Eltahir A. Musa and Mohammed Hamed Hassan Abubakr

Introduction

> *Since the war began in Sudan between the Security Committee led by General Abdel Fattah Al-Burhan and the Rapid Support Forces led by General Muhammad Hamdan Dagalo, we have said that it is a meaningless war.*
>
> *(Noureddine Madani in Pase & Pepa, 2023)*

On the April 15, 2023, the war in Sudan erupted, with little signs of the ongoing conflict's magnitude and the resulting devastation, which continues to cause civilian displacement, destruction of infrastructure and entire cities, sexual assaults and psychological trauma.[1] Since the war broke out between the Sudanese Armed Forces (SAF) led by Abdel Fattah al-Burhan and the paramilitary Rapid Support Forces (RSF) under General Mohamed Hamdan Dagalo, known as Hemedti, more than 8 million people have been forcibly displaced (UNHCR, April 2024). From the beginning, this war has received little media coverage, especially in Western outlets, and the news has slowly disappeared from the headlines. Similarly, the role of humanitarian aid interventions has been limited, and the majority of international organisations left the country or temporarily suspended their activities (World Food Programme (WFP), 2023).

Fortunately, mutual aid efforts led by the Sudanese people and local grassroots organisations (Carstensen & Sebit, 2023), the Emergency Response Room[2] and the Sudanese diaspora are sustaining the burden of this war (Shabaka, 2023).

The materials collected for this chapter were obtained from the fieldwork conducted in Sudan between January and March 2023 within the framework of a collaborative project undertaken by geographers from the University of Khartoum and the University of Padua. In that period, Khartoum was a vibrant city where the legacy of the 2019 revolution was still tangible[3] (Berridge et al., 2022; Khiar, 2022). We believe that it is crucial to talk about Sudan and narrate the discourses, practices, challenges and desires related to the transformation of the Sudanese irrigation spaces prior to the war, as well as to create visual archives of these transformations (Pepa, 2023).

DOI: 10.4324/9781003429111-15

This chapter is structured into five main sections. In the first section, a framework is offered to situate the research project and collaboration, as well as to navigate the diverse irrigation projects under investigation. In the second section, we contextualise the usage of water and land as fundamental geographical lenses through which we examine societal transformation in Sudan. In the third section, we delve into specific narratives characterising discourses surrounding water and land, focusing on the period immediately before the outbreak of the war. In the fourth section, we outline the changes, challenges and transformations experienced by the stakeholders involved in irrigation projects, as observed during the fieldwork conducted in Sudan between January and March 2023. Finally, an initial attempt at periodisation to understand water and land under siege is presented in the fifth section, along with a discussion of the war's impact on irrigation spaces.

Setting the context: Water and land usage as geographical lenses

For more than three decades,[4] geographers at the University of Padua have engaged in collaborative research with the Faculty of Geographical and Environmental Sciences at the University of Khartoum, focusing on the territorial impacts of mega-irrigation projects in Sudan (Bertoncin et al., 1995, 2017, 2019, 2022). The research group has been interested in tracing the historical trajectories—covering the period from the start of the Anglo-Egyptian Condominium established in 1899 to the present—of the relations between the state, capital investments and natural resources while examining the diverse ways in which these shifts have subsequently reshaped both human and more-than-human geographies (Bertoncin et al., 2019). The transformation of Sudanese drylands through mega-irrigation projects has prompted the following critical research questions that have been central to our investigation over the years: What development paradigms underpin these projects? Who is benefiting from these investments or promises of development? Are local communities compensated, displaced or adequately included in these new enclaves? What are the environmental, social and political consequences of these initiatives?

Our attentiveness to water and land usage and its political implications in Sudan (Egemi, 2006) is parallel to the rapid interest in securing water and land in Africa, which has resulted in what is known as the new scramble for Africa over the past decades (Obeng-Odoom, 2021). Since the early 2000s, Sudan has experienced a parallel dynamic of land and water grabbing, with a growing number of international investors controlling substantial land and water resources, facilitated by concessions granted by the state (Bertoncin et al., 2022; Elamin, 2018). These dynamics also extend to domestic land grabbing led by urban elites backed by the Sudanese government. This scenario contributes to heightened competition and conflicts over land resources (Sulieman, 2015).

In this frame, mega-development projects have been strategically used to control territories while reshaping rural geographies. Mega-projects tend to introduce a radical transformation of territories in both tangible (e.g., canals, pumping stations) and intangible ways (e.g., relation between farmers and land). The transformation of rural spaces typically adheres to a modernisation logic that often advocates a vision of agriculture without farmers (McMichael, 2014). In this respect, a mega-project acts as a 'laboratory of modernity' in which the reinterpretation of proximity relations is crucial (Bertoncin et al., 2014). Nonetheless, Sudan has a rich history of community resistance and opposition to large-scale development projects, as in the case of the Kabjar Dam in Northern Sudan (Abd Elkreem, 2018).

The transformation of Sudanese territories by mega-irrigation projects has evolved over time, yielding diverse territorial outcomes driven by the cultivation of different crops, such as cotton, sugar and alfalfa. These changes have been influenced by shifts in the relations between state, capital investments and natural resources, the introduction of new technologies and changes in geoeconomic and developmental priorities (Bertoncin et al., 2019). The research group has observed the evolution of various mega-irrigation projects implemented during different historical periods, including the Gezira Scheme in the colonial era, the Sudanese Sugar Belt in the postcolonial era and investments in central-pivot irrigation systems in the early 2000s.

Projects explored in the research

Colonial mega-scheme: The Gezira

The Gezira Scheme stands out as one of the world's largest irrigation systems, with its extension of 2.2 million *feddans*[5] inclusive of the Managil extension, which followed the construction of the Roseires dam in 1966. Originating in 1925 under the British colonial rule, the scheme was established primarily to supply cotton to textile factories in England, reflecting commercial and colonial interests rather than domestic economic imperatives. By the 1970s and the 1980s, mismanagement and declining performance led the Gezira to become emblematic of the failure of development promises (Ertsen, 2016). In the new millennium, a turning point in the management of the scheme was symbolised by the introduction of the 2005 Gezira Reform Act by the Fundamentalist Islamic regime of al-Bashir. The major changes prompted by the law were the freedom to plant crops, the decentralisation of irrigation management and the possibility for tenants to rent land for cultivation. Additionally, in 2015, the al-Bashir regime abolished the Farmers Union, which had long served as a platform for farmers' voices. A Khartoum University professor claimed that "under the regime, the Farmers Union was dismantled because farmers were powerful and represented a threat to the state" (interview, February 2023).

Following the 2019 revolution, the efforts to reinstate the Farmers Union and oppose the progressive liberalisation of land ownership gained momentum (Dabanga, 2022). The 2005 law was suspended in 2019, and new negotiations were ongoing before the start of the war, yet land conflicts remained one of the major issues (Vadot, 2021). Despite the scheme's evident degradation and the pressing imperative for rehabilitation, particularly in the management of the irrigation system and canals, the pumping system and drainage, the project has never been abandoned. Instead, it is undergoing a perpetual metamorphosis in which the future (past) of the project has continuously been reimagined by the Gezira Board, the state, local farmers, international organisations, domestic and international investors, consultants and researchers (Abdelhadi et al., 2004; Goelnitz & Al-Saidi, 2020; Vadot, 2021).

The 'Sudanese Sugar Belt'

Following Sudan's independence in 1956, the state was interested in expanding its agricultural potential beyond the Gezira Scheme. Particularly, numerous large-scale sugar irrigation perimeter and sugar factories were initiated all around the Gezira, which we referred to as the "Sudanese Sugar Belt"—comprising Kenana, Guneid, Assalaya, Sennar, New Halfa and, most recently, the White Nile (Bertoncin et al., 2017). Among these projects, Kenana Sugar Company, established in the early 1970s, stands out as the largest and most prominent, serving as Sudan's primary sugar producer. Its establishment occurred during the presidency of Jaafar Nimeiri, aligning with a broader narrative supported by the Gulf States that positioned Sudan as the breadbasket of the Arab world (Elnur, 2009), and as the "Cuba of Africa" in sugar production (El Nazir & Desai, 2001). The project encompasses an area of 168,000 feddans, with 83,000 feddans designated for commercial purposes. However, during our visit in February 2023, only 74,000 feddans were used in production. The project utilises an intricate system of canals and hydro-flume[6] technology to distribute water from the White Nile. Kenana Company operates as a public–private partnership, involving the Government of Sudan (35%), the Public Investment Authority of Kuwait (30%), the Government of Saudi Arabia (11%), the Arab Authority for Agricultural Investment and Development (6%), and other stakeholders. This diverse investor composition has facilitated continuous economic support, enabling the diversification of economic activities even under challenging political and economic circumstances.

The commercial success and dynamism of Kenana Company prompted the al-Bashir regime to engage the company in developing Sudan's sugar sector as a means to showcase to foreign investors the country's local capacity in the sugar industry (Turrini, 2019). This led to the establishment of White Nile Sugar Company, which commenced operations at the end of 2013. Covering a total area of 165,000 feddans, the project encountered various challenges from its inception, including soil salinity, inadequate drainage systems and operational inefficiencies, leading to underutilisation of its full capacity.

Different from Kenana Company, the Sudanese Sugar Company (SSC) was established in 1992 as a state-owned project, which also affected its economic possibility to operate over the years. The SSC comprises four sugar factories: Guneid, Assalaya, Sennar and New Halfa. In this chapter, we pay more attention to Guneid, which was originally part of the Gezira Scheme. In 1962, it shifted its production from cotton to sugar, retaining many of its old tenants. The peculiarity of Guneid is that the company works with farmers for the production of sugarcane from planting to harvesting. The company supplies the necessary inputs and, at the season's end, calculates the costs of inputs and distributes profits to the farmers. Since 1992, each farmer has held an individual account, which replaced the previous common account. Furthermore, in contrast to other factories under the SSC, Guneid allocates 2% of its annual profit (1% from farmers' profit and 1% from the company's profit) to socio-economic initiatives, including healthcare and education. Nonetheless, recent years have witnessed a decrease in profits, restricting the extent of these activities.

New wave of territorial investments: the circular green oasis in the desert

Water and land in Sudan are strategic assets that have attracted interest from the colonial era to the present militarisation of resources. Since the early 2000s, Sudan has witnessed a new wave of agricultural investments based on the introduction of new technologies, such as the central pivot irrigation system, also known as the 'green circles in the desert'. Compared to previous irrigation projects that implied a radical territorial transformation, such as the Gezira Scheme, pivot systems are characterised by lighter infrastructure (Fig. 10.1) and minimal intervention requirements (Bertoncin et al., 2017). The interest in securing vast tracts of land

Figure 10.1 Example of a pivot system operating under the Al-Waha project owned by DAL Group. During the war, the project was completely destroyed.

Source: Mariasole Pepa, 2023.

and access to water initially came from the Gulf countries, which were interested in Sudan's water for the cultivation of alfalfa (*Medicago sativa*), highly requested in the global market for its productivity and greater nutritional value compared to other forages. The alfalfa produced in Sudan was then exported especially to Gulf countries and also to China. For this reason, some interviewees claimed that through alfalfa, "Sudan exports its own water and land" (interview, February 2023). From the breadbasket strategy in the early 1970s to the pivot systems, there is a continuity of Arab countries' involvement in controlling land and water resources in Sudan (Bertoncin et al., 2022).

The neoliberal agenda, supported by the Sudanese government, aimed to attract investments, particularly after the secession of South Sudan in 2011 and the subsequent loss of oil revenue. The National Investment Encouragement Act of 2013 represented another turning point in the Sudanese government's attempts to attract investors, with Arab countries playing a prominent role in the agricultural sector (Bertoncin et al., 2022). The expansion of the pivot system has particularly interested the River Nile State; this rapid proliferation is visible from above.[7] Investors range from Arab countries to Chinese private investors, as well as joint ventures. Despite being lighter agricultural infrastructures, pivot systems exert significant pressure on land due to the requirements for vast tracts of land and substantial water usage, thereby increasing pressure on the Nile water resources and on unused land. Furthermore, these systems rely on technological management rather than on the work of farmers, raising further questions about the future livelihoods of peasants.

Narratives on water and land in Sudan

The way that narratives on water and land were framed during fieldwork in early 2023—just before the outbreak of the war—provides an informative insight into both national and international actors' vision for the future of agriculture In Sudan. In most meetings and interviews, there was a shared belief that "Sudan will become more important because of the war in Ukraine and the food crisis in Europe" (interview, February 2023). However, this interpretation of Sudan's distinctive role in addressing global food insecurity is a recurring theme throughout Sudanese history (Verhoeven, 2013).

For the Kenana Board, the food crises turned into a major issue of discussion to the extent that Arab Gulf states asked the company to produce wheat (interview, February 2023). Moreover, during our visit to the Gezira headquarters in Barakat in mid-February 2023, a member of the Gezira Board claimed that since Ukraine was involved in a war, countries such as Turkey, France and the UK, in addition to international companies and organisations, such as the World Bank, Food and Agriculture Organization (FAO) and the WFP, would turn to Gezira. A member of the Gezira Board stated, "There is a food crisis around the world, and Sudan has land and water". However, the Gezira Board members were pushing these discourses without bringing into the discussion the ways in which financial speculations in global food markets affected Sudan (Greco, 2023).

Showcasing Gezira as the answer to the world's food problems was expounded further with the title of the PowerPoint presentation the Gezira Board members shared with us: Gezira: The Gift—a document prepared to "encourage the private sector to invest in Gezira". This vision was also supported by members of the Agricultural Research Cooperation (ARC) in Wad Medani, who agreed that new irrigation technologies were needed since the "Gezira Scheme is not only important for Sudan but for the world" (interview, ARC, February 2023).

This narrative was reinforced also by international actors; for instance, in February 2023, a video message of the United Nations WFP Executive Director David Beasley was circulated on social media, such as WhatsApp groups. The video was recorded in Gezira by Dr Beasley, who showed wheat fields and argued for Sudan's potential to solve national and global food crises by bringing in the private sector and linking it with small-holder farmers. The scope of the WFP visit in Gezira indeed included introducing new agricultural companies to invest in Sudan's agricultural sector (World Food Programme (WFP), 2023).

Narratives about the potentiality of Sudan agricultural spaces were similar to dominant discourses constructed to legitimise the global land rush, built on the dichotomy of global scarcity and abundance of resources in Africa and on the power of technologies and large-scale investments to fix food insecurity (Scoones et al., 2019). Global food crises related to the war in Ukraine justified looking for agricultural space in Sudan, with its abundant water and land resources, without taking into consideration the intricate cultural, political and social entanglements characteristic of Sudanese land (Egemi, 2006). These descriptions perpetuate the idea that African countries have vast amounts of land that are unused and can be occupied, contributing to undermining the rights of rural communities. In different formal contexts (e.g., companies' headquarters, governmental offices), the portrayal narratives of Sudan as the land of '*unlimited resources*' (Fig. 10.2) were propelled without critically discussing the creation of a web of dependencies on new investors or traditional (previously colonial) donors and the risks of agricultural commercialisation (Martiniello & Azambuja, 2019). Farmers' agricultural experiences, like those of female farmers, and their vision for the future of the sector tended to be either silenced in most official contexts or excluded in the design of irrigation reforms (Goelnitz & Al-Saidi, 2020).

For example, the FAO reports that women comprise 49% of farmers in Sudan's irrigated sector and 57% in the traditional rainfed sector (UN Women, 2024). Despite the Sudanese women's significant contributions to agricultural work and unpaid care work, their perspectives and aspirations were not mentioned in the interviews we conducted. Notably, there was a glaring absence of discussion on women's contributions to agriculture, and attempts to address this matter were met

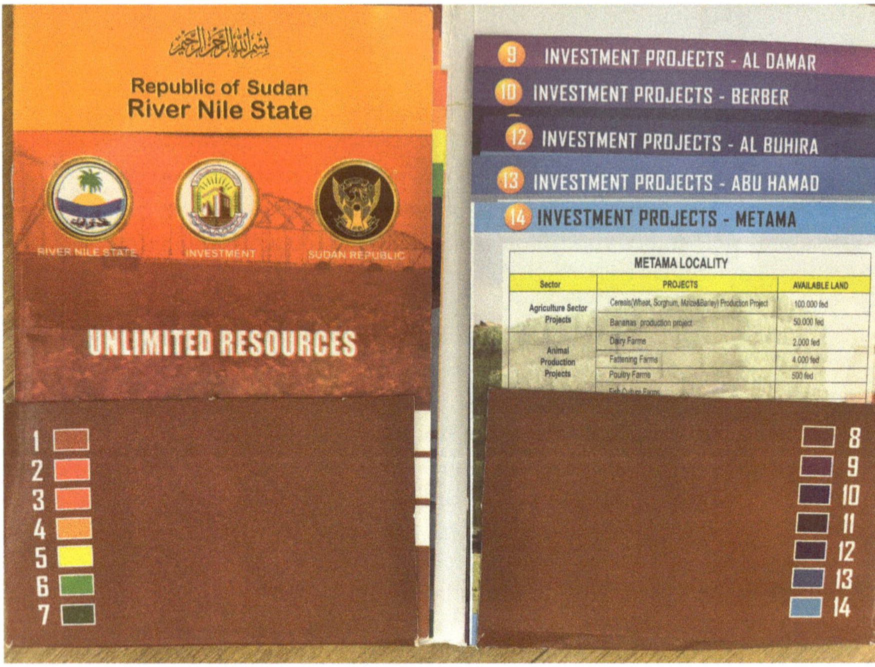

Figure 10.2 Brochure prepared for investors by the Ministry of Investment in River Nile State in 2022. Each card delineates the opportunities in various sectors, projects and land availability for different areas in River Nile State.

Source: Mariasole Pepa, 2023.

with disinterest. Nonetheless, it is also important to emphasise that 90% of the places we visited for interviews were male-dominated spaces, which also urges us to reflect on the predominance of masculine perspectives in influencing concepts surrounding development, agriculture, land and water management.

Exploring Sudanese mega-irrigation projects

The research visit started in Khartoum, where the headquarters of major companies and government offices are situated. From Khartoum, we then moved to the fields (Fig. 10.3). Visiting both locations allowed us to reflect on debates happening in the centre (Khartoum) and in the peripheries (irrigation spaces) and to rethink proximities and space fragmentation.

Perspectives from Khartoum

In Khartoum, we visited the corporate headquarters of GLB Invest, DAL Group, Kenana Sugar Company and the Chinese-owned Green Wall Company, alongside official meetings with representatives from the Ministry of Water and the Ministry

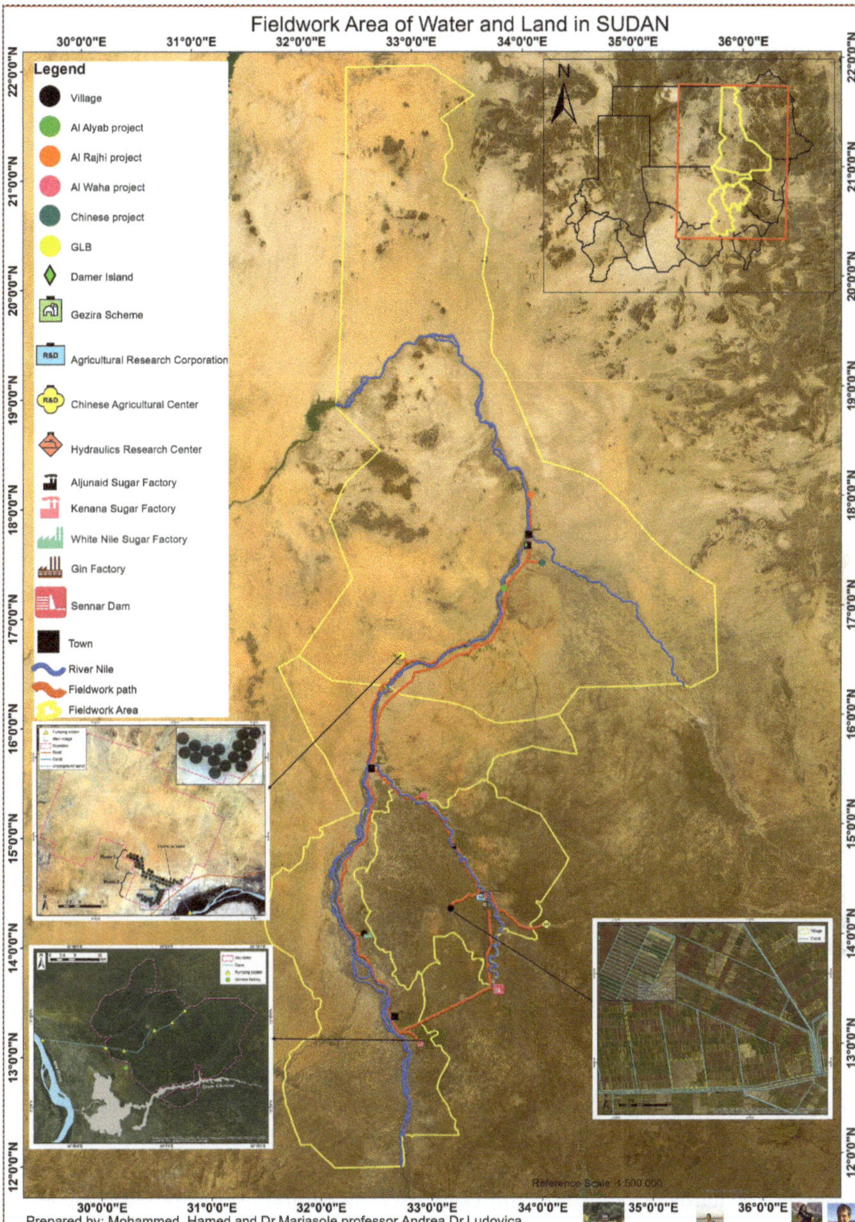

Figure 10.3 Fieldwork area.

Source: Mohammed Hammed Hassan, University of Khartoum, in the version presented at the Padua exhibition in April 2023.

of Agriculture. Additionally, several dialogues were held with representatives from academic institutions, including Khartoum University and the French Centre for Economic, Legal and Social Studies in Sudan (CEDEJ-Khartoum).

An overarching consensus emerged regarding significant challenges facing agriculture in recent years, primarily concerning energy and electricity, particularly the escalating cost of diesel.[8] This problem seemed to paralyse the agricultural activities of both national and international companies. For instance, GLB Invest, a major company in central-pivot irrigations, reported a halt in new pivot installations since 2014 due to the increasing cost of electricity.

However, GLB continued its production over the years to avoid the risk of losing control over the land if it is not under cultivation: “If we stop production, we lose the land” (interview, February 2023). The soaring diesel prices prompted widespread interest in investing in solar panels; at the same time, there was a shared concern about the lack of qualified workers: “All skilled people are outside the country” (interview, February 2023). Indeed, a vast number of Sudanese-educated people have emigrated from the country due to its political and economic instability, recurrent conflicts and the dearth of opportunities (Kurokawa, 2023). This exodus directly affected the agricultural sector, evidenced by numerous companies reporting difficulties in securing skilled labour in recent years. Various strategies have been employed to address this challenge; for instance, GLB initiated on-site training programmes for pivot management, while DAL Group planned the establishment of a training centre in Abu Hamad. Nevertheless, many mega-agricultural development projects continue to rely on expertise from abroad, particularly in hiring engineers from countries such as the Philippines, India, Egypt and Brazil, as observed in the case of Kenana Company. Interestingly, while we (Italian academics) were expected to also investigate the ways that COVID-19 impacted Sudan’s economic activities, biased by the effects in Italy, the role of the pandemic was never mentioned as a cause of impoverishment of the agricultural sector (see Chapter 16).

A noteworthy observation, which became apparent only after our fieldwork outside the capital, pertains to the information provided by the interviewees in Khartoum. They indicated a similarity in the land and the new investment landscape to those we had previously observed in our previous visit. Yet, driving around Sudan, we noticed a huge agricultural investment close to Atbara, financed by Zadna Company, a major agricultural and construction company that is 99% owned by the Special Fund for the Social Security of the Armed Forces. Zadna, as one of the military instruments used to control the Sudanese economy, reflects the strategic roles of land and water as critical political assets and in the militarisation of the economy (Thomas & el Gizouli, 2021). Following the outbreak of the war, the UK, the EU and the US imposed sanctions on several Sudanese companies with ties to military groups, including Zadna Company (Dabanga, 2024).

From Khartoum to the projects

The Sudanese Sugar Belt

WHITE NILE SUGAR COMPANY

The production of the White Nile Sugar Company (WNSC) has significantly diminished since its establishment in 2013. Indeed, in 2013, the total cultivated area amounted to 43,000 feddans, with a productivity of 20–25 t of sugarcane per feddan. Conversely, in 2023, the company planned to cultivate only 13,000 feddans for sugarcane production. The managers attributed the significant decline in production to various factors, including inadequate drainage design, challenges in procuring new machinery or locating spare parts, technical constraints resulting from US sanctions, and overall shortages in agricultural inputs, primarily financial resources (interview, February 2023). Moreover, the sugar factory was underproducing due to the lack of sugarcane. To keep the factory open, raw sugar from Brazil and South Africa was imported. Despite the managers' and the engineers' enthusiasm regarding the testing of new seed varieties, the overall capacity of the project was really limited compared to its full potential.

From the sugar factory, we moved to the villages within the perimeter. There were two types of villages: those accommodating displaced communities that had resettled outside the project's perimeter (totalling seven compounds) and one village that refused to be relocated. From the farmers' village that refused relocation, the company leased 33,500 feddans, albeit at a rate of 3500 SDG[9] per feddan annually. Then, 90% of this land was rented out by the White Nile Sugar Company to investors external to the project, usually coming from Khartoum; however, the villagers were not happy with these investors. Moreover, they were complaining about the lack of water for human consumption, a service that supposedly should have been provided by the company. One of the villagers walked us to the pump and showed us that there was no water (Fig. 10.4). In the village, grievances about the project were extremely polarised. On one hand, the WNSC managers reiterated the benefits that the villagers received for participating in the project; on the other hand, the villagers asserted the absence of tangible benefits beyond employment opportunities provided by the company.

A NEW VISION FOR KENANA SUGAR COMPANY

The situation in Kenana was completely different from the immobilisation of the White Nile Sugar Company, particularly in relation to the economic possibilities of Kenana's investors. Negatively affected by the decrease in sugar production due to political instability and international competition, the Kenana Company invested in economic diversification.

Since 2016, six new projects have been launched: compost, poultry for eggs, dairy house, horticulture, ethanol from sweet potato and animal feed. Among these,

Figure 10.4 Lack of water in the village.

Source: Mariasole Pepa, 2023.

the compost project stands out as the largest in Sudan, covering an area of 75 feddans. The composting process takes approximately 45 days for maturation, with an annual capacity of 200,000 m^3. Operations run continuously throughout the year, with tasks distributed between the dry and the rainy seasons. Its operations are centred on four materials: (1) filter mats derived from bagasse, (2) field trash (inclusive of shredded scrub cane), (3) manure from cows and chickens and (4) vinasse. Vinasse is a potassium-rich by-product of ethanol production, which comes from the factory pipeline directly to the compost turner machine. Previously considered a pollutant, vinasse is now utilised as a valuable additive in compost production. The project was managed by foreign experts, and the compost was mainly used internally in Kenana and for the local market.

The new projects centred around egg production and meat chicken breeds were unexpected in terms of their technological advancement and versatility. The agricultural machinery came from Valli, an Italian company, as well as the Agri World International, the society in charge of the projects. The projects, developed by an Italian engineer whom we met during our visit to Kenana, included three poultry houses for egg production and eight poultry houses for bred-chicken varieties. The egg production facilities could accommodate approximately 60,000 hens, yielding an output of 1800 trays per day, each containing 30 eggs, culminating in a daily production total of 54,000 eggs. The new complex for chicken breeds included eight buildings for chicken meat production, with 25,000 chickens in each building undergoing 40- to 45-day cycles. The markets for the products were mainly Rabak and Kosti, as well as Khartoum, Wad Medani and El Obeid. Alongside the establishment of these large new complexes, other initiatives focusing on horticulture were underway. Some areas within the perimeter were allocated to the cultivation of bananas, totalling 1000 feddans, with plans to expand production for the export market. The managers emphasised the importance of compost in banana production for enhancing productivity. An additional 400 feddans were dedicated to mango production, comprising a total of 27,000 mango trees, intended for export to Saudi Arabia and Egypt, and with the interest of exploring opportunities to enter new markets.

The economic diversification of Kenana Company gave rise to new narratives addressing 'green' issues and the circular economy model. Examples of these practices included the utilisation of compost as organic fertiliser and the implementation of bamboo filters for water treatment. These initiatives were framed in the context of contributing to the achievement of the UN Sustainable Development Goals. Among the companies we visited, Kenana stood out for its attention to green proposals, echoing broader Western narratives promoting the advantages of the circular economy. Nonetheless, as usual, a discrepancy emerged between the social responsibility discussions and the reality on the ground. In contrast to the 'greening' discourses framed by Kenana Company, perspectives from local communities around the project, such as those in Camp 5,[10] illustrated a gap in the narratives. This incongruence raises questions about the alignment of the company's green initiatives with the perceptions and needs of the communities directly affected by them.

Camp 5, situated in close proximity to the boundaries of Kenana, accommodates approximately 10,000 inhabitants, half of whom are employed by Kenana Company. The community voiced apprehensions regarding the efficacy of the water filtration provided by Kenana, citing instances of water scarcity persisting for more than three days. A primary source of discontent was the pollution originating from the sugar factory, which lacked a filtration system—"bagasse produce dust, waste hair, smell bad, there is too much pollution" (interview, February 2023). Residents asserted that this pollution has led to adverse health effects, particularly respiratory and ocular complications. Despite their concerted efforts to communicate these concerns and engage in protests to address the harmful impacts of pollution, these endeavours reportedly faced indifference from the company. Furthermore, some villagers raised objections concerning the low wages offered by Kenana for daily tasks, involving 9–10 hours of work with a remuneration of only 3500 SDG (approximately 5 euros at that time). Women were paid around 30,000 SDG monthly (less than 50 euros) for cleaning fields after sugar harvesting, enduring 12-hour shifts. Attempts to protest against the meagre wages resulted in the firing of the participants (Dabanga, 2022; interview, February, 2023). Against the backdrop of these labour-related grievances, several strikes were organised in Kenana Sugar Company in 2020 (The Middle East and North Africa Solidarity Network, 2020). The demands included calls for better wages, enhanced working conditions, fundamental trade union rights, and ameliorations in the education and health provisions for the families. Nevertheless, managerial representatives from Kenana contested these allegations, asserting the lack of occupational challenges.

SUDANESE SUGAR COMPANY: GUNEID

In Guneid, one of the factories of the SSC, we encountered similar challenges that deeply affected other companies in the sugar industry. Starting in 2013, there has been a continuous decline in production, which has intensified significantly over the past five years. Sugarcane output has plummeted, with productivity per feddan dropping from 45 t a few years ago to 28 t during the visit. In 2023, the company cultivated 13,000 feddans compared to 19,000 feddans in previous years, reflecting a substantial decrease in agricultural activity. Major challenges are related to the political instability and the financial constraints for the pump station rehabilitation and procurement of spare parts. For instance, in February 2023, only one of the four pumping stations from the Nile was operating in the scheme. Moreover, the decline in production was attributed to delays in land preparation, resulting in soil degradation, as well as the high costs associated with agricultural practices. A clear example of the financial difficulties faced under the scheme was related to the delay in the sugarcane harvest. Normally, harvesting was done in November, but due to financial difficulties, it was delayed until February, when we visited the project site. This delay was also closely connected to a lack of input for starting the agricultural season, which contributed to a chain of problems, including lower sugar maturation, decreased productivity and diminished quality. On the fields, the farmers claimed that the shortage of water and the poor water distribution due to the pump limitation reduced their production. The sharp decrease in productivity also has detrimental effects on the socio-economic component of the project. Indeed, 2% of the company profit is supposed to be allocated to the improvement of social infrastructures in Guneid, but over the past several years, it seemed that there was not enough profit for this allocation. Concerning this point, we attempted to visit the new Guneid Farmer Association to gather information. However, its members appeared reluctant to talk with us, possibly due to the minimal progress being made amid the ongoing crises.

Gezira: The gift from God

In 2018, during his fieldwork, Stefano Turrini[11] described the Gezira Board headquarters in Barakat as follows:

> After receiving a cordial greeting from the sleepy guards at the gates, one enters the courtyard, which appears deserted, as do most of the offices positioned on the ground and second floors of the large structure… the few computers are turned off, many are covered with dust and sand, an unmistakable sign of their limited service; the shelves contain no files, books or records.
>
> *(Turrini, 2019, pp. 151–152)*

While visiting Barakat in 2021, Guillaume Vadot, at that time a postdoctoral researcher (IMAF/CNRS) and a visiting researcher at CEDEJ Khartoum, asserted:

> The premises I saw had been completely repainted, the large courtyard was being rehabilitated, and above all, the whole place was bustling with activities. *(2021, p. 15)*

In February 2023, when crossing the Barakat gate, we immediately noticed the renovated buildings and multiple cars at the entrance—the first signs that something was happening. The situation encountered in Barakat was much more similar to the one described by Vadot. This also provided the opportunity to reflect on the continued metamorphosis of the project over time. According to some members, 2016[12] represented the lowest point for the Gezira Board, with between 70 and 100 members. However, after the 2019 revolution, the administration changed, and a 'new phase' began. Indeed, most of the conversations were interspersed with glorifying comments, such as "[The] Gezira Board has reshaped the modern Sudan" or "Gezira is a country; [it] is not a project". Despite this positive outlook for the future, the members appeared reluctant to delve into discussions about the situation in 2016.

The Gezira Board members were enthusiastic to engage in dialogue regarding the strategic plan being formulated for the scheme's rehabilitation, stressing the centrality of Gezira in mitigating both national and global food crises, as well as the necessity to increase connections with private investors and international organisations. In particular, the board was committed to increasing relations with the private sector for the latter's financial support of novel technologies and new crop varieties. Other challenges were related to institutional issues (e.g., the need for a new law to regulate the relations between farmers and companies) and the need for research in agriculture management and for new agricultural machinery. Despite the challenges, after the revolution, new initiatives were launched, particularly related to digital agriculture. In 2022, the renovation of the headquarters' offices was undertaken with the establishment of a new geographic information systems (GIS) laboratory.[13] This allowed for the implementation of a new mapping experiment in Gezira based on the use of QField, a mobile application for QGIS, which enabled the exchange of data between farmers and the central office, facilitating the monitoring of agricultural production. The pilot project started in 2022 in a limited area from a collaboration between the Gezira Board and the Ministry of Irrigation. The QField mobile app acts as an early warning system and at the same time provides farmers with opportunities to send messages to the central office and ask questions, for example, about the amount of pesticide to be used or the time for irrigation (Fig. 10.5). Due to the vastness of the Gezira Scheme, this attempt to remotely control agricultural production was regarded as a possibility for the future development of the scheme. However, there was a lack of ministerial funding to continue the project.

During this period, the Gezira Board faced a shortfall in government funding, leading it to primarily rely on revenue generated from farmers' fees and the sales

Figure 10.5 QField mobile app. The managers demonstrated the utilisation of the QField mobile app, which they used to map land parcels and associate them with the names of individual farmers. Using this data, the central office sends tailored Messages to each farmer based on the remote information collected.

Source: Mariasole Pepa, 2023.

of certified seeds to other agricultural schemes. Legal advisors informed us of the introduction of the "Law for Improvement and Transaction" in 2021, which allowed leasing smaller plots of land, reducing the previous allocation from 4 to 2 feddans. According to the legal team, the law was introduced to increase the number of smaller farmers and thus the revenue of the Gezira Scheme. The rent for one feddan per agricultural season was set at the rate of US$200–250, with individuals outside the Gezira area also permitted to lease land. For instance, Egyptians were allowed to rent land under the terms of the Four Freedom Agreement signed between Sudan and Egypt in 2004.[14] Furthermore, the legal advisor asserted that the Gezira Act of 2005 had been disruptive and that after the 2019 revolution, the process for a new Gezira Act started but had not yet been discussed with the government.

In Barakat, discussions revolved around the scheme's future prospects, including investment strategies, enhanced collaborations with international organisations and private sectors and the promotion of smart agriculture. However, moving from the renovated offices to the remote sites under the Gezira Scheme revealed disparate narratives and priorities. At Aggar Village,[15] around 3 hours from Barakat headquarters, we gathered in the shadow of the mosque with farmers from diverse social groups (e.g., owners of large plots of land, owners of agricultural machinery, farmers with limited finances).

One of the farmers immediately argued:

> There are three enemies in the Gezira:
>
> 1 the Gezira Scheme itself because they [the members of the management] do not do what they should do: supervision of extensions, advice, etc.;
> 2 farmers themselves because they do not do a good job, and they do not follow the recommendations;
> 3 leaders of farmer unions because they only pursue their interests; they do not develop new technologies.

The villagers agreed that the Gezira Scheme was non-participatory, and they mostly relied on their own resources. However, their agricultural production was impacted by: financial constraints, a dearth of expertise since most young people left, limited market access, poor land preparation and the high price of agricultural machinery. While some farmers could start the season by using their personal funds, often supported by remittances from diaspora family members, the majority relied on private companies or banks for loans. Despite banks offering loans at a 12% interest rate for a four-month term, farmers favoured private company loans, which carried interest rates of up to 28%. This preference stemmed from the perception that private loans were more reliable and accessible compared to bank loans, which were hindered by political instability. Private companies, referred to as 'contract companies', such as Magrub Brother and African Company, provided inputs to farmers at the outset of the season in exchange for future crop yields. For instance, a company would supply inputs for cotton production and agree on a predetermined price for the cotton harvest, which farmers would then repay with their harvested cotton. Other companies adopted a similar approach by offering agricultural inputs directly to farmers at the beginning of the season, always at inflated prices compared to local market rates. For instance, if the price of 50 kg of urea was 17 SDG at the local market, a company would ask 48 SDG for the same product. Villagers discussed several informal loan mechanisms that private companies activated within the scheme due to the absence of governmental funding, yet the relations between farmers and such companies were unregulated, allowing the latter to look after their own interests.

Given the Gezira Board's limited involvement in agricultural extension services, farmers turned to alternative networks for information dissemination. These included pilot farms established by the Agricultural Research Centre of Wad Medani, television and radio programmes and, increasingly, WhatsApp groups for sharing agricultural insights, particularly regarding cotton and soybean cultivation. Additionally, farmers noted a rise in labour costs, attributed to increased wages in the gold mining industry, which pushed agricultural workers to move into the gold sector (previously, the gold mining salary was 500 Sudanese pounds per day, but at the time of the interviews in February 2023, it increased to 3000 Sudanese pounds). However, on the way back to Wed Medani, still under the Gezira Scheme, we came across a large informal settlement of around 400 farmers from the Tuman tribe of North Kordofan. These farmers worked during the harvesting period from January to May and already have a recurrent agreement with the same farmers in Gezira every year. They received a fixed payment of 1500 SDG for each task; the wages were the same for men, women and children. Every year, these farmers travel for two days via lorries to Gezira, with each lorry accommodating approximately 30–45 individuals. During their travels, they transport all their necessary belongings, ranging from furniture to tableware, to sustain their stay on the site under the Gezira Scheme. After the harvesting season, they return to Kordofan for the rainfed agricultural season.

The rapid expansion of pivot irrigation systems in White River Nile State

From the second decade of the 2000s onwards, there has been a rapid proliferation of areas devoted to pivot irrigation systems in Sudan, with a notable expansion in River Nile State (see Chapter 8). This surge has attracted both domestic and international investors, primarily induced by the promising prospects offered by pivot systems, particularly for the cultivation of alfalfa, a highly demanded crop in the global market. According to the Ministry of Investment in Atbara, investors from 21 countries are present in the state but mostly from Saudi Arabia and the United Arab Emirates (Bertoncin et al., 2022).

Part of our fieldwork was dedicated to exploring the territorial transformation activated by investments in pivot systems, given their substantial land and water requirements.

Additionally, we were interested in the reason behind the rapid expansion of pivots in Sudan, in their futurities, as well as in the challenges faced by the companies amid the nation's political instability. Specifically, we visited four distinct project sites—Green Wall, GLB, Al-Rajhi and Waha Farm (south of Khartoum)—each varying in scale, objectives, production and ownership nationality.

Overall, the projects were adversely affected by similar challenges: political instability, fluctuations in diesel prices and the absence of government subsidies for diesel, electricity cuts, difficulties in procuring spare parts and the blockage at Port Sudan. Consequently, over the past two years, numerous companies ceased their operations. Investors have referred to the rising electricity costs and the volatile political environment in Sudan as significant deterrents, leading to increased perceived risks associated with agricultural investments in the country. Nevertheless, the Ministry of Agriculture and the Ministry of Investment in Atbara have been actively engaged in initiatives aimed at enticing new investors and facilitating their entry. At the Ministry of Agriculture, the General Director claimed that River Nile State had over 6 million feddans available for investors, with "the best locations, a lot of opportunities, good infrastructure, availability of electricity and water" (interview, February 2023). However, when we tried to explore the extent to which these 6 million feddans were truly free, the members of the office sniggered that there were no problems because people were not living on those lands. Similarly, the Ministry of Investment released a 'map of investment' in 2022 to attract investors (Fig. 10.2), supporting the projection that the future would be driven by pivots. However, both sites tried to avoid discussing the challenges faced by the local communities that daily endured the detrimental effects of these mega-projects, including displacement and impediments to accessing water resources, and the fact that community life was partially redrawn by these developmental trajectories.

In the Berber region, the Al Rajhi International for Investment (RAII), a subsidiary of one of the large Saudi companies, has implemented a project consisting of 132 conventional irrigation pivots and 33 solar-powered pivots, accompanied by 6 pump stations along the Nile. The majority of the pivots are under wheat cultivation; according to the manager, this choice depends on the company owner, considered one of the richest persons in the world and who intends the project to be 'for charity, not for profit'. Among the pivot project sites we visited in the country, RAII was the first and only company to introduce pivots operating with the use of solar energy. Over the last five years, the company introduced solar pivots as a way to contrast the volatility of diesel prices and electricity cuts. Given the considerable financial burden imposed by diesel fuel costs on agricultural ventures in the country, RAII asserts that various entities, including Zadna, have visited its installations to glean insights into solar pivot irrigation systems. Nevertheless, the comprehension of solar pivot management remains limited in the country.

Different from other projects, the private Chinese Green Wall Company that started its operations in 2017 was developed under the Almukabrab Agricultural

Scheme, located approximately 40 km from Atbara. The scheme was established in 2015 to provide land compensation to the people displaced by the construction of the Merowe Dam in Northern Sudan (Ahmed, 2019). However, resistance to the relocation of some communities within the scheme prompted the government to sign an agreement with the Chinese company for land allocation. The project covered an area of 7000 feddans and relied on 30 pivots. Interestingly, despite its location in the northern part of the country, approximately 50 workers from South Sudan were employed by the company. This highlights the role of irrigation projects as spaces of convergence for attracting workers from various geographic locations.

At the start of the project, the majority of the pivots were designed for alfalfa production, capitalising on environmental conditions conducive to high-quality yields compared to those of alfalfa cultivated in China. However, at the time of our field visit in 2023, a new ginning factory was under construction utilising second-hand machinery imported from China (Fig. 10.6). A Chinese engineer, specialising in ginning factory construction, was hired as a subcontractor, tasked with overseeing the construction site, which would include two production lines, one manual and the other with machinery. The engineer was dispatched from China for a limited duration (which was a usual practice in China) and was going to construct another factory in Wad Medani in June 2023. At that time, the aim of the Green Wall Company was to shift from alfalfa production to cotton production

Figure 10.6 Construction site of the new ginning factory, Green Wall Company.

Source: Mariasole Pepa, 2023.

because of the rising demands in China due to the high number of cotton factories. For this reason, the company was experimenting with new varieties of cotton seeds and testing the production of cotton under pivot irrigation systems. Overall, the company intended to train local farmers in cotton production to sell the resultant yield to the Chinese company as a contract farming practice. This arrangement could have allowed the maintenance of the 'new-second hand' ginning factory, which demanded a significant quantity of cotton for its operations. In the end, both alfalfa and cotton were shipped to China.

On the way back to Khartoum, we visited one of the largest centre-pivot irrigation farms for alfalfa, situated in the Wad Hamid area, owned by GLB Invest. In the previous months, the company shifted from being a Sudanese/Lebanese joint venture to being fully owned by the HKS-United Arab Company. The project extended over 872,000 ha of land, which had been secured through a 99-year renewable lease agreement with the Sudanese government. Moreover, the manager told us that GLB was the only company that acquired the Nile River water rights due to its acquisition of a 300 m × 200 m stretch of riverbank, where the pumping station was located. This made a big difference between other companies that had rights over land and GLB, which also had guaranteed water rights. However, this situation also created conflicts with local communities that needed access to riverbanks for water. The project began in 2011, and the production started in 2014 with the launch of the first development phase, consisting of 63 pivots designed for alfalfa production. In 2018, 90% of alfalfa production was exported to the Chinese market, yet the introduction of double compression[16] increased the possibilities for entering new markets. What was curious and at the same time triggered doubts was that from our last visit in 2016 to the one in 2023, the project remained stalled at the first development phase and no new pivots were installed despite the huge amount of land available, considering the final capacity of 1000 pivots. The managers claimed that this was related to the previous owners' reluctance to inject new money and due to the cost of electricity. The plan was indeed to install 150 pivots by 2032. However, there were also rumours that this represented another case of land grabbing.

Recent trends and discourses on irrigated spaces

On the March 19, 2023, at CEDEJ Khartoum, our research group held a seminar on the recent fieldwork conducted in Sudan with the inauguration of the photo exhibition "Endlessly Changing Irrigation Spaces" (Pepa, 2023). The seminar represented the opportunity to discuss the manifold transformations affecting irrigation projects as well as their challenges and futurities—without even imagining the violence of the war that would unfold in just a few weeks.

With the exception of Kenana Company, which was experiencing significant economic injection, most irrigation projects appeared to be on the brink of collapse, yet always finding a way forward. The case of Gezira is an example; projects that seemed abandoned in 2016, with just a few members operating, were rehabilitated in 2023, renewing national and international interests and offering hope for future schemes. Yet these transformations need to be contextualised within the framework of the geopolitical stop/go—embargo, coup d'état and revolution—which halted long-term planning and created temporal development zones with alternating periods of anticipation for the future of projects and disillusionment. For instance, following the overthrow of al-Bashir in 2019, negotiations with international organisations resumed; however, immediately following the October 2021 coup d'état, international donors halted their disbursements (Abdelaziz, 2021). The geopolitical stop/go has directly impacted the futurities of irrigation projects, considering the prolonged time needed by international agencies, like the World Bank, spanning the period from project evaluation to implementation. The case of Sudan exemplifies the myriad ways in which the humanitarian aid system consistently falls short in addressing multiple crises, necessitating prompt and timely responses instead of years of negotiation processes. The inadequacies in the current humanitarian aid responses to the Sudanese war serve as another illustration highlighting the imperative to not only rethink aid itself but also to reassess the timing of aid delivery, which often proves too sluggish to effectively respond to the humanitarian crises in Sudan (Aly & Fundira, 2023).

Despite the challenges encountered, significant changes and shifts have been observed compared to the situation in 2016. The interest in digital agriculture, especially in the use of remote sensing and GIS, and the growing focus on solar energy due to the rising cost of electricity have been particularly noteworthy. The emphasis on digitalisation strategies in agriculture aligns with a global shift towards smart farming, with digital technologies permeating the rural space contributing to the emergence of new digital geographies (Zhang et al., 2023). Moreover, climate-smart agriculture (CSA) is spreading within African governments and international political circles as an effective way to combat climate change and improve food security, despite its constraints (Autio et al., 2021). In the case of Sudan, discourses were oriented to an increase in productivity through new digital technologies rather than also considering the sustainability factor. However, CSA and digital agriculture raise similar concerns about the depoliticisation of agricultural interventions, which tend to oversimplify the complexities of agriculture, such as issues related to land tenure, class dynamics and gender inequalities, in favour of technological solutions (Taylor, 2018). Such technocratic views may obscure relevant issues when considering Sudanese agriculture, including the diverse meanings attached to land by communities (Elamin, 2018), corruption in land negotiations, the colonial legacy of land and water management systems (Taha, 2016) and diverse social groups of farmers, as we observed in Gezira.

Frictions arose between the different perspectives of the Gezira Board members, who were excited about the new GIS laboratory and the use of the QField app, and the priorities of farmers under the scheme. On one hand, the QField app offered the promise of facilitating Gezira's management, especially considering the challenges associated with accessing remote areas. On the other hand, with remote management there were risks of imprecision and data sovereignty (Fraser, 2018), as well as rendering invisible its inherent complexities. Indeed, while the Gezira Board viewed GIS as holding the potential to revolutionise the scheme management, most farmers participating in the scheme were struggling even to start the agricultural season because of financial constraints. This situation exacerbated the proliferation of informal credit systems, primarily towards contract companies that capitalised on land and water resources, thereby entangling farmers in a cycle of dependency on agricultural inputs. Nevertheless, it is important to note that these dynamics are not uniformly applicable to all farmers across the Gezira Scheme because of class stratification. For instance, during the interviews, we met farmers with available capital to start the season, particularly due to the remittances sent from the Sudanese diaspora in the United Arab Emirates. In this way, these farmers may have access to more capital, thus offering them the possibility to purchase their own agricultural machinery and rent it out to 'smaller' farmers. Additionally, some farmers exercise control over the land without directly cultivating it, instead employing seasonal workers, such as those from Kordofan and South Sudan. The interviews conducted in Gezira revealed an inconsistency (OR an incongruence) between the discourse and future visions for scheme development advanced by the Gezira Board and international organisations, on the one hand, and the everyday needs and priorities of farmers, on the other hand. This disconnection was also recurrent in the headquarters of the companies in Khartoum (the centre), the on-site headquarters (peripheries) and farmers (representing the ultra-peripheries).

In this context, the global food crisis has prompted another geopolitical stop/go, leading to renewed strategic national and international interests in the financialisation of Sudanese water and land resources. The prevailing 'development' perspective appears rooted in an extractive economic model that attracts foreign investments and facilitates the introduction of neoliberal mechanisms into rural areas through practices such as contract farming and the allocation of vast tracts of land. This vision prioritises the utilisation of Nile water resources for pivot systems aimed at exporting commodities outside Sudan, thereby fostering farmers' dependency on national contract companies or foreign institutions. This 'development' paradigm regards land and water primarily as commodities for exploitation, disregarding alternative ways of understanding, relating to and utilising water and land resources (see Chapter 17).

Water and land under siege: Mega-irrigation projects after the 15th of April[17]

One year after the onset of the war, an attempt was made to delineate a threefold periodisation of how the conflict has harmful effects on water and land resources. It is important to note that the situation is continuously changing while this book is being written, and the aim is not to rigidly define temporality but to examine the conflict through the lenses of water and land. The first phase commenced with the fall of Khartoum, the primary epicentre of the war. This event turned into the looting of companies' headquarters in the capital, such as the plundering of Kenana Company's facilities, including the theft of its warehouses' inventories and sugar stocks. From April 15 to mid-December 2023, the conflict predominantly affected Khartoum and Western Sudan. During this period, companies and the government relocated their operations to Wad Medani while still maintaining a minimal level of governance over the land. The second phase, which marked a crucial turning point, occurred in mid-December with the fall of Wad Medani, the heart of Gezira, the Sugar Belt and the centre of Sudan's main seed bank (Elsafy, 2024). This forced institutions to move to Port Sudan while the residents sought refuge in villages. Officials and technicians departed from Gezira, leaving the heart of the country's agriculture without management. The ongoing third phase involves the retreat to Port Sudan. However, it is impossible to control remote irrigation projects from a distance, resulting in the government's loss of effective control over the country's irrigation production. The control of the RSF in Gezira risks exacerbating famine in Sudan due to the impossibility of initiating the 2024–2025 agricultural season. Nonetheless, the Sudan Platform for Agriculture and Food Security are making an effort to open a dialogue with the RSF and the SAF, aiming to halt the conflict in Gezira and prevent incursions into irrigation zones, given the severe repercussions of starvation.

In this scenario, we are imagining how to sustain our collaboration and research project. First of all, we strongly believe that it is crucial to continue conducting research and not give up because now more than ever, it is even more important to keep our eyes on Sudan. We have accomplished this by staying in contact with our Sudanese colleagues and collecting stories (Pase & Pepa, 2023) with WhatsApp enhancements, exchanging photos and videos, phone calls[18] and remote interviews. We are putting into practice what we learned during the COVID-19 pandemic by undertaking research 'from below and from above' as a means of adapting to the evolving circumstances (Chapter 16). Moreover, after the displacement of hundreds of thousands of Sudanese in Egypt and the relocation to Cairo by some of the authors of this chapter, we have decided to continue our research on Sudan from Egypt with the support of CEDEJ Khartoum in Cairo, which provides a space for dialogue. Undoubtedly, this situation continues to pose everyday questions, including the following: How can research on water and land be conducted in the midst of this war? What are the applicable ethics (if any) when carrying out research during an ongoing conflict? How do we deal with the emotional burden of the people we interview? More simply, what can we do as academics? We have agreed to continue doing what we know: undertaking research, asking questions, creating bridges and weaving alliances while elaborating on creative strategies.

Several noteworthy insights emerged from the remote interviews conducted by Prof. Ganawa and Dr Musa with members of irrigation projects teams from August to December 2023, as well as from interviews conducted by Pepa, Pase, Ganawa and Osama in Cairo between January and April 2024. All interviewees and interviewers emphasised the necessity of having the war end to pave the way for future imaginaries of water and land usage. The war has profoundly affected all irrigation projects, with those that continue operations facing multiple challenges. Residual and punctual forms of agriculture remain amidst widespread difficulties stemming from input scarcity, a notable 30–50% increase in input costs, obstacles in accessing seeds and spare parts and electricity shortages. Furthermore, security concerns have impeded diesel transportation from Port Sudan to farms, exacerbating import–export difficulties. Companies also asserted that the departure of foreign workers from Sudan adversely impacted project management.

In River Nile State, 70% of pivot irrigation projects have ceased operations, primarily due to political and security concerns or the lack of input. In all four factories of the SSC, the financial situation is dire, the wartime instability and insecurity have disrupted markets, making it difficult for farmers to sell their products. Rising fuel prices have also increased the operational costs of agricultural activities, further compounding the challenges faced by farmers. In Gezira, farmers have adopted different strategies in response to the challenges posed by the war and escalating fuel costs. These strategies include implementing reduced tillage practices, utilising traditional techniques and transitioning to alternative crops that require less fuel—such as legumes or vegetables—to meet the needs of the mass flow of people to Gezira state and its surrounding areas, at least before the fall of Medani.

In February 2024, a seminar was organised at CEDEJ Khartoum in Cairo to explore the repercussions of the war on irrigation projects and foster dialogue among Sudanese professors, researchers, experts and representatives of agricultural companies. What emerged from the discussion was the centrality of restarting the agricultural sector in the postwar, as well as the importance of rethinking the decentralisation of infrastructure and activities beyond Khartoum, in other words, reimagining the relations between the centre and the peripheries. The seminar participants articulated various reflections on the significance of rethinking development approaches in Sudan, reconsidering a bottom-up approach, the role of Indigenous knowledge and the priority of land rights. At the same time, other views supported the need to attract foreign companies to invest in agriculture since most of the national companies lost their capital. Finally, during the discussion, a participant raised this crucial question: "What will happen to the agricultural sector if the war stops today?" This left us with old (new) questions—*Which kind of development? For whom? Who will benefit?*—and with the necessity of critically addressing postwar agricultural imaginaries.

Acknowledgement

We thank the University of Khartoum, Faculty of Geographical and Environmental Sciences for their kind support and help.

We are grateful to Marie Bassi, the Coordinator of CEDEJ Khartoum, for her support.

Notes

1 Different testimonies and stories of displacement within and beyond Sudan can be found at Peace of Sudan (https://www.peaceofsudan.space/blog?lang=en).
2 https://khartoumerr.org/.
3 Revolutionary Graffiti in Khartoum. To see a digital collection of mural art from the 2019 Sudanese Revolution, created by Josepha Wessels and Ahmad al Naji, visit https://storymaps.arcgis.com/stories/7f8b62d66c4243a9b93cd5656cd525f5.
4 Before the 2023 fieldwork, the latest research visits occurred in 2010, 2014, 2016 and 2018.
5 1 *feddan* is the unit of measure used in Sudan = 0.42 ha.
6 The hydro-flume technology entails a system of PVC flexible tubes that transport water to the fields. The pipes can be easily moved, and during periods of non-use, the tubes remain empty on the ground.
7 In AtlaSahel, it is possible to access a temporal and spatial analysis of pivot systems in Sudan in 2005, 2016, 2018 and 2021 (https://atlasahel.it/atlasahel/?locale=en/#/viewer/openlayers/220).
8 "In 2019, for 2 tanks of 48 gallons, each was 3000 SDG; today, with the same price, you pay for 1 gallon" (interview, February 2023).
9 When we left the project, we visited the local market where 1 kg of meat cost 3000 SDG, a scandalous figure compared to the meagre compensation the farmers received for their land.
10 The village is known by two names. Originally designated as Camp 5 under the Kenana Project, it is now situated beyond the project's boundaries. In 1983, its name was changed to Alijraa Village. However, both names are still in use today.
11 Stefano Turrini is an Italian geographer who conducted research in Sudan as part of his PhD dissertation at Padua University. Turrini was previously part of the research group.
12 One of the years when the research group visited the project site
13 During our visit in February 2023, more than once, we tried to ask questions about the source of funding for these renovations and new projects, but the responses were vague. Apparently, after the 2019 revolution, US$3 billion had been allotted to the Gezira Scheme by international organisations for rehabilitation but their funding stopped with the October coup d'etat.
14 The Four Freedom Agreement is based on the freedom of movement, residence, work and property ownership between Sudan and Egypt. However, starting on the June 10, 2023, Egypt limited the free movement of Sudanese entering the country. A visa in Port Sudan or Wadi Madi is now required before crossing the Egyptian border. With the ongoing war in Sudan, entering Egypt has become particularly challenging, and new smuggler routes have been incentivised as well as deportations of Sudanese refugees (Creta & Khalil, 2024).
15 The village was previously visited in 2016.
16 Alfalfa double compression is a process that compresses alfalfa and can save quality and space. Many countries only accept double-compressed alfalfa because it makes alfalfa much more advantageous to export. For example, previously, a single container could accommodate 14 t, but with the implementation of double compression, it can now hold up to 25 t.
17 This section was written in March 2024 in Cairo.
18 Following February 4, 2024, Sudan experienced a telecommunication shutdown, which made it even more difficult to understand the situation in the country (Fikra, 2024).

References

Abd Elkreem, T. M. A. (2018). *Power relations of development: The case of dam construction in the Nubian homeland, Sudan*. Lit Verlag.

Abdelaziz, K. (2021, October 27). *World Bank halts Sudan operations in blow to coup leaders, strike calls gain support*. Reuters. https://www.reuters.com/world/africa/sudan-state-oil-workers-join-civil-disobedience-movement-association-2021-10-27/
Abdelhadi, A. W., Adam, H. S., Hassan, M. A., & Hata, T. (2004). Participatory management: Would it be a turning point in the history of the Gezira Scheme? *Irrigation and Drainage*, *53*, 429–436. https://doi.org/10.1002/ird.139
Ahmed, I. A. I. (2019). The Merowe Dam in Northern Sudan: A case of population displacement and impoverishment. In J. D. Schmidt, L. Kimathi, & M. O. Owiso (Eds.), *Refugees and forced migration in the Horn and Eastern Africa. Advances in African economic, social and political development*. Springer. https://doi.org/10.1007/978-3-030-03721-5_8
Aly, H., & Fundira, M. (2023, October 19). Rethinking humanitarianism. How mutual aid in Sudan is getting international support [Audio podcast]. *The New Humanitarian*. https://www.thenewhumanitarian.org/podcasts/2024/04/11/rethinking-humanitarianism-mutual-aid-sudan-groups-receive-international-support
Autio, A., Johansson, T., Motaroki, L., Minoia, P., & Pellikka, P. (2021). Constraints for adopting climate-smart agricultural practices among smallholder farmers in Southeast Kenya. *Agricultural Systems*, *194*. https://doi.org/10.1016/j.agsy.2021.103284
Berridge, W., Lynch, J., Makawi, R., & de Wall, A. (2022). *Sudan's unfinished democracy. The promise and betrayal of a people's revolution*. Hurst.
Bertoncin, M., Bicciato, F., Bonollo, L., Croce, D., Faggi, P., Mariani, L., Minoia, P., & Pase, A. (1995). Irrigazione, stato e territorio in Sudan: Il gioco della posta in gioco. Terra d'Africa, 1995, 8–40.
Bertoncin, M., Pase, A., & Quatrida, D. (2014). *Geografie di prossimità. Prove sul terreno*. Franco Angeli.
Bertoncin, M., Pase, A., Quatrida, D., & Turrini, S. (2017). Territory dello zucchero: Le dinamiche della prossimità nella sugar belt Sudanese. *Rivista Geografica Italiana*, *125*, 93–114.
Bertoncin, M., Pase, A., Quatrida, D., & Turrini, S. (2019). At the junction between state, nature and capital: Irrigation mega-projects in Sudan. *Geoforum*, *106*, 24–37. https://doi.org/10.1016/j.geoforum.2019.07.002
Bertoncin, M., Pase, A., & Turrini, S. (2022). Water, land and Arab investments in Sudan. In J. N. Bach (Ed.), *Routledge handbook of the Horn of Africa*. Routledge.
Carstensen, N., & Sebit, L. (2023, October 11). *Mutual aid in Sudan: The future of aid?* https://odihpn.org/publication/mutual-aid-in-sudan-the-future-of-aid/
Creta, S., & Khalil, N. (2024, April 25). How Egyptian authorities are deporting thousands of Sudanese refugees. *The New Humanitarian*. https://www.thenewhumanitarian.org/investigations/2024/04/25/exclusive-inside-egypt-secret-scheme-detain-deport-thousands-sudan-refugees
Dabanga. (2022, November 14). Farmers of Sudan's El Gezira Agricultural Scheme discuss unionising. *Dabanga*. https://www.dabangasudan.org/en/all-news/article/farmers-of-sudan-s-el-gezira-agricultural-scheme-discuss-unionising
Dabanga. (2024, February 2). New round of US sanctions on companies 'perpetuating violence' in Sudan. *Dabanga*. https://www.dabangasudan.org/en/all-news/article/new-round-of-us-sanctions-on-entities-perpetuating-violence-in-sudan
Egemi, O. (2006). *Sudan: Land and peace processes in Sudan*. Accord Conciliation Resources. https://rc-services-assets.s3.eu-west-1.amazonaws.com/s3fs-public/Accord18_18LandandpeaceprocessesinSudan_2006_ENG.pdf
El Nazir, O. A., & Desai, G. D. (2001). *Kenana, kingdom of green gold. Grand multinational venture in the desert of Sudan*. Kegan Paul.
Elamin, N. (2018). 'The miskeet tree doesn't belong here': Shifting land values and the politics of belonging in Um Doum, central Sudan. *Critical African Studies*, *10*(1), 67–88. https://doi.org/10.1080/21681392.2018.1491803
Elnur, I. (2009). *Contested Sudan. The political economy of war and reconstruction*. Routledge.

Elsafy, M. (2024, January 29). Under siege: The race to save Sudan's crucial gene bank amid devastating civil war. *Ecomedia Africa*. https://ecomedia.africa/under-siege-the-race-to-save-sudans-crucial-gene-bank-amid-devastating-civil-war/

Ertsen, M. W. (2016). *Improvising planned development on the Gezira Plain, Sudan, 1900–1980*. Palgrave Macmillan. https://doi.org/10.1007/978-1-137-56818-2_1

Fikra. (2024, February 11). Complete shutdown of telecommunications services in Sudan. *All Africa*. https://allafrica.com/stories/202402110252.html

Fraser, A. (2018). Land grab/data grab: Precision agriculture and its new horizons. *The Journal of Peasant Studies*, *46*(5), 893–912. https://doi.org/10.1080/03066150.2017.1415887

Goelnitz, A., & Al-Saidi, M. (2020). Too big to handle, too important to abandon: Reforming Sudan's Gezira Scheme. *Agricultural Water Management*, *241*, 106396. https://doi.org/10.1016/j.agwat.2020.106396

Greco, E. (2023). Keeping eyes on Sudan – Keeping eyes on austerity. *Review of African Political Economy*, *50*(75), 1–8. https://doi.org.10.1080/03056244.2023.2240675

Khiar, Y. (2022). Les fresques murales à Khartoum pendant la révolution soudanaise de décembre 2018: Généralisation et politisation d'un phénomène artistique. *Manazir Journal*, *4*, 17–33. https://doi.org/10.36950/manazir.2022.4.2

Kurokawa, C. (2023). Sudan's brain gain: What motivates the intention of skilled Sudanese in Japan to return and contribute to their home country. *Journal of International Cooperation in Education*, 25(2), 221–235. https://doi.org/10.1108/JICE-07-2022-0021

McMichael, P. (2014). Historicizing food sovereignty. *The Journal of Peasant Studies*, *41*(6), 933–957. https://doi.org/10.1080/03066150.2013.876999

Martiniello, G., & Azambuja, R. (2019). Contracting sugarcane farming in global agricultural value chains in Eastern Africa: Debates, dynamics, and struggles. *Agrarian South: Journal of Political Economy*, *8*(1–2), 208–231. https://doi.org/10.1177/2277976019851955

Obeng-Odoom, F. (2021). Land grabs, imperialism, and anti-imperialism in Africa. In I. Ness & Z. Cope (Eds.), *The Palgrave encyclopedia of imperialism and anti-imperialism*. Palgrave Macmillan. https://doi.org/10.1007/978-3-030-29901-9_236

Pase, A., & Pepa, M. (2023). L'inferno all'improvviso. Testimonianze dal Sudan. Una città, 22–29. https://www.unacitta.it/it/sommario/313-una-citta-nr-293-giugno-luglio-2023

Pepa, M. (2023). Endlessly changing irrigation spaces: A multi-sited exhibition to reimagine proximities between Khartoum and Padua. *J-Reading*, 103–116. https://doi.org/10.4458/6083-08

Scoones, I., Smalley, R., Hall, R., & Tsikata, D. (2019). Narratives of scarcity: Framing the global land rush. *Geoforum*, *101*, 231–241. https://doi.org/10.1016/j.geoforum.2018.06.006

Shabaka. (2023). *Delivering localisation: Lessons from diaspora humanitarians*. Shabaka. https://shabaka.org/wp-content/uploads/2023/09/Delivering-Localisation-Lesson-from-Diaspora-Humanitarians_Aug-2023-.pdf

Sulieman, H. M. (2015). Grabbing of communal rangelands in Sudan: The case of large-scale mechanized rain-fed agriculture. *Land Use Policy*, *47*, 439–447. https://doi.org/10.1016/j.landusepol.2015.04.026

Taha, M. (2016). *Land use, ownership and allocation in Sudan: The challenge of corruption and lack of transparency*. Sudan Democracy First Group's Sudan Transparency Initiative. http://www.dehai.org/archives/dehai_news_archive/2016/nov/att-0830/Land-Use-Ownership-and-Allocation-in-Sudan.pdf

Taylor, M. (2018). Climate-smart agriculture: What is it good for? *The Journal of Peasant Studies*, *45*(1), 89–107. https://doi.org/10.1080/03066150.2017.1312355

Thomas, E., & el Gizouli, M. (2021, November 11). Creatures of the deposed: Connecting Sudan's rural and urban struggles. *African Arguments*. https://africanarguments.org/2021/11/creatures-of-the-deposed-connecting-sudans-rural-and-urban-struggles/

The Middle East and North Africa Solidarity Network. (2020, September 21). *Urgent call out for solidarity with Sudan sugar workers*. https://menasolidaritynetwork.com/2020/09/21/urgent-call-out-for-solidarity-with-sudan-sugar-workers/

Turrini, S. (2019). *Cotone, zucchero, Alfalfa. Il grande progetto tra sviluppo e "sradicamento" territoriale.* [Doctoral dissertation, University of Padua.] https://www.research.unipd.it/retrieve/e14fb26f-b2e3-3de1-e053-1705fe0ac030/turrini_stefano_tesi.pdf
UNHCR. (2024, April 28). *Situation. Sudan situation.* UNHCR Data Portal. https://data.unhcr.org/en/situations/sudansituation
UN Women. (2024, March 7). Enhancing the capacities of women in agriculture in Sudan. *UN Women Africa.* https://africa.unwomen.org/en/stories/news/2024/03/enhancing-the-capacities-of-women-in-agriculture-in-sudan
Vadot, G. (2021). *Revolution, back to the future.* CEDEJ Khartoum. https://cedejsudan.hypotheses.org/3704
Verhoeven, H. (2013). Sudan and its agricultural revival: A regional breadbasket at last or another mirage in the desert? In J. A. Allan, M. Keulertz, S. Sojamo, & J. Warner (Eds.), *Handbook of land and water grabs in Africa. Foreign direct investment and food and water security.* Routledge.
World Food Programme (WFP). (2023, April 4). *WFP Sudan Country Brief, February 2023 – Sudan.* Relief Web. https://reliefweb.int/report/sudan/wfp-sudan-country-brief-february-2023
Zhang, Q., Webster, N. A., Han, S., & Ayele, W. Y. (2023). Contextualizing the rural in digital studies: A computational literature review of rural-digital relations. *Technology in Society 75*, Article 102373. doi: 10.1016/j.techsoc.2023.102373

11 Traces of irrigation

Some recent developments in the Chadian region

Andrea Pase, Aboukar Mahamat, Dobah Marsala, Marina Bertoncin and Mariasole Pepa

Introduction

In this chapter, we aim to describe some recent developments in irrigated agriculture in the Chadian region. We do so by following its traces, that is, signals, sometimes barely visible, to be deciphered in complex contexts where irrigation is only one of the ways of exploiting water and land resources. The term "traces" takes on three different meanings. The first indicates what remains: the afterlives of interventions or the little that is left of these schemes. These large-scale irrigation schemes suffer from numerous difficulties, which drastically diminish their potential and sometimes place them on the edge of collapse. The second meaning refers to the visible signs on the ground of transformations linked to irrigated agriculture: new perimeters, rehabilitations, and initiatives by private individuals or local communities. Finally, the third meaning deals with the possible future paths that are under discussion in feasibility reports and in conversations with heads and technicians of government agencies and ministries regarding new interventions (or the hopes for new developments) in irrigated agriculture or otherwise related to the management of water and land resources. This chapter's cognitive framework is derived from the research on irrigated agricultural modernisation projects that Marina Bertoncin and Andrea Pase (2012), geographers at the University of Padua, conducted in the Chadian region (see Fig. 11.1). However, over the past decade, the desire of the research team to continue this research has clashed with the extreme difficulty, if not impossibility, of reaching the study areas, particularly the most remote ones, first due to security issues and later the movement restrictions caused by the COVID-19 pandemic (see Chapter 16). For this reason, two researchers who resided in the region joined the research team. Aboukar Mahamat, an environmentalist and expert in integrated water resource management, has extensive experience in action research aimed at development and conflict-resolution initiatives in the Waza Logone region in Cameroon's Far North, led by the NGO he directs—Alliance Citoyenne pour le Développement et l'Education à l'Environnement (ACEEN). Dobah Marsala is a Chadian political scientist, a PhD candidate in International Environmental Law at the University of Youandé II, interested in decolonial studies and critical development analysis, and the founder of the Chadian NGO Espoir. Between September 2020 and August 2022, the two researchers wrote eight reports based on their fieldwork in the

DOI: 10.4324/9781003429111-16

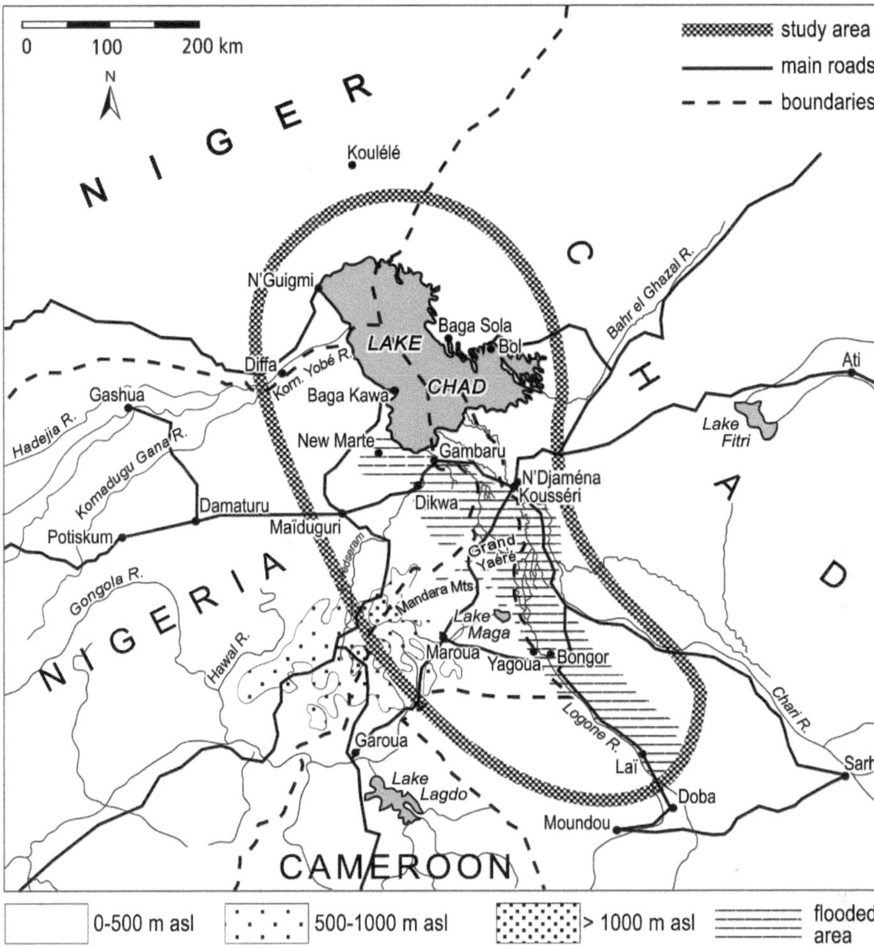

Figure 11.1 Lake Chad region and study area.

Far North of Cameroon and in Chad, along the Logone River, and in Bol, on the northern shore of Lake Chad, respectively. Mariasole Pepa remotely coordinated the research through Zoom meetings and WhatsApp conversations, which facilitated the exchange of ideas to think together the steps of the research process.

We continue this chapter with a brief section on the historical-geographical framework of irrigated agriculture in the Chadian region. This is followed by two sections where we report the information gathered by Mahamat and Marsala on the ground during the period 2020–2022, with some updates up to August 2024. In the final section, we hypothesise possible interpretations of current processes regarding irrigated agriculture in the study areas and propose some lines of interest for future research on ongoing socio-territorial transformations.

Times and spaces of the spread of modern irrigated agriculture in the Chadian region: An overview

The Chadian region straddles four states—Nigeria, Niger, Cameroon and Chad—whose borders are legacies of colonial partition (Bertoncin & Pase, 2012, pp. 90–91). From an administrative standpoint, the study areas of the research group focus on some regions of Chad, particularly Lac, Hadjer-Lamis, Chari-Baguirmi, Mayo-Kebbi East and Tandjilé. The administrative subdivision of the other countries is much more centralised: the territories of interest are essentially in the Borno State (Nigeria), the Diffa Region (Niger) and the Far North Province (Cameroon). The major irrigation schemes are concentrated along the main rivers and around Lake Chad. Logone, the main tributary of Chari River, which it joins at N'Djamena, is the first river placed under the irrigated agriculture project, since the 1950s (Brunet-Jailly, 1981; Levrat, 1975; Roupsard, 1987), and the most intensively affected by the spread of modern irrigation schemes. In Cameroon, the Société d'Expansion et de Modernisation de la Riziculture de Yagoua (SEMRY) has played an essential role in this regard. Three areas have been affected by the Société's initiatives over time: SEMRY I at Yagoua, with four pumping plants from the Logone River (5500-ha potential), built in the mid-1970s, SEMRY II at Maga (6000 ha), which involved the construction of a retention basin (Maga Lake) at the end of the 1970s, fed by the river flood and the *mayo*, seasonal watercourses descending from the Mandara mountains, and, finally, SEMRY III in the northernmost Department of the Cameroon region, the Logone and Chari Rivers, with ten pump-fed medium perimeters (1927 ha in total) at the turn of the 1970s and the 1980s (Beauvilain, 1989; Roupsard, 1987, 2000). SEMRY I and II remain active today (although with difficulties, as discussed later in this chapter). On the Chadian bank of the Logone River, the first irrigation pilot test, in the 1950s, was casier A at Bongor, devoted to rice cultivation and based on an embankment that allowed flood control; however, by the 1970s, the infrastructure began to deteriorate, leading to the project's failure (Bertoncin & Pase, 2012, p. 320). The more significant projects are casier B at Bongor (800 ha, partly with flood control and partly by pumping, active since the second half of the 1960s; the part in controlled submergence was extended in the early 2000s by Taiwanese cooperation) and the perimeter at Satégui Deréssia (1500 ha in flood control, established in the second half of the 1970s) (Bertoncin & Pase, 2012, pp. 190–200). Casier C (about 250 ha) is located in Nya, near Doba. Along the Chari River to the north of N'Djamena, seven small perimeters (30–40 each) for pumping from the river were built between the late 1980s and the early 1990s, at the stimulus of European cooperation (Bertoncin & Pase, 2012, pp. 291–292). Finally, it is worth mentioning that along the Chari River but south of Sarh, close to the border with the Central African Republic and therefore outside the limits defined for the Sahel (see Chapter 5), the irrigation scheme for sugarcane cultivation (3500 ha) that feeds the Banda sugar refinery dates back to the 1960s and is now managed by the Compagnie Sucrière du Tchad.

Another river along which irrigation schemes have been established is the Komadougou Yobé, which flows into the northern basin of Lake Chad (Bertoncin

& Pase, 2012, pp. 201–205). In the terminal section of the course, numerous initiatives have been implemented over time on both the Niger and the Nigerian shores. In Niger, many small perimeters (ranging from 8 to 160 ha) take water from the river or the water table; they were designed and built in the second half of the 1970s. Earlier, on the Nigerian side, three perimeters with flood control or pumping from the river, totalling 1700 ha, have been built since the late 1950s. However, in Nigeria, the most important projects are upstream, away from the lake area and outside the study area, along the tributaries of Komadougou, Hadejia and Jama'are Rivers: the Kano River Irrigation Project (13,300 ha), which takes water from the Tiga dam, and the Hadejia Valley Irrigation Project (2200 ha), which draws water from the Hadejia barrage. Other smaller irrigation schemes are connected to secondary dams along the Komadougou's tributaries. Upstream withdrawals take most of the water resources away from the final stretch of Komadougou River.

Finally, Lake Chad itself is a water reservoir that can be drawn on for irrigation. On the northeastern shore of the lake, in the Bol and Baga Sola area, the system of modern polders (i.e., with pumping from the lake; 2500 ha) or semi-modern polders, which receive water by gravity through bulkheads on the dams separating from the lake itself (6000 ha in 2012; Mugelé, 2015, pp. 110–111), is used, inspired by tried and tested practices in the area (Bouquet, 1990, Vol. II, pp. 19–24; see Chapter 1 of this volume). Built between the late 1970s and the early 2000s, three modern polders are concentrated around Bol: Guini (350 ha), Bérim (800 ha) and Mamdi (about 1600 ha) (Bertoncin & Pase, 2012, pp. 294–296; Bouquet, 1990, Vol. II, pp. 49–82; Mugelé, 2015). Large-scale irrigation schemes were devised and initiated on the Nigerian shore, which were supposed to take water from the lake through intake canals and pumping systems; the most important was the South Chad Irrigation Project (SCIP) in the Marte area, and the second (smaller) one was the Baga Kawa Polder (Bertoncin & Pase, 2017; Blench, 1997). While the former witnessed its greatest extension in the early 1980s, with 7000 ha in production (out of the potential 49,000), the latter operated for a short time, on a few hundred hectares and only with temporary systems. Conceived in the 1970s before the great Sahelian droughts, they underwent a crisis in 1984 when the lake level dried up the intake canals.

Compared with other areas of the Sahel (see Chapter 1), the development of irrigated agriculture in the Chadian region has shown specific characteristics: it arrived later, in the second half of the 20th century, and with a reduced scale of intervention, compared in particular with giants such as Gezira or the Office du Niger. It also has an even more pronounced history of failures and repeated crises. The late arrival of hydro-agricultural modernisation and the frequency of failures in the Chadian region can be attributed to two elements that further complicate the already difficult picture of the Sahelian context. The first is enclavement, that is, the distance from the sea ports on the Atlantic: the problem has been highlighted since colonial times and is a recurring theme in all economic development planning in the area. It was, and is, difficult and expensive to both bring in machinery and materials and export produced goods. The friction of distance slows down any kind of intervention. The second aspect concerns the peculiarities of the hydrographic basin, that is, being endoreic, mountainous only on the edges but with very low slopes throughout its central part. The floods of the Chari and Logone Rivers seasonally inundate large areas; Lake Chad itself is a thin lens of water, with its extension and banks constantly being redefined. This situation is reflected in a difficult predictability and an even more challenging control of water resources. In this respect, an example is the complete failure of the SCIP and the Baga Polder in Nigeria. These extreme situations—enclavement and hydrographic variability—paradoxically triggered the hypotheses about decidedly challenging and improbable technical solutions, such as the transfer of water from Ubangi River to the Chari River, that is, from the Congo basin to the Lake Chad basin, which in the initial versions also included the opening of a navigable channel to Benue River and thus to Nigeria (see Chapter 2).

Due to the multiple crises that affected these territories, in the following sections, we focus only on the areas where it was possible for Mahamat and Marsala to conduct fieldwork. The research activity consisted of the collection of grey literature; interviews with institutional decision-makers, as well as technical officials and farmers; field trips to verify the situation and document it with photos and videos; and the selection of information circulating in the media, both in the press and on social networks. The collected materials were sent during the drafting of the research reports and on a daily basis as news emerged, especially in the press and on the internet. The corpus is both substantial and diverse. We thus obtained first-hand, timely and localised information, containing facts, sometimes minute details, and accounts of events that, while they may appear specific to situations delimited in space and time, are capable of rendering the materiality and concreteness of daily life in irrigation projects. In the following sections, we attempt to summarise these rich materials, highlighting the information that seemed most significant for the purposes of this volume. The researchers encountered similar difficulties in the Far North of Cameroon and in Chad. Restricted access to information, coupled with burdensome application procedures, inadequate preservation of project reports and refusals to participate in interviews—especially when recording or videotaping is involved—pose significant challenges. These difficulties are compounded by frequent power outages, limited access to the internet and mobile phone networks, and, at times, unfeasible mobility conditions. Additionally, researchers affiliated with Italian universities have reported being regarded with suspicion or perceived as "spies" by the local staff. Despite these many hardships, the gathered information is consistent and of high quality.

Far North of Cameroon: News from the field

Aboukar Mahamat pursued three fundamental targets: the verification of the conditions of SEMRY I and II (Yagoua and Maga), the state of implementation of some new irrigation-related projects and, finally, the spontaneous spread of small-scale irrigation, with particular reference to the Grand Yaéré wetlands (i.e., the large flood expansion zone on the hydrographic left of the Logone River). The survey zone is more concentrated than the dispersion over vast areas of the points of interest in Chad. For this reason, the different investigated dimensions present many points of contact and intersection. Over the last decade, water management has been severely affected by interventions and projects stemming from two emergencies: the first is that of the great floods of 2010 and especially 2012, which subjected all the hydraulic infrastructures (particularly the Maga dam and the embankments along the Logone River) to intense stress that even led to fears of their collapse. The second is related to the perception of increasingly irregular rainy seasons, with, on the one hand, prolonged dry periods and delays in the arrival of rainfall and, on the other hand, extreme events with intense and destructive rainfall.

The first emergency is linked to not only the hydrology of Logone river but also anthropic interventions. For instance, the construction of an 80-km embankment on the right bank of the river between 2012 and 2015 (which should have prevented floods from spreading on that side), promoted by the Chadian government and financed by the African Development Bank (ADB), increased the pressure on the Cameroonian side. When visiting the areas affected by the 2012 floods, the President of Cameroon promised the construction of a levee topped by a 330-km road—a major work capable of securing the entire province. In anticipation of this intervention, the Projet de Lutte contre les Inondations (PULCI) was promoted, which first allowed the Maga dam and the river banks from Yagoua to Téléké to be reinforced and raised, and then the primary irrigation channels of SEMRY II to be rehabilitated.

To contrast the effects of irregular rainfall, the Cameroonian government planned to intervene with a project comprising eight dams in the Mayo Tsanaga basin, which descends from the Mandara Mountains onto the Diamaré plain (Mission de Développement Intégré des Monts Mandara [MIDIMA], 2016). The project is part of the Plan d'Urgence pour l'Accélération de la Croissance Économique (PLANUT), launched by the President of Cameroon on 9 December 2014. The *volet* of the plan is intended to meet the local residents' needs for drinking and irrigation water, as well as to feed the fishing areas downstream. The total capacity of the dams, which vary greatly in size, would be 43,012,631 m^3. ACEEN's (the local NGO led by Aboukar) analysis of the feasibility studies reveals numerous violations of environmental impact assessment procedures. The public consultations carried out with local communities also broke the rules. Indeed, the participants did not have enough information to discuss the projects. Moreover, local authorities have opposed the project, fearing its effects on local production systems. For example, the mayor of the municipality of Bogo, who is also the *lamido* (customary chief), has repeatedly fought against the Mokong dam project since it could have serious impacts on the availability of water in the area downstream of the reservoir.

Turning now to the second area explored by Aboukar—SEMRY—the overall picture is one of progressive worsening of a scheme already in a precarious situation (Bertoncin & Pase, 2015). SEMRY employees constantly suffer delays in the payment of their salaries (up to seven months) and are sometimes paid in kind, with sacks of husked rice. The scarcity of available earthmoving and agricultural machinery (e.g., only five were operational in 2020) considerably slows down mechanised ploughing in the two perimeters, so much so that the allottees often resign themselves to doing it manually. SEMRY I, which obtains its water from four electrified pumping stations, is adversely affected by the province's energy crisis, with frequent cuts in electricity coming from the Lagdo dam hydroelectric plant, itself in a precarious condition. Hundreds of hectares are lost each dry season due to lack of water. SEMRY I is always on the verge of complete cessation of activities and revolt of the allottees. In SEMRY II, PULCI-related interventions have led to an improvement in water distribution. In parallel with the raising of the dam and the rehabilitation of the main canals (which took place between 2013 and 2019), fields outside the perimeter, which were illegally drawing water from the canals, were destroyed. The maintenance of the secondary and tertiary canals has been transferred to the Water Users Association, but they lack suitable machines because the existing ones are old and almost all out of service. Rice has reached a particularly high selling price of up to 54,000 CFA francs per 100-kg bag. For the same reason, buyer-traders from Nigeria have reappeared since 2020. Lake Maga is overexploited in fishing, and the population of hippopotamuses, which are particularly aggressive, has substantially increased.

A key issue is land tenure insecurity, as is the case in many other irrigation schemes in the Sahel, which would require land reform. In theory, the land belongs to the state and is entrusted to the SEMRY, which allocates the plots to the allottees in exchange for an annual fee (51,000 CFA francs). In practice, however, the allottees consider the plots their properties, so much so that there is an informal land trade, with land being sold at prices ranging from 200,000 to 300,000 CFA francs per plot; others rent plots from season to season. In the framework of the Programme de la Valorisation des Investissements Agricoles dans la Vallée du Logone (VIVA Logone; this is phase II of the PULCI), financed by the World Bank, land reform is planned. This sensitive issue had already led to a violent revolt against the Director General of the SEMRY in Maga in 2008 (Bertoncin & Pase, 2012, pp. 185–187). According to Aboukar's latest updates, at the end of August 2024, news of the activation of a topographical survey campaign aimed at the "regularisation of the SEMRY's land titles" caused the people of Maga to rise up against the proposal. The Association pour le bien-être des populations du fleuve Logone had called for a protest march on 6 September 2024, later banned by the authorities for fear of violence. The allottees are afraid that the SEMRY will claim back the lands that they now consider their own properties. Moreover, the VIVA Logone plan's main focus is on privatisation in areas such as rice production and ploughing. In fact, there were also expressions of interest from foreign investors. One example is the 2013 visit of some 30 Turkish business people, who were, however, attracted by the lands outside the SEMRY perimeters; the insecurity situation in the region blocked any initiative.

A significant new development over the past decade is the launch of two new large irrigated extensions, both of which aim to cultivate around 10,000 ha of rice for an expected 15,000–19,000 allottees. The first is in Zina, some 50 km north of Maga, and the second will be in Karam, even farther north. These are therefore projects that would each approach the total potential extension of SEMRY I and II (which together amount to 11,500 ha). Both are framed in the PLANUT. Work began on the Zina perimeter in 2017, but due to both shortcomings in the design and lack of financial means, only 20% of the planned total land is currently prepared. The project takes water from the Logomatya, the watercourse that drains the Grand Yaéré, through diesel-powered motor pumps. The second project, farther downstream along Logone River, about 50 km from Zina, in the Arrondissement of Logone Birni, is still at the feasibility study stage, entrusted to an Indian design firm (PKS Infra Engineers Pvt. Ltd. of New Delhi; Ministry of Economy, Planning and Regional Development [MINEPAT], 2019).

Focusing on the Zina perimeter, the only one partly realised, it must first be mentioned that it is in the heart of the Waza-Logone plain, in a delicate area already crossed by considerable tensions between and within the different user groups, farmers, fishers (especially with the proliferation of fishing canals) and transhumant breeders (see Chapter 15). ACEEN conducted two studies that showed that the project was designed without any participation of the so-called beneficiaries and without carrying out adequate environmental and social impact research. In particular, the project team did not take into account the delicate conflict management mechanisms that are essentially based on customary regulation, which is also under considerable strain as the pressure on resources increases. The irrigation scheme, although still far from being completed (the hectares are expected to be 9521 in total), has already taken land away from neighbouring villages and has altered the hydrological functioning of the plain in a major way, making some 50 villages even more exposed to flooding. The local population's demands for compensation were not heeded. On the contrary, while the irrigated areas were initially intended to be allotted to local farmers, organised in a management committee, in 2020, taking local actors by surprise, the Ministry of Agriculture and Rural Development allocated two-thirds of the perimeter lands to private operators, namely 1000 ha to the Société Camerounaise Production du Riz SARL (Société à responsabilité limitée) of Yagoua, 3521 ha to Afrifood of Yagoua and 2000 ha to Agroinvest of Yaoundè. The remaining one-third of the lands were granted to the inhabitants of the 10 neighbouring villages, divided among some 3100 households. ACEEN tried to highlight to the ministerial delegation the many violations of the rights of the communities whose lands have been requisitioned. While the Zina project had initially raised hopes and even enthusiasm among the communities involved, these new developments have created deep disillusionment and an atmosphere of opposition, if not open rebellion. Construction work on the protective dam resumed in February 2022, with the aim of completing it by 2025. However, at the time of writing this chapter (August 2024), only about 1000 ha are under cultivation.

The spread of small irrigated perimeters has increased in the Logone plain since 2008, as an alternative to rainfed rice cultivation. ACEEN estimates the dry-season-irrigated areas to be between 1200 and 1500 ha. The farmers make use of motor pumps by taking water from rivers or from agricultural drills they have built themselves or with the support of NGOs and the Lake Chad Basin Commission through the Programme de développement durable du bassin du lac Tchad. There are approximately 700 agricultural wells for irrigation in the plain. Irrigating depressed areas has very good yields, between 6 and 8 t/ha of paddy (Fig. 11.2). In other areas, such as in Zimado along the Logone River, irrigation is mainly for horticulture, with produce sold in Kousséri and exported to Chad and Nigeria. On the two banks of Logone River in Cameroon and Chad, it is estimated that 5000 ha are irrigated. If this spread is the manifestation of a spillover of hydraulic knowledge from projects such as the SEMRY and of the farmers' adaptability, it has consequences on the other uses of the plain: grazing areas are reduced, thus increasing conflict, and the use of chemicals (fertilisers and pesticides) can pollute water and soil.

Chad: News from the field

Dobah Marsala's focus on Chad followed a double track. On the one hand, he visited some of the most significant "historical" irrigation projects to verify their situation, particularly casier B in Bongor and the modern polders in Bol on several occasions. On the other hand, the second direction taken aimed at recent or

Figure 11.2 Small family-run irrigation scheme in the village of Manjour (canton of Mazera, arrondissement of Zina), rice cultivation.

Source: Aboukar Mahamat, 2016.

minor realities or possible future projects. These research opportunities have thus illuminated situations such as those of the irrigation perimeter of Boumou, a small Taiwanese perimeter on the Chari River and possible initiatives for the hydro-agricultural enhancement of Batha Province and, in particular, the Lake Fitri area. Key references for the researcher were the managers (in the different study areas) of the Agence Nationale d'Appui au Développement Rural (ANADER), created in 2016 by bringing together three structures that previously dealt with agriculture in the country, namely the Office National de Développement Rural, the Société pour le Développement du Lac (SODELAC) and the Programme National de Sécurité Alimentaire (PNSA). In 2020, SODELAC became an autonomous body again.

Casier B, Bongor

After rehabilitation and new extensions implemented between 2002 and 2004 by the Taiwanese cooperation, Casier B now has 500 ha irrigated by pumping and 2000 ha in flood control (Abakar et al., 2020; Bertoncin & Pase, 2012, pp. 198–200). There are currently more than 1000 allottees, who are required to pay a fee of 45,000 CFA francs per year, which includes the cost of ploughing as well as water. The project management consists of ten people, including technicians and supervisors. The agreement signed in 2008 with Brasserie du Tchad for the purchase of the produce means that there are no problems with unsold produce. The interviewees have expressed ambivalent opinions. On the one hand, Marsala has collected positive testimonies from allottees, for example, "The income from rice has allowed me to send my two children to school, one in N'Gaoundéré and the other in Yaoundé (Cameroon). It has also allowed other families to build permanent homes, to buy medicines and means of getting around". Some interviewees have been able to buy a motorbike or resume their studies. The financial autonomy of women has also improved, according to the testimony of a female farmer. On the other hand, there are many difficulties. No major maintenance work has been done for more than ten years; the irrigation network, the tractors and the pumping station are in alarming conditions. The canals are often obstructed by weeds or clogged with sediment, so much so that part of the perimeter served by the pumping station cannot function during the counter-season. Only one tractor is currently operational; the others are out of order. Finally, the pumping station is described as being "in agony". A witness says:

> Of the station's four power units, only two are operational. The two groups work alternately so as not to overheat. Every ten minutes, water has to be poured over them to cool them down. These are still the engines brought by the Taiwanese at the end of the 1990s. Both engines, received as a gift from PNSA in the 2010s, are blocked.

The allottees who do not receive water because of the blocked channels are protesting and demanding a refund of the fee. For years now, there has been no outside help, neither from the state nor from NGOs. For the future, there is hope

Figure 11.3 Embodying hope: Allottees from casier B in Bongor. The photo highlights the determination, courage and, above all, the resilience of these people who, despite a hostile, turbulent environment and intricate crises, continue to fight with all their might.

Source: Dobah Marsala, 2021.

for a new rehabilitation of all hydraulic works and the purchase of new pumping systems (Fig. 11.3).

Polders, Bol

The fundamental change that has had negative impacts on the north shore since the 2010s is the instability related to terrorism from Nigeria. Among other things, the arrival of refugees has led to an increase in cereal prices. The cultivated areas in the three modern polders of Bol comprise 1540 ha in Mamdi for 4276 allottees, 300 ha in Berim for 608 allottees and 212 ha in Guini for 568 allottees. Management is entrusted to SODELAC. Management committees represent the allottees. Maintenance work is minimal because SODELAC does not currently have adequate machinery. Of the five pumping stations in the Mamdi Polder, two are out of order. The poor condition of the facilities causes some allottees to refuse to pay the annual fee. A major problem is the entry of livestock into the irrigated areas, which can devastate entire plots. On the one hand, some pastoralists believe that the land occupied by the polders is theirs and that it is therefore their right to graze their animals. On the other hand, for herders arriving on the shores of the lake in the dry season, finding pasture is a "matter of life and death". Moreover, often, individuals are both farmers and herders, bringing their animals onto their plots. In addition to the obsolescence of tractors and pumps, the lack of maintenance and the entry of animals, there are other problems such as the low water level of the lake, especially in June, which prevents the pumps from drawing water, the insufficient number of agricultural advisors, the theft of Polyvinyl Chloride pipes that bring water to the plots, the invasion of weeds in the drainage channels and the consequent salinisation of the soils. Despite these weaknesses, many local residents are still demanding the allocation of plots. The hope is that an overall rehabilitation of the polders (particularly Guini and Berim) can begin and that new tractors, heavy-duty engines and pumping systems will arrive. An agronomist engineer with long experience in the region notes that the soils of the lake are fertile, rich in organic materials, and they do not require fertilisers. The Mamdi Polder has been active for 20 years and still does not need fertilisers. There is a lot of potential but the works to exploit it are expensive, especially the modern polders with pumping. The so-called semi-modern polders, where sluice gates are opened seasonally to moisten the soil by controlled submersion, are more sustainable. This option has also found support in more recent plans, such as the Projet de Relance du Développement du Lac (PROLAC), launched in 2019, financed by the World Bank as support for the populations most exposed to destabilisation due to terrorism. In the opinion of the same agronomist engineer, the government was right to make SODELAC autonomous again because the situation of the lake is too specific and requires a dedicated structure capable of preparing the necessary projects to obtain funds from donors:

> You have to write a 200-page book to propose a project and then be able to negotiate with donors. Otherwise, there is nothing left but to make room for private investors. However, as long as the situation in the region remains so unstable, it will be difficult for big investors to arrive, be they domestic or foreign.

Boumou irrigation scheme

The Boumou irrigation scheme is located between Laï and Djoumane, in the same plain where the Satégui Déréssia project was realised in the 1970s, and follows the same principle of operation, that is, through intake structures on Logone River to capture floodwaters and direct them to the irrigation network. The preparation of the hydro-agricultural works lasted from 2010 to 2013, but the project was completed in 2015. Initiated by the PNSA, the project aimed to increase national paddy production in quantity and quality and to combat poverty. The feasibility studies, which began in 2007, were conducted by Bureau DIC, while the work was carried out by a Chadian company, Geyzer SA (https://geyser-sa.com/), under the control of another national company, AgriTchad SA. The two lots, both divided into five hydraulic units of varying sizes, cover a total of 3139 ha, including both hydraulic works and cultivable land (2740 ha). The first 200 ha have been cultivated since 2010, gradually increasing the cultivated areas as the project progresses. The plots entrusted to allottees each have a surface area of 1 ha. Some allottees have more plots, which are dedicated to sharecropping, although this practice is theoretically prohibited in the perimeter. The PNSA has provided allottees with tractors and, for seven farming years, mineral fertiliser at subsidised prices and improved seeds free of charge. However, since the 2017–2018 crop year, there have been signs that crop yields have fallen to 2.5 t/ha of paddy, much lower than expected (25 t/ha). According to the ANADER sector leader, the primary reasons lie in the imperfect execution of certain hydro-agricultural projects. This issue has been acknowledged by various technical missions, though no solutions have been implemented. As a result, the allottees themselves often make improper modifications to these works. Tensions have also arisen among the allottees, with some displaying aggression towards their neighbours, occasionally escalating to acts of vandalism. Additionally, fertilisers have also been lacking over time, as well as adequate technical-agronomic support for the growers. Water shifts and the agricultural calendar are not respected in the way necessary for good project management.

A small irrigation scheme on the banks of the Logone River, near N'Djamena

The focus of Marsala's fieldwork in this case is a small irrigation perimeter created by the Taiwanese cooperation in 1999, which also took care of training the farmers. It involves five villages: Blabli, Jardin présidentiel, Zaraf, Seyaba and Bourmaneyé, about 12 km from N'Djaména. The 20 ha are cultivated with rice and legumes. Two crops are cultivated per year: rice in the wet season and tomatoes, aubergines, onions, peppers, courgettes and other vegetables in the hot season. The income depends a lot on the market performance of the produce, which widely varies, according to the season (the tomato case can range from 1000 to 3000 CFA francs). Sometimes, the market does not even absorb the produce, which has to be thrown away. There are 80 plots and about 100 allottees because some plots are divided among members of the same family. The allottees are organised into a Regroupment that manages the pumps, the maintenance of the canals and the sale of the produce, retaining the fee (which is 100,000 CFA francs per campaign/ha). The Regroupment has a current account and a savings account in the bank and is independent of state structures. Self-management has been profitable; the Regroupment even succeeded in buying a car to transport the produce to the capital. There is a waiting list of people who want to enter the perimeter as allottees; it can happen that someone leaves or cultivates poorly and is then replaced. The major problem faced by the allottees is the theft of their produce, which has led to the need for a fence to deter both thieves and stray goats. "When we catch goats, we take them to the village headman, who returns them to their owners upon payment of a fine". Just beyond their fields, a perimeter is held by the family of the President of the Republic. The whole area is characterised by irrigated perimeters. However, these schemes belong to private individuals, who can also have large areas, up to 35 ha.

The projects for Batha Province and Fitri Lake

Batha Province is located in the west-central zone of Chad, outside the areas affected by development projects related to irrigated agriculture. However, the endorheic Lake Fitri, which is seasonally fed by Batha River and other smaller temporary streams, is situated in the province. This lake, similar to Lake Chad, is shallow, so that its surface area can reach 1300 km^2, but in dry seasons and in the absence of Batha River's supply, it can dry up completely (Lemoalle & Hourtal, 1996, p. 27). Nevertheless, it is a valuable wetland for agriculture, fishing and animal husbandry. The head of ANADER, interviewed in Batha, pointed out that the hydro-agricultural potential of the province has not been exploited so far. In fact, no irrigation projects have been implemented. Studies by AgriTchad SA, initiated by the World Bank, have shown at least four sites where large-scale hydro-agricultural projects could be established, drawing water from the water table or seasonal flooding from Lake Fitri. Each site would serve four or five neighbouring villages. The sector head stated that the development of large-scale irrigation schemes to enable cultivation in the dry season would be essential to strengthen the province's food self-sufficiency.

Tricky issues, tricky traces

Regarding agriculture, the challenges facing the Chadian region are well indicated by Magrin (2023), who emphasises above all the demographic variable (with the population expected to increase from 50 million in 2015 to 130 million in 2050) in the context of the weakening state and customary frameworks for managing tensions, particularly land disputes. The constraints of global warming, with higher temperatures making rainfed cultures more fragile, "will give wetland enhancement and irrigation development an even more central place" (Magrin, 2023, pp. 17–18). Irrigated agriculture is therefore far from losing its weight, quite the contrary. However, the troubled history of modern irrigation in the region warns that without addressing crucial issues, the risks of outcomes that fall short of expectations and above all, of the enormous needs, are high. In this section, we focus on three of them, highlighting aspects that recur frequently in the research reports of Mahamat and Marsala: the enclavement, the approach with which climate and hydrological conditions are viewed and narratives about the failures of irrigation development projects. The first two aspects, as mentioned above, have an important bearing on the third, the problematic outcomes of irrigation schemes.

Enclavement, corridors, circulations

At the height of the security crisis in the 2010s, the perception of a worsening isolation of the region became even more acute. Although road insecurity is a long-established phenomenon, linked to brigandage (e.g., think of Maroua-Kousséri Road; Issa, 2004), terrorism has certainly added another layer of danger. Beyond the security problems, the issue of the distance of the Chadian area from the southern and coastal areas of Nigeria and Cameroon is always topical. Taking the Far North of Cameroon and the fundamental roads that cross it as a reference, not only for the Cameroonian province but also for Chad, the strategic sections of the Ngoundéré-Garoua and Mora-Kousséri highway are at a standstill. Likewise, the promise of the President of the Cameroonian Republic (made in Maroua during his visit on 29 September 2018) to build that Ngoundéré-Maroua-Kousséri-N'Ndjamena railway, which would give vast breathing space to the entire region, has not been fulfilled.

Therefore, if the enclavement has not yet found adequate infrastructural solutions, it is also true that something from the Chadian region does reach and, even more, can reach the ports on the coast. It is oil, and not by chance, if it is true, as the Iranian philosopher R. Negarestani states, that "the cartography of oil as an omnipresent entity narrates the dynamics of planetary events. Oil is the undercurrent of all narratives, not only the political but also the ethics of life on earth" (2008, p. 19). A 1070-km pipeline, built between 2000 and 2003, allows oil extracted from the Doba field in southern Chad to reach the port of Kribi in Cameroon (Magrin, 2015). In March 2024, the pipeline connecting Koulélé in Niger to the offshore terminal at the port of Sèmè in Benin was completed. This will allow Niger to export crude oil extracted from the Agadem Rift Basin field, located near the border with Chad. One of the longest oil pipelines in Africa (1950 km—1275 in Niger and 675 in Benin), it was built by the China National Petroleum Corporation. Work started in 2019 and was slowed down, first by the COVID-19 pandemic and then by the tensions related to the coup d'état in Niger, with the latter still ongoing. Furthermore, in June 2024, the Nigerian rebel group, Patriotic Liberation Front, attacked the Chinese project, threatening further assaults if the deal with China would not be terminated (Asadu, 2024). An alternative route, through Chad to join the Doba-Kribi pipeline, had at first been envisaged, but insecurity conditions made it opt for the much longer route through Benin. Thus, if the Chadian region in many respects remains isolated, this is not the case for the export of oil. It can be interpreted as the continuation of the "logic of corridors and tunnels" that the colonisers had imposed to connect "centres of extraction to ports of exportation by the quickest route possible" (Mbembe, 2023, pp. 206–207). These "extroverted transportation plans put in place during the colonial period are in fashion once again" (Mbembe, 2023, p. 207), even if today, many other actors are at play, and among them, as in the case of the Niger-Benin pipeline, China is in the foreground (Schritt, 2016). The logic of "circulations" is quite different, understood as "a series of complex operations by means of which a society invents, through movement and exchange, a vital equilibrium with its environments, or is able to give form to such environments and to link them to one another" (Mbembe, 2023, p. 209). This was the situation before colonisation when vast regional African spaces, such as the Chadian one, were characterised by "a complex series of vertical corridors, lateral axes, and networks that were often mutually imbricated according to the *principle of intertwining and multiplicity*" (Mbembe, 2021, pp. 181–182, italics by the author). It is precisely a recovery of this logic of circulations, "by intensifying mobilities, and developing interconnections between places", that could lead to the "creation of an Afropolitan public space", which could transcend national frontiers (Mbembe, 2023, pp. 216–217). From this point of view, some developments appear interesting, with particular reference to the Far North of Cameroon. Traffic has picked up again after the difficulties linked to Boko Haram. Municipalities have been rehabilitated or opened rural tracks, due to funds received from the government through programmes such as the PROLAC, the Projet de Développement Rural Intégré du Chari-Logone and the Programme National du Développement Participatif, which are, in turn, financed by the ADB, the Islamic Development Bank and the World Bank. Portions of national roads (Maroua-Magada, Maroua-Mora and Magada-Yagoua) and regional roads (Maroua-Bogo et Bogo-Guirvidig) are also being rehabilitated. The Guividig-Pouss Regional Road is being paved. The improvement of the connection between the Far North and neighbouring countries is also important. Of particular note are the completion of a large bridge over Logone River connecting the towns of Yagoua (Cameroon) and Bongor (Chad), which were previously precariously connected by a ferry service, and the construction of a crossing between Limani and Amchidé, which greatly facilitates the movement of goods and people between Cameroon and Nigeria via the Mora-Banki-Maiduguri corridor.

Approaches to the water

The second relevant aspect highlighted in the reports is the approach to the peculiar hydrographical conditions of the Chadian basin and the outcomes of climate change. It showed what solutions were envisaged to cope with the problems in the Far North of Cameroon, with exceptional flooding on the one hand and water scarcity on the other hand. Both the proposed responses to the two emergencies, that is, a clearer separation of Logone River from the flooding areas and the massive spreading of dams in the Mayo Tsanaga basin, reflect the state government's policy to intervene in the territory through operations of deep modification of environmental equilibrium and massive transformation of territories and their spatial relations, which have already demonstrated their short-sightedness and ineffectiveness over time. The complexity and, at the same time, the fragility of the water use systems of the communities living in the Diamaré plain and the value of flood zones, such as the Grand Yaéré, is inadequately assessed. Furthermore, the needs and grievances of the affected villages are frequently overlooked or ignored.

In numerous instances, as exemplified by ACEEN, local associations and NGOs are capable of proposing alternative approaches that are deeply embedded in the local context, focusing on affordable, incremental changes rather than aiming for radical transformations of territories. Scott (1998) refers to "thick spaces" as those constructed by local communities through an ongoing relational process involving both human and non-human entities, resulting in a fluid and adaptive management of resources. These approaches do not depend on large-scale territorial transformations but leverage local knowledge and technologies. Through the active involvement of inhabitants, they generate practical innovations that are easily adaptable because they are grounded in the specificities of the local context. Mahamat and Marsala's reports have highlighted the need to reconsider interventions designed with and by local communities, rather than imposing top-down projects. This perspective aligns with critiques of development that emphasise modernisation through large-scale infrastructure. It points to the necessity of embracing and listening to plural ways of practising "development" through transformative initiatives (Kothari et al., 2019).

Narratives on blame

There is one point on which the interviewees and the reports repeatedly insist, namely the search for a scapegoat for the troubled progress of irrigation projects.

The fault-finding moves in the interviewees' narratives on different scales. It may be useful in this sense to adopt the multiscalar analysis of forms of blaming proposed by the anthropologist T. H. Eriksen, with particular reference to the "blame" of climate change (2016, pp. 138–145). The starting point of his reflection is the path-breaking work of anthropologist M. Douglas (1992), who, regarding the sudden and mysterious death of a woman in a village, identifies three forms of blaming: "first, the woman might have been blamed herself: she could have offended the ancestors or broken a taboo"; "second, the blame could be attributed to an individual adversary or a competitor within the community"; and "finally, the cause of the woman's death might be located outside the community" (Eriksen, 2016, p. 141). This tripartition can also extend to other, more complex contemporary phenomena such as unemployment, mass immigration or climate change itself:

> When something is seen to go wrong, it could ultimately be the victim's fault, it could be the result of conflicts, contradiction or competition within society, or it could be the work of insidious outside forces, often seen as some version of "the global" or its permutations.
>
> *(Eriksen, 2016, pp. 141–142)*

Note that in contemporary multiscalar societies, blame can be attributed to all three levels simultaneously. This analytical scheme can also be applied, even in more detail, to narratives about the causes of failures or otherwise difficulties in irrigated perimeters. The crises can thus be attributed to the allottees themselves: "They are lazy; they do not respect the rules; they do not do maintenance; they do not follow the water shifts; they do not pay the fee; they quarrel among themselves; they trade illegally the fertilisers they receive". Otherwise, it can be imputed to the organisation; it is the fault of those who designed the irrigation scheme ("planning errors"), of the project managers or technicians or government ("they do not listen; they do not understand the needs; they do not care"). A finger can also be pointed at those who live around the perimeters (theft of water and produce or cattle entry) or to the major traders, who take the bulk of the profit at the expense of the producers. Alternatively, the blame can be placed on a widespread social propensity for the "predation" of available resources and thus on corruption (many cases are reported along these lines). Finally, the blame can be directed externally, at the dynamics of economic globalisation (e.g., cheaper imported rice or the rising costs of seeds, fertilisers and pesticides) or at the unfair division of political and economic power on a global scale that penalises African countries, particularly the Sahelian ones, or even at the paternalistic logic of international donors. An interesting example is the reflection of an agronomist, interviewed by Marsala, who interprets the great droughts of the 1970s and the 1980s as the first signs of the ecological crisis linked to global warming:

> If the rich countries had taken charge of the problem at that time, much damage could have been avoided. They would have had the means. We,

> African countries, are poor and nature started to attack us first. The origin of these attacks is first in the behaviour of rich countries. Today we know that it is the effect of carbon dioxide, but where does this carbon dioxide come from? Unfortunately, only today are people starting to react. If the rich countries had reacted earlier, perhaps many solutions could have been found and concrete results achieved.

Some or even all of these "attributions of blame" can be intertwined and overlapping in explaining the outcomes of specific situations. Each indicates, by cast, possible directions of actions for the future of irrigated agriculture in the Chadian region.

Conclusion

In the scenario of irrigated agriculture, which in the Chadian region is marked more by problems than successes, reports and interviews nonetheless reveal several reasons for hope. The first reason involves young people—many of them have a great desire to get out of the dire straits in which they find themselves. However, governments cannot continue to propose to them the tiring and unprofitable agriculture of the hoe. Agricultural modernisation can instead arouse interest. There is no shortage of skills either, given the thousands of young people trained in polytechnic schools. Training should extend to all players in irrigated agriculture, disseminating innovative experiences and skills, with a focus on consolidating a maintenance culture. Finally, it is important to strengthen an endogenous socioeconomic fabric; its signs are the proliferation of national engineering and construction companies, as well as investors potentially interested in the agricultural sector. Of course, there is no dearth of potential social conflicts triggered by a consolidation of the private sector, at the expense of collective forms of resource management. Even more positive is the emergence of associations capable of contesting political choices and building networks of relations in the area, among local actors, to bring out their knowledge and needs and externally, even in an international dimension, dialoguing as equals or, in any case, competently with development promotion agencies or those holding possible financing funds.

References

Abakar, H., Ourdé Ousta, H., & Adoum, A. (2020). *Rapport de mission des périmètres aménagés des Casier A, B, et C de Nya*. Agence Nationale d'Appui au Développement Rural (ANADER).

Asadu, C. (2024). *Coup-hit Niger was betting on a China-backed oil pipeline as a lifeline. Then the troubles began*. AP News. https://apnews.com/article/niger-benin-oil-pipeline-china-agadem-coup-01891707e30450361eeaedbb3daf7bce

Beauvilain, A. (1989). *Nord-Cameroun. Crises et peuplement*. T. I–II. C. Bellée.

Bertoncin, M., & Pase, A. (2012). *Autour du lac Tchad. Enjeux et conflits pour le contrôle de l'eau*. L'Harmattan.

Bertoncin, M., & Pase, A. (2015). Progetti di sviluppo e trasformazioni territoriali: L'esempio di una zona umida saheliana (Camerun). *Rivista Geografica Italiana*, *CXXII*(1), 15–32.

Bertoncin, M., & Pase, A. (2017). Interpreting mega-development projects as territorial traps: The case of irrigation schemes on the shores of Lake Chad (State of Borno, Nigeria). *Geographica Helvetica*, *72*, 243–254.

Blench, R. (1997). The history and future of water management of the Lake Chad basin in Nigeria. In H. Jungraithmayr, D. Barreteau, & U. Seibert (Eds.), *L'homme et l'eau dans le bassin du lac Tchad* (pp. 143–166). ORSTOM.

Bouquet, C. (1990). *Insulaires et riverains du lac Tchad: Une étude géographique* (Vols. I–II). L'Harmattan.

Brunet-Jailly, J. (1981). L'introduction de la riziculture irriguée en pays Mousgoum. *Revue de Géographie du Cameroun*, *II*(2), 71–95.

Douglas, M. (1992). *Risk and blame. Essay in cultural theory*. Routledge.

Eriksen, T. H. (2016). *Overheating: An anthropology of accelerated change*. Pluto Press.

Issa, S. (2004). L'embuscade sur les routes des abords sud du lac Tchad. *Politique Africaine*, *94*, 82–104.

Kothari, A., Salleh, A., Escobar, A., Demaria, F., & Acosta, A. (2019). *Pluriverse: A post-development dictionary*. Tulika Books.

Lemoalle, J., & Hourtal, D. (1996). Ressources en eau du Bassin Conventionnel. In I. de Zborowski (Ed.), *Atlas d'élevage du Bassin du Lac Tchad* (pp. 23–28). CIRAD.

Levrat, R. (1975). *Une expérience de développement rural au Cameroun: La SEMRY*. Fac. des Lettres et Sciences Humaines, Université de Yaoundé, Département de Géographie.

Magrin, G. (2015). Le lac Tchad à l'épreuve de l'ère pétrolière. In G. Magrin, J. Lemoalle, & R. Pourtier (Eds.), *Atlas du lac Tchad* (pp. 166–168). Passage.

Magrin, G. (2023). Agriculture, crise climatique et violences: Du nouveau sous le soleil du Sahel? Retour sur un vieux débat à partir de la région du lac Tchad. *Bulletin de l'association de géographes français*, *100*(1), 6–19.

Mbembe, A. (2021). *Out of the dark night*. Columbia University Press.

Mbembe, A. (2023). Circulation. In K. Botanova, Y. Kamara, & Q. Latimer (Eds.), *Sahara. A thousand paths into the future* (pp. 204–219). Culturescapes.

Ministry of Economy, Planning and Regional Development (MINEPAT), Republic of Cameroon. (2019). *Detailed project report. The development of 10,000 ha of hydro agricultural area in the locality of Logone Birni, Logone, Birni Subdivision, Logone and Chari Division, Far-North-Region*. PKS Infra Engineers Pvt. Ltd.

Mission de Développement Intégré des Monts Mandara (MIDIMA). (2016). *Etudes de faisibilité des retenues d'eau de: Mizil, Adatz (Baldama), Maboudji, Tchevi, Yola Maliki, Manguirlda, Soukkoungo, Makilingaï*. ETS MARADUR.

Mugelé, R. (2015). Les polders de la SODELAC. Résilience et défis. In G. Magrin, J. Lemoalle, & R. Pourtier (Eds.), *Atlas du lac Tchad* (pp. 110–113). Passage.

Negarestani, R. (2008). *Cyclonopedia. Complicity with anonymous materials*. Re.press.

Roupsard, M. (1987). *Nord-Cameroun: Ouverture et développement*. C. Bellée.

Roupsard, M. (2000). Production rizicole. In C. Seignobos, & O. Iyébi-Mandjek (Eds.), *Atlas de la province Extrême-Nord Cameroun* (pp. 94–97). IRD.

Schritt, J. (2016). The petro-political configuration: Entanglements of Western and Chinese oil zones in Niger. *ZFW – Advances in Economic Geography*, *60*(1–2), 40–56.

Scott, J. C. (1998). *Seeing like a state. How certain schemes to improve the human condition have failed*. Yale University Press.

12 The production of land insecurity and conflict in the Sourou valley (Burkina Faso)

Moumouni Ouande, Lassane Yameogo and Francesco Staro

Introduction

The Sourou valley, located between the Seno-Mango and the Mondoro plains in northern Burkina Faso, has been a major site in the country's economic development strategies since the 1960s. This was the period when the construction of water infrastructures changed the natural hydraulic regime of the Sourou and Mouhoun Rivers in order to increase the floodable area and water storage and thus to accommodate agricultural development projects.

The Sourou region has been transformed by hydraulic interventions over the past 50 years. Since 2020, the valley has been in the spotlight due to a new wave of national and foreign investments and land deals to further develop irrigated schemes. This new impetus for agricultural development projects has been undermined by the emergence of different forms of armed violence and rural insecurity, targeting irrigation systems and infrastructures.

In this chapter, we discuss how these critical events are the results of a long-term process of land tenure insecurity and environmental regulations that have restricted access to key natural resources and exacerbated tensions and competition among different groups of users.

Methodology and evolution of the research questions

This chapter is the result of the fieldwork carried out by Ouandé Moumouni under the supervision of Lassane Yaméogo, both geographers, using a mixed method, including semi-structured interviews, household surveys, field observations, documentary analysis and the geo-referenced databases of the Institut Géographique du Burkina Faso. Agricultural statistics from local institutions, data on internally displaced persons (IDPs) from CONASUR (Conseil National de Secours d'Urgence et de Réhabilitation) and health statistics about COVID-19 cases from the Centre des opérations de réponses aux urgences sanitaires (www.corus.gov.bf) were also used. The data was collected over a three-year period (in 2021, 2022 and 2024) in the localities of Di, Lanfiéra and Tougan. The qualitative survey of local

DOI: 10.4324/9781003429111-17

stakeholders took place in January–February 2021 and January–February 2024. Semi-structured interviews were conducted with 27 stakeholders, including 8 cooperative organisations and 19 state and private structures, non-governmental organisations and civil society organisations. The household survey took place from May to June 2022, and involved 218 individuals, including farmers and herders, government employees and IDPs. The KoboCollect app was used during the field surveys. The data was later analysed and reorganised by Francesco Staro, an anthropologist, in the framework of the AtlaSahel project.

The *burkinabé* researchers experienced several difficulties during the fieldwork phase due to rural violence and insecurity, including roadblocks, targeted kidnappings and violent attacks that resulted in internal displacements. Following the declaration of the COVID-19 pandemic, the government's restrictions limited internal movements and accessibility to rural areas and made fieldwork even more challenging. Because of these constraints, the research methodology was adapted by conducting telephone interviews and using secondary data.

Beyond logistical and security constraints in reaching the valley and carrying out fieldwork, other issues during the research process were related to the overall approach to the analysis of the ongoing dynamics of economic development and rural insecurity in northern Burkina Faso.

At first, our research focused on the impacts of COVID-19 on Burkina Faso, the armed attacks and the internal displacements that have occurred in rural areas of the Boucle du Mouhoun region since 2014. As our study proceeded, it became clear that it was impossible not to take into account how these recent events were deeply related to the local history of natural resource management and land tenure insecurity in the context of large-scale agricultural investments.

To address this issue, we question the narrative that considers this series of critical events as a "threat" to the project of economic development in Sourou Valley. This is the mainstream discourse in the reports and in the grey literature on agricultural development schemes; according to such studies, projects to extend irrigated areas are being suspended and investments are being frozen because of rural violence and insecurity (Souroupole, 2022, 2023). By focusing on the impacts of external factors on local dynamics, this representation de-historicises the political and economic relations among the local, national and global actors involved in agribusiness and large-scale agricultural schemes.

The COVID-19 pandemic also played a role in legitimising the dominant understanding of the socio-economic crisis affecting Sourou Valley as a "local consequence" of an external problem. From this perspective, the risks related to a virus arriving from abroad are evaluated by considering the impacts of the health crisis on local communities, depending on their capacity to limit the infection rate. This approach is problematic because it implies a passive status for local populations presented as victims of "natural" phenomena and fails to acknowledge that the capacity to cope with critical events depends not only on the social and economic characteristics of local communities, considered distinct and separate entities but also on the changing relations and balance of power among different interest groups and different levels of interaction.

Similarly, it is necessary to avoid viewing rural insecurity as simply the perverse result of development projects that would have compromised the local traditions regulating natural resource management or as the result of a neo-liberal scramble for land over local institutions. Instead, it is essential to consider how agricultural investments are hijacked and become elements of local dynamics (Wolford et al., 2013). This approach also implies an adaptation of the research focus: rather than considering the criteria for qualifying a development project as "successful" or "unsuccessful", the aim is to investigate the practices that the development discourse contributes to legitimise (Ferguson, 1994; Mosse, 2004).

Against this background, we decided to study current events in Sourou Valley, not in terms of the "impacts" of critical events at the local level but rather by taking into account the interplay of external and internal factors at different scales of analysis, in a regional situation hit by multiple and overlapping crises.

Development of irrigated agriculture in Sourou Valley

Sourou River is emblematic of the interaction between ecological and socio-economic dynamics related to the Sahelian natural resource management and access to such resources. Before the Léry dam was built in 1976, Sourou River naturally functioned as a reservoir, helping regulate the flow of the Mouhoun River downstream (Zoungrana et al., 2009). Sourou River acted as a buffer or an expansion tank, storing high water levels and supporting low water levels, depending on its natural regime (Bethemont et al., 2003). During the rainy season, floodwaters from Mouhoun River (around one-fifth of the total flow, i.e., 1,500 million m^3) entered Sourou River at the end of July and flowed slowly upstream for almost 125 km to the north, up to Baï, in the Malian territory. During that season, Sourou River appeared to be defluent. As a result, 25,000 ha of the area was flooded (Koussoubé, 2010). During one dry season (starting in December), Sourou River reversed its course and returned some of the stored water to Mouhoun River (a volume of around 70 million m^3 for 200 million m^3 of inflow), while the rest was eliminated by evapotranspiration (Leprun, 1979). Natural factors may explain this unusual flow regime in Sourou Valley.[1]

These hydrological dynamics in Sourou Valley exemplify the seasonal complementarity of wetland and dryland resources in the Sahel region. In this changing and moving environment, the succession of the rainy season and the flood season and their interdependence illustrate the local temporalities of the natural resources used by different groups throughout the year, thus allowing complementarity among different productive systems (García et al., 2022).

Over time, Sourou River (including its water resources) has been an essential watering place for livestock during the dry season, and the Boucle du Mouhoun region has historically been an area of interactions between sedentary and nomadic populations due to the importance of the water resources and grazing land available during the dry season. In this regard, it is worth recalling the strategic role of *bourgoutières*, natural areas that provide green grass and rich pastures as the water gradually recedes from the flooded areas (Toutain, 1987). These shared resources are pivotal in the organisation of the seasonal mobility of people and herds on a regional scale and allow the economic and ecological complementarity among transhumant Peul pastoralists, sedentary Mossé herders and mainly farmers and fishers from Bobo, Bwa and Samo groups. Sourou Valley is also a favourable area for the development of numerous fish species. Fish breeding has long been influenced by the flooding of the Mouhoun-Sourou water streams and the seasonal migration of fish to the flooded plains. Two types of fishing are practised here: capture fishing (mainly customary and collective) and fish farming. In particular, capture fishing is practised in rivers, flood plains and reservoirs.

The 190 million m^3/year dam at Léry was built in 1976 to increase the maximum submersion time and ensure irrigation of the rich lands of the Sourou plain. Subsequently, the growing demand for water-irrigated plots, accentuated by irregular rainfall, led to the construction of a 700 m long, 25 m wide and 9.5 m deep canal in 1984 to deviate the waters of Mouhoun River. This increased the flooded plains by 10,000 ha and their storage capacity by 603 million m^3/year.

Since the construction of the first 144-ha scheme in 1966–1967, the development of irrigated agriculture has continued, reaching 460 ha of irrigated land in 1981. Starting in 1985, most of the hydro-agricultural development works were carried out in the framework of the Maîtrise d'Ouvrage du Sourou (MOS) project, which in 1986 became the Autorité pour la mise en valeur de la vallée du Sourou (AMVS). Under AMVS management, a dozen schemes were built until 2005 (around 3,818 ha). Moreover, the construction of 500 ha of irrigated land at Di in 2005 and 2,240 ha of irrigated land as part of the Millennium Challenge Corporation – Millennium Challenge Account (MCC-MCA) Compact project[2] in 2014 resulted in an additional 6,558 ha of irrigated land (see Table 12.1), that is, 21.86% of the irrigable potential (Somda et al., 2010; Souroupole, 2022).

As a result of these projects, cultivated areas and annual production increased, with an average production of 101,773 t between 2014 and 2022. This income for producers was estimated at CFAF 15 billion for the 2021–2022 season (Souroupole, 2023). Due to the good quality of the soil, several crops can be grown: cereals (maize, rice, wheat and sorghum), vegetables (onion, tomato, green bean, potato, garlic, lettuce, carrot, cabbage, etc.) and fruits (papaya, banana and citrus). Cereals such as maize and double-cropped rice, with yields of 3 and 5 t/ha, respectively, are the dominant crops, covering 63% of the total sown area. These are followed by fruit crops (banana and papaya) and vegetables such as onions, tomatoes and chillies, which account for 36% of the total cultivated area (Souroupole, 2023).

Table 12.1 Summary of developments in Sourou Valley from 1967 to 2014.

Irrigated areas	*Year*	*Funding*[3]	*Irrigation system*	*Surface area (ha)*	*Type of farm*
Guiédougou1	1967	CCE	Semi-Californian	144	Countryside
Guiédougou2	1980	CCE	Semi-Californian	144	Semi-Californian
Di 1	1977	National budget	Spraying	210	Régie paysannat
Di 2	1986	National budget	Spraying	210	Régie paysannat
Niassan (Coopérative Agricole de Niassan-CANI)	1986	FED	Gravity	50	Countryside
Niassan (Coopérative Agricole et Pastorale du Sourou-CAPSO)	1987	CNSS	Gravity	140	Countryside
Niassan 3	1990	CNSS and IDB	Gravity	475	Mixed
Niassan (Coopérative Agricole du Périmètre Irrigué de Niassan-CAPIN)	1993	FED	Gravity	500	Countryside
Niassan (Société coopérative simplifiée BADENYA)	1993	FED, BOAD	Gravity	200	Mixed
Niassan (Coopérative Agricole de Débé-Société Coopérative Agricole Maraîchère de Débé-CAD/SOCAMAD)	1996	BID, BADEA, FSD and OPEP	Gravity	925	Countryside
SONO	1997	BADEA	Spraying	210	Private
Débé plain	2005	BID	Gravity	610	Mixed
Di 500 ha Pivots	2005	National budget	Spraying	500	Mixed
Di	2014	MCA	Gravity	2,240	Countryside

Data source: AMVS report (2014).

Large-scheme agricultural development as the target of rural violence and attacks

In 2020, the AMVS started to be reorganised into a semi-public company called Société de développement intégré du Pôle de croissance de la vallée du Sourou. At that time, the region was already fragile and unstable due to rural insecurity. Since 2018, various incidents and terrorist attacks (destruction of assets and buildings, roadblocks, kidnappings and murders) have occurred in Sourou Valley. The Lanfiéra police post was attacked in 2018 and 2021. In Dî, national security forces (Groupement des Compagnies Républicaines de Sécurité [GCRS]) and the border police station were attacked in October and November 2021. As we shall demonstrate throughout this chapter, these terrorist groups seized the opportunity to assert their power by taking advantage of the fragile institutional setting and the situation of land insecurity.

Table 12.2 summarises the information concerning the projects that have been cancelled or suspended following the deterioration of the security situation in Sourou Valley.

The 2,000-ha Dangoumana Agricultural Development Project (Projet de développement agricole de Dangoumana, PDAD) and the Agricultural Resilience

Table 12.2 Agricultural development projects that have been suspended due to insecurity.

Project name	*Cost (million CFCA)*	*Potential partners*	*Remarks*
2,000-ha Dangoumana Agricultural Development Project (PDAD)	45,000	To be found	The initial funding was lost when the projects (PDAD and PReCA) were relocated outside Sourou Valley due to insecurity
1,500-ha Hydro-agricultural perimeter at Bissan	30,000	To be found	
2,000-ha Hydro-agricultural development project for wheat growing in Sourou at Dî	3,701,988	Chinese company: China Machinery Engineering Corporation (CMEC)	In negotiation with CMEC
2,000-ha Hydro-agricultural perimeter at Sono Kouri	3,701,988	Saudi and Kuwaiti funds	22 billion, excluding the national counterpart (4.019 billion) has already been the subject of financing agreements with the Abu Dhabi Fund
Project to develop 1,000 ha at Koumbara	11,901	To be found	No technical and financial partner
Total: 8,500 ha	**162,140,76 million CFCA**		

Source: Souroupole (2023).

and Competitiveness Project (PReCA) for the development of the 1,500-ha hydro-agricultural perimeter at Bissan have been suspended. Depending on available funding, three projects included in the National Economic and Social Development Plan will enable the development of an additional 5,000 ha. These comprise the 2,000-ha Di Pivot Wheat Project, the 2,000-ha Sono Kouri Project and the 1,000-ha Koumbara project.

Water infrastructures and agricultural development projects have often been the main targets of the attacks. There have been targeted kidnappings, such as the abduction of three workers from the Souroupole national agency on 30 September 2020 and of three drivers (from a company in charge of the rehabilitation of irrigated areas) on 19 October 2021. In addition to these events, a warehouse in Sono was set on fire on 23 December 2021 and machinery for asphalting the Tougan-Lanfiéra-Dî road was destroyed in Doussoula and Kassoum on 9 September 2021. Shops, buildings and telephone networks were destroyed as well.[4] Thefts of vehicles and motorbikes were also reported by local newspapers.

In the institutional discourse on Souropole, the decision to suspend agricultural investments and extensions of irrigated perimeters is mainly attributed to rural insecurity. This explanation may hide other factors, such as the high costs of maintenance of hydraulic infrastructure (which includes the Léry dam, the dam gates, the Sourou Valley reservoir and the flood control infrastructure) and the renovation of irrigation perimeters. These costs can be high due to flood episodes that can damage these infrastructures and limit crop productivity. In 2022, heavy rains and the rising waters of Sourou River in the irrigated plots led to huge losses at the end of the agricultural season and the failure to meet the 2022–2023 production target. In total, 1,160 ha of crops were destroyed by the floods, including 886 ha of maize and 274 ha of rice, out of a total of 6,423.21 ha. These data show the structural limits of the Agropole economic model in a variable and unpredictable climatic environment, as well as the difficulties in ensuring the economic viability of agricultural investments.

In the next sections, we trace the recent history of irrigation projects, paying special attention to the reconfiguration of land access rights, which is a crucial destabilising factor that prompted the current situation of rural insecurity.

Agricultural investments and environmental protection initiatives and the emergence of land insecurity

The development of irrigated agriculture in Sourou Valley started in 1966 when 148 families of agricultural settlers of the former Office du Niger were relocated from Mali in the Guiédougou irrigated areas. Since the very first phases of hydro-agricultural projects, the local populations' internalisation of external objectives had been problematic because agricultural schemes in many cases prioritised cash crops over food production. The pilot project in Guiédougou achieved limited results due to the high maintenance costs of the irrigation system and also because of the lack of motivation of agricultural settlers who opposed the project's economic objectives that focused on the production of industrial crops, mainly cotton (Bertoncin & Pase, 2000). The overlap between economic development goals and local actors' interests has to be taken into account in the analysis of recent episodes of violence and rural insecurity.

The weak consensus among the agricultural settlers was also related to land tenure insecurity. In 1984, the Agrarian and Land Reorganisation (Réorganisation Agraire et Foncière) did not grant full ownership rights to the farmers because, in principle, the state remained the owner of all lands and had to grant ultimate land rights by certification, thus negatively affecting farmers' commitment to agricultural productivity. It should be noted here that the argument of the unproductive farmers was used in the national debate in Burkina Faso to support the arrival of new private players capable of exploiting irrigated plots in a more profitable way (Dialla, 2002).

Other destabilisation factors were related to internal migrations and the growing competition over natural resources. From the 1970s onwards, other Mossé farmers migrated from the central plateau of the country to settle close to the irrigated land in the valley. This influx affected the organisation of livestock movements at the regional level, further reducing pastoral areas and resulting in increasing land pressure. The arrival of agricultural migrants transformed the access rights to common-pool resources, which typically attribute control and precedence use to the first settlers while providing secondary access to latecomers according to local arrangements and cultural norms and permissions (Haller, 2016).

These dynamics were particularly strong in the lowlands used by transhumant pastoralists as important grazing areas during the dry season (Benoît, 1978) and lasted for several decades. It has been observed that the banana plantations along the river, which are detrimental to livestock farming, are mostly owned by agricultural migrants who arrived in the 1990s from the north and centre-north regions (Pellerin, 2021).

Land pressure was also exacerbated by the shift from nomadic to sedentary pastoralism among Peuhl migrants from northern Sahel and the consequent increase in the concentration of livestock in the Mouhoun region as part of the process of conversion to agro-pastoralism accelerated by the Sahelian droughts of the 1970s and the 1980s (Gonin & Tallet, 2012).

The Sourou water development projects were therefore implemented in the context of emerging competition over land resources. In 1978, the situation was described as follows:

> The right bank of the Sourou (…) is entirely occupied by Peuhl herds; on the left bank, which is densely populated, the Peuhls still have access to water and 'bourgoutières'. The Léri dam on the Sourou, which has now

> been completed, will have many repercussions on pastoral activities. The 'bourgoutières' will be submerged and therefore inaccessible to the herds.
> *(Bellot & Bellot-Couderc, 1978, p. 285)*

Since 2014, the situation of land insecurity has led to the emergence of armed self-defence militias at the village level, particularly in the northern regions. The aim of these fragmented and poorly coordinated groups, described in terms of a brigandage problem (Crisis Group, 2020), is to react to the government's inaction in the context of growing violence related to land access, pitting farmers against herders in many cases.

Other episodes of land insecurity and competition over key resources have been reported in Sourou Valley following the creation of biodiversity reserves that excluded groups of users. These incidents have occurred since 2009 after three sites in Sourou Valley were included in the list of the Ramsar convention.[5] It should be noted that these initiatives have been aimed at preserving and protecting a natural area that is socially constructed; the valley's biological diversity has been enhanced over time by the increase in water levels as a result of successive hydraulic infrastructures and human-made works.

Alongside biodiversity preservation,[6] these environmental protection measures (e.g., the restoration of the riverbanks) and the restricted access to protected sites were partly introduced to mitigate the effects of the rise in water levels in Sourou Valley. In fact, the construction of dams has led to the shrinking of key pasture areas and the transformation of the traditional transhumance system into the sedentarisation of mobile pastoralists. In this context, the preservation of wetlands imposes further constraints on local populations, in addition to the rigidity of hydro-agricultural projects.

Restoration projects and environmental protection initiatives have thus been implemented in the context of changing patterns of natural resource management and access to such resources, in turn contributing to heightened local tensions (Hubert, 2021). This point has been raised in recent research in other regions of north-west Burkina Faso, where protected natural areas have become inaccessible to local populations, forcing farmers, herders and fishers to support armed groups in order to regain access to the resources that they used to live on (Pellerin, 2021). In some cases, armed groups have forced farmers to allow livestock to graze on their cultivated fields. Other claims concern the access to crop residues for livestock after the harvest. In the Boucle du Mouhoun region, "systematic corruption" is said to be practised today to enable herders to settle for several months in protected forests during the rainy season (Pellerin, 2021, p. 112).

The Agropole model: From tenure insecurity to dispossession

The institutional reforms that led to the reorganisation of AMSV in Souroupole, a mixed public-private company, followed the 2008 price crisis and the risks of food insecurity due to the dependence on global food markets. In this context, Western

African countries introduced new policies to support local production of staple crops, mainly rice, as part of the public-private partnership programmes, such as the Feed Africa Strategy of the World Bank and the African Development Bank, to develop the industrial agricultural sector (Tyrou et al., 2023).

In Burkina Faso, the creation of the Souroupole followed the case of another semi-private agency named Bagrépole SEM, established in 2012 to manage a project worth US$133.7 million (Hauer & Østergaard Nielsen, 2020; Tyrou et al., 2023; Yaméogo, 2016). A similar process of economic reorganisation has also been observed in the fishing activity. New economic players such as the Périmètre aquacole d'intérêt économique and the Périmètre halieutique d'intérêt économique were constituted in 2004 and 2012, respectively, to develop commercial and small-scale fishing in Sourou Valley.[7]

The Agropole economic model was based on Perroux's (1950) model of "growth poles" and the assumption that industrial infrastructure located in one place and provided with adequate connections with regional and international markets can reduce production costs, foster agricultural productivity and scale up local economies. These industrial clusters or agglomerations can attract private investments and foster an economy of scale.

The economic development model, grounded on the reindustrialisation of European countries in post-World II was thus re-adapted to raise the competitiveness of African countries following the 2008 food crisis. The Agropole thus appears as a type of "travelling model", referring to the standardisation of development interventions that are duplicated from one country to another and are supposed to have "intrinsic properties independently of the various contexts where they are implemented" (Olivier de Sardan & Vari-Lavoisier, 2000, p. 10). This transfer of economic development models also responds to the interest in creating new investment opportunities for local and transnational actors and occurs in regions such as African semi-arid areas that were previously on the periphery of global markets (Gargallo et al., 2022).

This point is confirmed by an analysis of the economic development policies adopted by some national states to support the creation of new agropoles. In Burkina Faso, the reorganisation of AMSV in Souroupole was coupled with economic measures and favourable regulations, such as a tax and customs benefit; a special dispensation from the public procurement code for spare parts and consumables; and the application of preferential energy tariffs because of the high cost of the energy used in the irrigation system (Autorité de mise en valeur de la vallée du Sourou (AMVS]), 2018).

In the case of the Bagrepole mentioned above, a specific measure to attract foreign investments is the 100% exemption of corporate taxes over a 7-year period, even if the duration of the exemption may differ from site to site, ranging from 5 to

30 years. The targeted actors are mainly agribusiness firms because these measures favour external investors to the detriment of local farmers (Tyrou et al., 2023).

Other regulations include land tenure reforms aimed at easing both farmers' and agricultural enterprises' access to land. In the case of Sourou Valley, this process started in 2011 with the ratification of a governmental decree[8] that declared the investment areas as Zone d'Utilité Publique (ZUP). According to national laws, all irrigated areas are registered as state property, and farmers and agricultural enterprises are granted land titles and long-term leases, in turn enabling them to secure their plots. In March 2022, 1,498 land titles and 1,933 long-term leases were issued and handed over. However, a large part of the work of land demarcation and registration remains to be done to ensure compliance with the national regulations. These delays particularly concern the farmers affected by the planned developments of irrigated plots totalling over 2,240 ha as part of the MCC/MCA-BF project; the farmers' rights have been cleared while waiting for compensation based on the land-for-land principle (Souroupole, 2022). As of 10 June 2024, there has been no update on this situation. The failure to complete the process of land demarcation and registration can be explained not only by the inadequacy of the financial resources allocated by the state but also by the budgetary austerity imposed by rural insecurity, which has forced the state to redirect investments to other priority areas.

The same procedure of land dispossession was already applied in 1998 when the area around the Bagré dam was declared a ZUP. In this case, 1,800 ha of the first irrigated schemes were allocated to smallholders, while 1,500 ha of the first extension were allotted to agro-investors to ensure that their contribution would cover construction and maintenance costs (Coulibaly, 2019). According to a World Bank (2010) report on the Bagré economic pole, the legislation regulating land deals within a ZUP does not oblige the state to pay compensations to expropriated farmers because the land can be allocated to any investor "without further negotiations with/compensation to local populations" (Report Number AB5668). However, land conflicts and protests against the allocation of land to large-scale investors led to a change in this plan, and only 300 ha were allotted to agro-investors (Hauer, 2021). A new approach was introduced in 2011 – based on the diversification of cropping cycles and the extension of irrigated plots on higher lands that were reserved for agro-investors – because they could only be irrigated by pumping. This project became a "vanguard for further growth pole projects across the country" (Coulibaly, 2019).

In the Sourou valley, episodes of tensions and violence related to land dispossession in the framework of large-scale agricultural investments have been reported in Béléhédé village (commune of Tongomayel, Soum province), where agricultural schemes to develop the rice industry attracted Mossi and Peuls migrants from 2013 onwards. As a result, dozens of Peuls were expropriated from their land without compensation. Starting in 2016, terrorist groups took advantage of the economic and socio-political dissatisfaction of the Peuls to recruit from within this community and in particular among the expropriated individuals (Crisis Group 2020).

These are sedentary pastoralists who have experienced the expropriation of land by newcomers, the obstruction of transhumance routes, the cultivation of traditional pastures and the difficulty of accessing water courses and watering points for their herds. It is important to note that pastoralists are not the only group that has decided to protest in order to defend their rights and denounce this situation of oppression and injustice; other ethnic groups have been involved in clashes. For example, another study carried out in the region refers to ongoing land-related disputes between Mossé farmers and local authorities (Pellerin, 2021).

These considerations point out the role of large-scale agri-food investment schemes in causing land tenure insecurity, which may result in tensions and conflicts. The Agropole model should be re-evaluated in light of recent evidence (provided by scholars) on how titling programmes feed land conflicts and dispossession (Tyrou et al., 2023).

Another problematic aspect of the Agropole that contributes to exacerbating tensions at the local level is that this model fosters the concentration of economic activities within specific geographical boundaries. This approach echoes the process of rigidisation of resource use that characterises large-scale irrigation projects in wetland areas of the Sahel (García et al., 2022). For example, this process has been observed on the shores of Lake Chad in northeastern Nigeria, where the South Chad Irrigation Project fostered a new geographical organisation of productive activities, completely erasing the previous multiplicity of resource uses and introducing modern technologies, institutions and techno-scientific knowledge to control what entered and left its economic perimeter (Bertoncin & Pase, 2000). These measures to govern geometrically partitioned spaces stand out in contrast to the flexibility of resource management and access that is suitable for accommodating multiple resource uses and users at different times and seasons and is adapted to the Sahel's climate variability and the ecological dynamics characterising its wetland areas.

In the case of the Agropole model, the rigidity of territorial boundaries also contrasts with the permeability of borders in arid and semi-arid environments, where seasonal mobility across different ecological zones is a key economic strategy to make use of unevenly distributed resources and irregular rainfall patterns.

These considerations are completely neglected in the narrative of agropoles since the territorial separation of the irrigated perimeters from the surrounding environment is considered a necessary condition to increase agricultural productivity. In this regard, a World Bank (2011) document on Burkina Faso refers to the Bagrépôle project as an "island of success and promising cluster of initiatives and enclaves that can (…) pull along the rest of the economy". The Agropole model is thus considered an adequate strategy to shift "from low-input/low output subsistence farming to diversified and high income-enhancing agriculture" (World Bank, 2011, p. 33).

This problematic approach undermines the viability of the Agropole economic model, whose success is highly dependent on economic corridors connecting the production sites to regional markets. As the case of the Souroupole shows, this has proven to be a major weakness of this development model in the context of rural insecurity and violence.

The economic logic of land investment or how land becomes a resource

In Sourou Valley, the livestock farming system is generally extensive and practised by sedentary communities (Dafing, Samo, Mossi, Pana, etc.). It involves a few cattle (draught animals), small ruminants (sheep and goats) and poultry. Animals are generally left to roam during the dry season. During the rainy season, the children look after their family's livestock, which is only entrusted to a third party (mainly Peuhl herders) when the herd becomes large (Fig. 12.1).

Livestock farming is integrated with transhumant pastoralism as herds cross the region to reach watering points and pastures during the dry season (January–February). At this time, Peul herders from the interior of the country (the communes of Kassoum and Barani on the left bank of Sourou River and localities to the north of Ouahigouya on the right bank) as well as from outside the country (Mali) travel

Figure 12.1 Sheep and cattle looking for grazing resources in the Di perimeter.

Source: Moumouni Ouande, 2021.

with their herds (mainly cattle) to the Sourou River, the Mouhoun dykes and the Léry bridge for watering and grazing. These pastoralists return to their places of origin at the start of the rainy season.

Today, livestock is watered directly from surface water (rivers, lakes, ponds and dams) or from wells and boreholes. However, farmers and livestock herders in the valley are increasingly faced with problems of water quality due to surface water pollution caused by chemical fertilisers that are mostly used by farmers cultivating irrigated areas. Moreover, expanding farmlands results in the lack of grazing areas and the obstruction of transhumance corridors. Consequently, the majority of transhumant herders now access water points through fallow lands or fields after the harvest. These practices often lead to conflicts between farmers and herders because of the crop damage caused by the animals.

The social and economic viability of agricultural development interventions could have been strengthened through initiatives aimed at preserving the complementarity among different productive specialisations and reducing land pressure and competition over shrinking resources. For example, a crucial issue is the availability of fodder, which in Sourou Valley is of good quality but limited in quantity, particularly at the end of the dry season. Irrigated agriculture could provide fodder resources after the harvest, but this would require an economic model that considers the development of livestock farming and irrigated agriculture as mutually dependent.

Unfortunately, these considerations remain understated in the economic policy for Sourou Valley. According to some authors (e.g., Bin, 2009), water development projects and irrigated agriculture contributed to the eradication of the Sahelian tradition of economic complementarity and synergy between sedentary and nomadic people. This is even more striking in a region of Burkina Faso where pastoral areas have historically represented "a kind of connective tissue linking the major agricultural areas" (Izard, 1988, p. 316). These studies highlight the incompatibility between agricultural development schemes and the seasonality of agro-pastoral productive systems, particularly during the dry season, when the herds' need for water and grazing is greatest. The sustainability of these agricultural development projects has thus been jeopardised by economic policies that have marginalised livestock farming in irrigated areas, particularly by abandoning fodder crops.

This complementarity among different groups of users – based on the seasonal mobility of humans and livestock and social institutions regulating access to shared resources – continues to be marginalised in large investment schemes, following the constitution of Souroupole. These investments focus exclusively on the quantitative assessment of irrigated land, which is considered a major indicator of agricultural development, whereas little or no attention is paid to existing arrangements and local patterns of resource use among different groups.

In this regard, previous research on the drylands already pointed out the "discrepancies and contradictions in quantitative reporting" of land-related investments in arid and semi-arid environments (Pase et al., 2022, p. 78). Studies such as this cited one rightly consider the lack of detailed data about land allocation as an interesting piece of information in itself because it shows that the legitimation of agribusiness initiatives is grounded on the marginalisation of qualitative information about land use patterns and systems of land rights. In this way, the land is considered a homogeneous and undivided space, thus imposing a classification of types of lands that overlooks local knowledge and taxonomies (Pase et al., 2022).

The targeting of areas for investments is based on the evaluation of the economic potential of agricultural land in a given time and place, but this leaves behind multiple and overlapping rights among different groups of users. This is because different stakeholders, such as pastoralists who enjoy seasonal and intermittent rights over resources, are directly concerned (Haller, 2016; Pase et al., 2023). This issue concerns more particularly the changing and "fluid" dimension of wetlands and semi-arid lands, where ecological rhythms of floods and rainfall patterns shape productive activities and seasonal relationships among different groups over time. It illustrates the limit of a bi-dimensional representation of land census and the importance of considering the temporal dimension of land use patterns in wetlands and semi-arid environments, where resident agricultural communities share land with other groups at different times of the year. In these moving and changing socio-ecological environments, shifting and seasonal bundles of rights are key features that govern common-pool resources. For this reason, there is the need for a "topology" more than a "topography" of spaces in drylands to ensure "a plural, open, and ever-changing view of what is 'local'" (Pase et al., 2022, p. 88).

The marginalisation of local logic and practices of land use reflects the over-simplification of socio-ecological interactions and processes and is one of the pillars used in modernist ideology to make local systems more "readable" (Scott, 1998). In this way, land *becomes* a resource as it is bound to a fixed representation of time and space. This concept contrasts with the idea of locality as a socially constructed and changing condition that is not taken for granted but is constantly redefined by its relational and contextual dimensions (Appadurai, 1996).

In Sourou Valley, this simplification is also evident when one looks at the relationships between mobile pastoralists and sedentary farmers, as well as other groups of users, namely fishers. In fact, alongside local arrangements between farmers and herders to negotiate their access to pasture and water sources, land rights are subject to customary institutions regulating collective fishing practices and techniques in Sourou River and its tributaries. Available research (Ouedraogo de Charles and Somé, 2003) points out the cultural and social values regulating the interactions among local communities, particularly the importance of customary fishing as a means of asserting the authority of local leaders during sessions of collective fishing, and forging socio-economic relations among villages. This example shows the diversity of actors, temporalities and social logics governing the management of and access to shared resources.

Old and new patterns of vulnerability

The impact of the COVID-19 crisis in Sourou Valley was modest compared with urban centres, as verified by the former's lower contamination rate. Health risks in the region are mostly related to the transmission of diseases linked to water as a vector of pathogenic organisms. Villages located along Sourou River and in flood-prone areas, such as Toma-île, Wèrè, Touroukoro and Yayo, generally have the highest prevalence of intestinal parasitosis compared with villages located farther from the water streams (Oué and Sono) or in areas where irrigated fields have been implemented, such as Guiédougou and Niassan (Dianou et al., 2004).

However, during the COVID-19 pandemic, the government's restrictive measures delayed the delivery of agricultural inputs in rural areas (fertilisers, seeds and tools) and spare parts for machinery. The ban on mass gatherings and the social distancing measures also hampered the support provided by agricultural extension agents to local farmers. This situation exacerbated existing problems, such as the failure to respect the irrigation calendar and water turns in the irrigated perimeters, and caused a drop in agricultural yields.

The mobility of people and goods was severely affected by government restrictions. According to the data collected by O. Moumouni, the rural areas of Dî, Lanfiéra and Tougan registered an increase in transportation costs (according to 26.7% of the respondents) and a significant reduction in travel (55.2%) and in the flow of goods to markets (44%). The restrictions disrupted the export and import networks, and this impact was particularly severe for those markets of the valley (such as Di, Gouran, Niassa and Tougan) that were mostly dependent on Ouagadougou and Bobo Dioulasso trading centres. As Compaoré et al. (2022) pointed out in their study on two fisheries in Burkina Faso (Samendéni and Sourou Valleys), the government's measures to curb the spread of COVID-19 affected fishing activities by reducing catch quantities and fishing seasons by 1–6 months, leading to unemployment and loss of income for 22.16% of the fishers surveyed.

Agricultural production was also affected by the rise in oil and transport prices because of the need for oil that powers the motor pumps used to irrigate the plots. It should be noted that this is not a new problem. Increasing prices of oil and other external inputs, such as fertilisers, have been a major issue for Sourou's farmers since the development of agricultural schemes focusing on the production of rice, wheat and other varieties, replacing millet and maize, in order to meet the growing demand from national and international urban markets (Bin, 2009). Thus, local economies were further impoverished and weakened by the pandemic, but these are long-term processes that cannot be attributed exclusively to exogenous factors.

Before the pandemic, restrictions on mobility were caused by the insecure situation in rural areas. Due to roadblocks, local farmers were obliged to purchase agricultural inputs and supplies from local traders at higher prices or from those in Bobo Dioulasso and Ouagadougou. This problem, coupled with the lack of technical supervision by extension agents who were unable to reach the rural areas, contributed to a fall in agricultural yields between 2017 and 2021. From an estimated total agricultural production of 115,334.38 t in 2018, the production in Sourou Valley dropped by 8,834.98 t (−7.7%) in 2021, according to the official figures (Souroupole, 2022).

The livestock sector has also been negatively affected by rural insecurity and the increase in the cost of transporting animals for export, which in the Mouhoun region has risen by 250–300%, depending on the traded animal (Pellerin, 2021). In addition to the soaring oil prices, this trend reflects a broader situation of rural insecurity and the rise in the prices of other goods and services needed for livestock farming (animal feed and vaccinations).

In this context, older livestock keepers had been hit hard by the restrictions imposed on transhumance during the pandemic, while younger generations had been able to get around these rules more easily. Several factors may have contributed to the emergence of these practices. Among them, we can mention the crisis of social mobility and the high unemployment rate in rural areas. This has contributed to making young people more inclined to challenge the restrictions imposed by the ruling and elderly classes, which are often held responsible for the current precarious situation. The impact of restrictions on herd mobility has also been less severe in localities where the interdependence between farmers and herders remains strong, such as in the village of Tierkou, where "most farmers entrust their draught oxen to Peulhs pastoralists between January and April and sometimes until May, in return for being allowed to graze their own cattle on the fields after the harvest" (Pellerin, 2021, p. 154). Thus, following the COVID-19 pandemic, local farmers reactivated the local strategies they had previously adopted in the context of rising prices and structural constraints to livestock mobility. These practices are part of the relations of economic cooperation between sedentary farmers and mobile pastoralists.

Thus, in Sourou Valley, the recent crisis of the global pandemic, rural insecurity and terrorist attacks are amplifying and making more visible old patterns of vulnerability. This consideration also applies to the analysis of internal migrations and displacement. Following the abductions of workers, the destruction of equipment and the departure of the national police and GCRS officers from Dî in November 2021, many people left the communes of Dî, Niassan and Lanfiéra and moved to the cities of Tougan, Dédougou and Bobo-Dioulasso. By 30 April 2022, 80,938 IDPs had been registered in the Boucle du Mouhoun region.

At the same time, rural emigration to urban centres is not the simple result of terrorist attacks or land insecurity. Internal displacement is part of local mobility patterns due to the meagre agricultural income and limited yields of dry farming, partly caused by the excessive fragmentation of agricultural plots. During the dry season, seasonal migration (an average of one to two persons per family) to plantations in southern states, particularly the Ivory Coast, is a well-known phenomenon (Bertoncin & Pase, 2000).

These processes have been exacerbated by the recent scaling down of agricultural investments. The unemployment crisis brought about by the closure of development schemes is a complementary destabilising factor in Sourou Valley. Here, as well as in other areas in the Sahel region, the lack of economic opportunities, coupled with the situation of land insecurity, is likely to foster various forms of violence, including those committed by terrorist groups that exploit the uncertainty and precarity of rural populations (Olivier de Sardan, 2023).

Conclusion

In Sourou Valley, rural insecurity is challenging the existence of large-scale agricultural development projects. At the same time, these interventions have contributed to the fragilisation of rural economies and to the exacerbation of existing tensions and competitions among different groups. In this chapter, we have attempted to analyse this multilayered crisis by focusing on three issues.

The first issue concerns land insecurity, which is central to understanding the region's instability and uncertainty. Whereas economic policies have weakened the local institutions regulating access to land since the early development of irrigated perimeters, this problem also intersects with longstanding local processes related to agricultural migrations and changing demographic balance at the regional scale.

The second issue involves the inefficiency of large-scale agricultural investments, related not only to the high maintenance costs of irrigated perimeters and hydraulic infrastructures but also to the dependence on external inputs that was particularly striking during the COVID-19 crisis. In this context, economic policies in favour of private operators aiming to attract foreign investments have further contributed to the dispossession and pauperisation of local farmers, to the point of becoming the material and symbolic target of local protests against socio-economic inequality and injustice.

The third issue relates to the divide between agriculture and mobile pastoralism since it is key to understanding the factors leading to the spread of rural insecurity in Sourou Valley. The Agropole economic model amplifies the process of land fragmentation and is based on a representation of natural resources that fails to accommodate seasonal mobility as a key economic strategy in semi-arid environments and does not value the interdependence among productive specialisations to ensure peaceful coexistence among different groups.

Notes

1 The natural hydrological regime of Sourou River, working as a tributary defluent, is not an isolated case in Africa. Guiers Lake in Senegal and the Zambezi River crossing Namibia, Zimbabwe and Mozambique are similar cases (Koussoube, 2010, p. 92). Like the Sourou River, they are all characterised by a flat topography in the catchment area and a reversal in the direction of the river flow, depending on whether it is low or high tide.
2 The MCC-MCA is an American aid agency created in 2004.
3 CCE: Caisse Centrale de Coopération Economique (CCE), now the Agence Française de Développement (AFD); FED: Fond Européen de Développement; CNSS: Caisse Nationale de Sécurité Sociale; IDB: Islamic Development Bank; BOAD: Banque Ouest Africaine de Développement; BADEA: Banque Arabe pour le développement économique; BID: Banque Islamique de Développement; MCA: Millenium Challenge Account, FSD: Fonds Saudien de Développement; OPEP: Organisation des Pays Exportateurs de Pétrole.
4 In the context of armed confrontations and attacks on hydro-agricultural infrastructures, chemical fertilisers have been used to manufacture improvised explosive devices. As a result, the Ministry of Agriculture was obliged to suspend the distribution of fertilisers (Source: fieldwork of O. Moumouni).
5 These sites are: (1) the Sourou floodplain (21.157 ha), a 120-km cross-border site between Burkina Faso and Mali, which was classified in 2009; (2) the Mouhoun-Sourou confluence zone (23.3 ha), classified in 2017; and (3) part of the Forest Corridor (134.553 ha), classified in 2017.
6 In 1999, the inventory carried out by the Ministry of the Environment and Water identified 23 endemic plant species in the Sourou Valley. The floodplain is home to numerous animals, including 31 fish species and 60 species of wild birds.
7 The valley has a catch potential of 802 t per year or 10% of the national production-generated turnover of 545 million CFA francs per year. This makes it Burkina Faso's third-largest fishery centre, after the Kompienga and Bagré fisheries in the east (Somda et al., 2010; Sow, 2015).
8 Decree No. 2011-425/PRES/PM/MAHRH/MEF/MATD of 24 June 2011

References

Appadurai, A. (1996). *Modernity at large: Cultural dimensions of globalization*. University of Minnesota Press.

Autorité de mise en valeur de la vallée du Sourou (AMVS). (2018). *Étude de faisabilité pour la transformation du statut de l'AMVS, Mai 2018*. CGIC/Afrique International.

Bellot, J.-M., & Bellot-Couderc, B. (1978). Pour un aménagement du cours moyen de la Volta Noire et de la vallée du Sourou. *Cahiers d'outre-mer*, *123*, 271–286. https://doi.org/10.3406/caoum.1978.2866

Benoît, M. (1978). Pastoralisme et migration. Les Peul de Barani et de Dokui (Haute-Volta). *Études rurales*, *70*, 9–49. https://doi.org/10.3406/rural.1978.2390

Bertoncin, M., & Pase, A. (2000). Le projet hydraulique dans la vallée du Sourou. In P. Faggi & P. Mozzi (Eds.), *La territorialisation hydraulique dans la vallée du Sourou (Burkina Faso). Lignes pour la recherche* (pp. 29–38). Materiali/Dipartiment di Geografia, Università di Padova.

Bethemont, J., Faggi, P., & Zoungrana, T. P. (2003). *La Vallée du Sourou (Burkina Faso). Genèse d'un territoire hydraulique dans l'Afrique soudano-sahélienne*. L'Harmattan.

Bin, S. (2009). La vallée du Sourou (Burkina Faso), entre agrobusiness et autonomie locale. *Géocarrefour*, *84*(1–2) (Les grands projets hydrauliques et leurs dérives), 73–82. https://doi.org/10.4000/geocarrefour.7241

Compaoré, I., Sanogo, S., Djerma, A., Zongo, B., Nacro, H. B., & Kabre, T. A. (2022). Étude comparative des effets du COVID-19 sur les activités des pêcheurs dans deux pêcheries au Burkina Faso. *International Journal of Biological and Chemical Sciences*, *16*(2), 798–811.

Coulibaly, N. (2019, December 1). *Burkina Faso: Mise en service du barrage hydroélectrique de Samendéni*. Jeune Afrique. Retrieved May 31, 2024, from https://www.jeuneafrique.com/863168/economie/burkina-faso-mise-en-service-du-barrage-hydroelectrique-de-samendeni/

Crisis Group. (2020, February 24). *Burkina Faso: Sortir de la spirale des violences*. Rapport Afrique N° 287.

Dialla, B. E. (2002). La sécurité foncière sur les périmètres irrigués du Burkina Faso: Le cas de la vallée du Sourou. *Africa Development*, *27*(1–2), 62–83.

Dianou, D., Poda, J.-N., Savadogo, L. G., Sorgho, H., Wango, S. P., & Sondo, B. (2004). Parasitoses intestinales dans la zone du complexe hydroagricole du Sourou au Burkina Faso. *Vertigo*, *5*(2).https://doi.org/10.4000/vertigo.3369

Ferguson, J. (1994). *The anti-politics machine. Development, depoliticization and bureaucratic power in Lesotho*. Cambridge University Press.

García, A. K., Pase, A., Haller, T., Artur, L., Lisboa, S. N., Mętrak, M., & Suska-Malawska, M. (2022). Wetlands in drylands: Large-scale appropriations for agriculture, conservation, and mining in Africa. In A. K. García, T. Haller, H. Van Dijk, C. Samimi, & J. Warner (Eds.), *Drylands facing change* (pp. 53–72). Routledge.

Gargallo, E., Haller, T., Chatty, D., Weissman, S., Meessen, H., Giger, M., & Chkhobadze, N. (2022). The new green grabbing frontier and participation: Conserving drylands with or without people. In A. K. García, T. Haller, H. Van Dijk, C. Samimi, & J. Warner (Eds.), *Drylands facing change* (pp. 132–252). Routledge.

Gonin, A., & Tallet, B. (2012). Quel avenir pour l'élevage dans le bassin cotonnier de l'Ouest du Burkina Faso? Dynamiques agro-pastorales et recompositions territoriales. *Autrepart*, *4*, 95–110. https://doi.org/10.3917/autr.060.0095

Haller, T. (2016). Managing the commons with foods: The role of institutions and power relations for water governance and food resilience in African foodplains. In T. Terje, O. Terje, & B. Jostein (Eds.), *Water and food: Africa in a global context* (pp. 369–397). *A history of water: Water and climate change* (Vol. 3). I. B. Tauris.

Hauer, J. (2021). Future-making in Burkina Faso: Ordering and materializing temporal relations in the Bagré Growth Pole Projec. *Geographica. Helvetica*, *76*, 163–175.

Hauer, J., & Østergaard Nielsen, J. (2020). Making land use change and markets: The global-local entanglement of producing rice in Bagré, Burkina Faso. *Geografiska Annaler: Series B, Human Geography*, *102*(1), 84–100.

Hubert, N. (2021). The nature of peace: How environmental regulation can cause conflicts. *World Development*, *141*, 1–12.

Izard, M. (1988). Youssouf Diallo, Les Fulbe du Boobola. Genèse et évolution de l'Etat de Baroni. *Journal des africanistes*, *68*(1–2), 315–317.

Koussoubé, Y. (2010). *Hydrogéologie des séries sédimentaires de la dépression piézométrique du Gondo (bassin du Sourou): Burkina Faso/Mali* [thèse de doctorat unique en Hydrologie, Université Pierre et Marie Curie – Paris VI].

Leprun, J.-C. (1979). *Les cuirasses ferrugineuses des pays cristallins de l'Afrique occidentale sèche* [Thèse de doctorat, Université de Strasbourg].

Mosse, D. (2004). Is good policy unimplementable? Reflections on the ethnography of aid policy and practice. *Development and Change*, *35*(4), 639–671.

Olivier de Sardan, J.-P. (2023). *L'Enchevêtrement des crises au Sahel: Niger, Mali, Burkina Faso*. Karthala Editions.

Olivier de Sardan, J.-P., & Vari-Lavoisier, I. (2000). Introduction: Pour une approche comparatiste des modèles voyageurs. *Revue internationale des études du développement*, *248*, 7–28. https://doi.org/10.4000/ried.280

Ouedraogo de Charles, F., & Somé, H. (2003). Lutte contre la vulnérabilité alimentaire par l'appropriation de l'eau au Burkina: Cas de Ramigtenga. *Cahiers CERLESHS, 2e numéro spécial*(Septembre), 94–107.

Pase, A., Hofman, I., García, A. K., Haller, T., Cirillo, D., Giger, M., & Bertoncin, M. (2022). Large-scale agricultural investments in drylands: Facing some blind spots in the grabbing debate. In A. K. García, T. Haller, H. Van Dijk, C. Samimi, & J. Warner (Eds.), *Drylands facing change* (pp. 73–92). Routledge.

Pellerin, M. (2021). *Entendre la voix des éleveurs au sahel et en Afrique de l'ouest: Quel avenir pour le pastoralisme face à l'insécurité et ses impacts?* Réseau Billital Maroobé et partenaires.

Perroux, F. (1950). Economic space: Theory and applications. *The Quarterly Journal of Economics*, *64*, 89–104.

Scott, J. (1998). *Seeing like a state: How certain schemes to improve the human condition have failed.* Yale University Press.

Société de développement intégré du pole de croissance de la vallée du Sourou. (2022). *Etat de mise en oeuvre du pole de croissance du Sourou, rapport annuel d'activités,* Souroupole Sem.

Société de développement intégré du pole de croissance de la vallée du Sourou. (2023). *Rapport de mise en œuvre des activités 2022 de Souroupole au 31 décembre 2022, rapport annuel d'activités*, Souroupole Sem.

Somda, J., Zonon, A., Ouadba, J.-M., & Huberman, D. (2010). *Valeur économique de la vallée du Sourou: Evaluation préliminaire.* Ouagadougou, Burkina Faso, Bureau Régional de l'UICN.

Sow, J. (2015). *Genre et pêcherie au Burkina Faso: Cas des villages de Di, Niassan et Gouran dans le Sourou, Université Professeur Joseph Ki-Zerbo (ex Ua).* Université Nazi Boni.

Toutain, B. (1987). Potentialités pastorales du Nord Burkina Faso. In *Élevage et potentialités pastorales sahéliennes. Synthèses cartographiques* (pp. 1–10). Burkina Faso, CIRAD.

Tyrou, E., Soullier, G., & Coulibaly, M. (2023). Unpacking policies for the development of agricultural growth poles in West Africa. *Canadian Journal of Development Studies/ Revue canadienne d'études du développement*, *44*(4), 554–574.

Wolford, W., Borras, S. M., Jr, Hall, R., Scoones, I., & White, B. (2013). Governing global land deals: The role of the state in the rush for land. *Development and Change*, *44*(2), 189–210.

World Bank. (2010). *Burkina Faso – Bagre Growth Pole Project (English).* World Bank Group. Retrieved May 31, 2024, from http://documents.worldbank.org/curated/en/774001468019806803/Burkina-Faso-Bagre-Growth-Pole-Project

World Bank. (2011). *Project appraisal document on a proposed grant in the amount of SDR 71 million (US$115 million equivalent) to Burkina Faso for a Bagré Growth Pole Project.* Retrieved May 31, 2024, from http://documents.worldbank.org/curated/en/581451468020361287/Burkina-Faso-Bagre-Growth-Pole-Project

Yaméogo, L. (2016). *Territorialisation hydraulique et développement local autour du lac de Bagré (Burkina Faso)* [Thèse de doctorat, Université de Padova, Italy].

Zoungrana, T. P., Bethemont, J., & Faggi, P. (2009). De l'échec d'un projet à l'émergence d'un territoire: La vallée du Sourou (Burkina Faso). In H. Ayeb, & T. Ruf (Eds.), *Eau, pauvreté et crises sociales* (pp. 458–475). IRD Editions.

Part IV

Connections, Rights and Conflicts

13 Land and water rights in the Inner Niger Delta

Between negotiation and conflict

Luca Pes and Ibrahima Poudiougou[1]

Introduction: A fluid environment

The area commonly known as the Inner Niger Delta (hereafter, the Inner Delta) is a flat territory covering a substantial portion of central Mali, characterised by a highly dynamic and ever-changing environment, where the interactions between land and water shape the landscape and the livelihood of the people inhabiting its territory.

From an ecological viewpoint, the most prominent feature of the Inner Delta is its seasonal flooding cycle. During the rainy season, which typically occurs from June to October, heavy rainfall in the upper Niger River basin causes the Niger and Bani Rivers to swell, overflowing their banks and inundating the floodplains. This flooding transforms the delta into an expansive wetland, covering vast areas with water and creating a mosaic of channels, lakes and marshes. As the dry season progresses, water levels gradually recede, exposing fertile soils and allowing livestock grazing and agricultural activities to take place.

This is a "fluid" landscape for two reasons: it is dominated by the water of the river (the fluid substance par excellence), and the main "natural resources" employed in productive relations in the locality are under constant change and transformation. Spaces such as fishable waters, grazing land and farming land have no fixed and stable consistency.[2] Productive relations among specialised groups of fishers, herders and farmers vary accordingly. Despite the western and developmentalist jargon of "natural resources", for the inhabitants of the Inner Delta, both land and water are far from being objects of the natural world.

First of all, the exploitation of water and land requires mobilisation of work – which is an inherently social object – involving relations among people rather than a "thing" (Mundy & Pottage, 2009). This fact is particularly evident in social settings characterised by low intake of technology, such as the Inner Delta.[3]

Secondly, farmland, grassland and fishable waters do not belong to the same kind of objects, as the term "resource" implies in western economic discourse, because they represent a symbolic dimension in social relations that is peculiar to the

DOI: 10.4324/9781003429111-19

environment of the Inner Delta. This characteristic has particular relevance in the domain of political relations, where control of land and water stands as the symbol of legitimate authority (Pes, 2016), and in the religious sphere, where land and water are imagined as containing invisible forces (*djinns*, ancestors, genies of the bush, etc.). From this perspective, land and water represent sacred objects and the means of access to the supernatural world.

In the particular landscape of the Inner Delta, farmland, grassland and fishable waters are extremely *fluid* objects, not fixable in any permanent way in space and time. The authority of customary chiefs over such resources and, more generally, the overall legal order that we are setting out to explain in this chapter take a correspondingly fluid character. The extension of available grassland (just like the quality of the grass and its endurance in time) depends on the many and highly variable environmental factors influencing the behaviour of the flood during the dry season, when the withdrawal of waters fosters the growth of fresh grass. If the flood has not been particularly abundant, more farmland rather than grassland will be available; this will change the opportunities for the authorities in charge of allocating rights of access to these resources, as well as the power balance among groups practising different forms of livelihood.

The fluidity of the objects of the political authority of customary chiefs is particularly apparent when fishing communities are observed. The image is conveyed vividly in French, where the term "fluidity" (*fluidité*) bears the same etymology as *fleuve* (river), the same root as the verb "to flow" in English. In fact, the most fishable areas are bends of the river, ponds and flooded plains that dry gradually during the season and convert to grazing or agricultural land, depending on environmental factors. Hence, rights on the fishable waters of the river (*fleuve*) are *fluid* by definition. In a metaphorical sense, this is also true for farmland and grassland.

The holistic authority of customary chiefs and the collection of "access fees"

Historically, three figures of customary chiefs have been central to the control of land and water in the Inner Delta (Gallais, 1984).[4] Far from being a thing of the past, these extremely resilient institutions have been reconfigured and redeployed over time, adapting to the socio-legal registers created by successive forms of political domination: from the Macina Empire in the 19th century, through the French colonisation, to present-day independent Mali.

The first of these authorities is called *dugukolo-tigi*. This is the chief of the land in communities whose livelihood is based on semi-subsistence farming, growing mainly dry crops (finger millet, maize and sorghum) and flooded crops (mainly rice). Signalling the relevance of farming as the material basis of the very existence of the community, his name translates as the master (*tigi*) of the bones or "skeleton" (*kolo*) of the village (*dugu*).[5] His office normally coincides with that of the head of the community of people inhabiting the same village (*dugu-tigi*).[6] He is also the master of the sacrificial practices (*murukala-tigi*, literally the holder of the sacrificial knife) honouring the sacred pact between the first settlers and the genies of the bush, on which the community is founded. Sacrifices ensure the fertility of the land and protect it from drought, harmful rains and attacks by crop-destroying insects.

In the first half of the 19th century, under the political domination of a precolonial state called Macina,[7] the *dugukolo-tigi* lost some of its land-controlling powers, to the advantage of another character called *jooro* (pl. *joroo'en*). In fact, the ruler Sékou Amadou (1776–1845) had divided the territory of the Macina into agro-pastoral units called *leydi*[8] by appointing noblemen from the sedentary Fulani communities of the Inner Delta as local authorities responsible for the distribution of both pastoral and farming lands. Significantly for the future development of land relations in the locality, this aristocracy comprised literate men claiming direct descent from the Prophet (*Sharif*).[9]

The counterpart of the land chief for a cattle-herding community is the *sila-tigi*, literally the master of the transhumance path (*sila*). An institution typical of semi-nomadic Fulani communities, the *sila-tigi* is an elderly man leading a group of cattle-herders at the start of transhumance. He possesses detailed knowledge (both spatial and social) of transhumance paths and corridors. He is also the minister of sacrifices necessary for the opening of transhumance and other pacts that bind the community of herders and their cattle to the spirits of the bush (Hampaté Bâ, 1991). As a master of the bush, the knowledge of the *sila-tigi* includes the fauna and flora, as well as the invisible forces populating the bushland. He protects the herders and their cattle from wild animals by leading them to safe corridors, rich in grass and water.

Finally, among the fishing communities inhabiting the Inner Delta,[10] control of the waters and the distribution of fishing rights are the responsibilities of a third authority called *ji-tigi*, literally the master of waters. A character corresponding to the customary chiefs of farmlands and transhumance paths, the *ji-tigi* is an elder representative of a community practising fishing; he is credited with knowledge about both territorial and social boundaries of such economic activity in the Inner Delta. His powers include negotiation of the relationships with neighbouring communities and the distribution of fishing rights, both within the community and among foreigners who can be "hosted" (i.e., allowed to fish) in varying numbers, depending on the season and the level of the flood. Similar to the other authorities, the *ji-tigi* also plays an intermediary role between the visible and invisible forces inhabiting the river.[11]

Overall, the legitimacy of customary chiefs lies in a kind of "holistic" knowledge of both the material and the mystical worlds. In fact, chiefs are called to master intertwining dimensions of social relations involving the management of natural resources (ecology), the relationships with other groups (politics) and the settlement of disputes (law). In this respect, land and water rights must be understood as a hybrid set of rules, not only confined to the settlement of disputes, but also extending to ecological, mystical and socio-political knowledge.[12]

The authority of customary chiefs is substantiated in their exercise of two main powers. On one hand, they are in charge of the distribution of rights, held collectively by the community, to the various lineage-based subgroups and eventually to foreigners: groups sensed as not permanently bonded to the community, mainly on the premise of common descent.[13] On the other hand, customary chiefs have the power to impose monetary fees in return for the exercise of such rights, especially when these rights are allocated outside the community. The *jooro* is an emblematic figure in this regard.

Joroo'en collect payments, called *conngi*, from transhumant herders grazing their cattle on the territory of the *leydi* (Barrière & Barrière, 2002), as shown in Fig. 13.1.[14] The amount paid per herd is set by the *jooro* and can vary significantly according to objective criteria, such as the quality of the grass between different grazing areas and moments of the season. Crucially, however, it also reflects subjective considerations, including the quality of the party, as some groups will be treated as closer than others. This may well be combined with the personal strategies and private interests of individual *jooro'en*. Originally paid in kind,[15] the *conngi* has been monetised as a result of processes of commodification tracing back to colonial times. For some *joroo'en* today, these payments clearly constitute a means of personal enrichment, in complicity with state agents (Jourde et al., 2019).[16]

A closer observation of payments made to water chiefs, in parallel to the above-mentioned *conngi*, provides a better understanding of their function in contemporary Inner Delta society.

Besides its mystical role of maintaining the pact with the water spirits and its economic function of distributing water rights to the various subgroups in the community, the *ji-tigi* also bears the more mundane office of collecting payments called *manga-ji* from foreign fishers (Fay, 1989). These should not be misinterpreted as "access fees" in a strictly economic sense ("access to fish" in return for money or kind) because the payment of *manga-ji* engages exchanges of a more complex kind.

The negotiation of the *manga-ji* clearly involves the symbolic domain of political relations, whereby a group of foreigners recognises the political authority of the "autochthonous" group (that is *different* from saying that it recognises the "ownership" or control of the resource). For both groups, the negotiation of the fee bears an importance greater than the price.[17] In fact, it is the guarantee of a lasting relationship, able to adapt to changing ecological, as well as socio-political circumstances. It provides the opportunity for both parties to mobilise unwritten, cultural codes of behaviour fostering a sense of common identity, renewing their alliance and commitment to future exchanges. Overall, the relationship between the payers and the recipients of these payments is more of a mutual and cooperative kind, rather than exploitative and marked by conflict; it is part of an exchange, bearing both material and symbolic implications (Godelier, 1984).

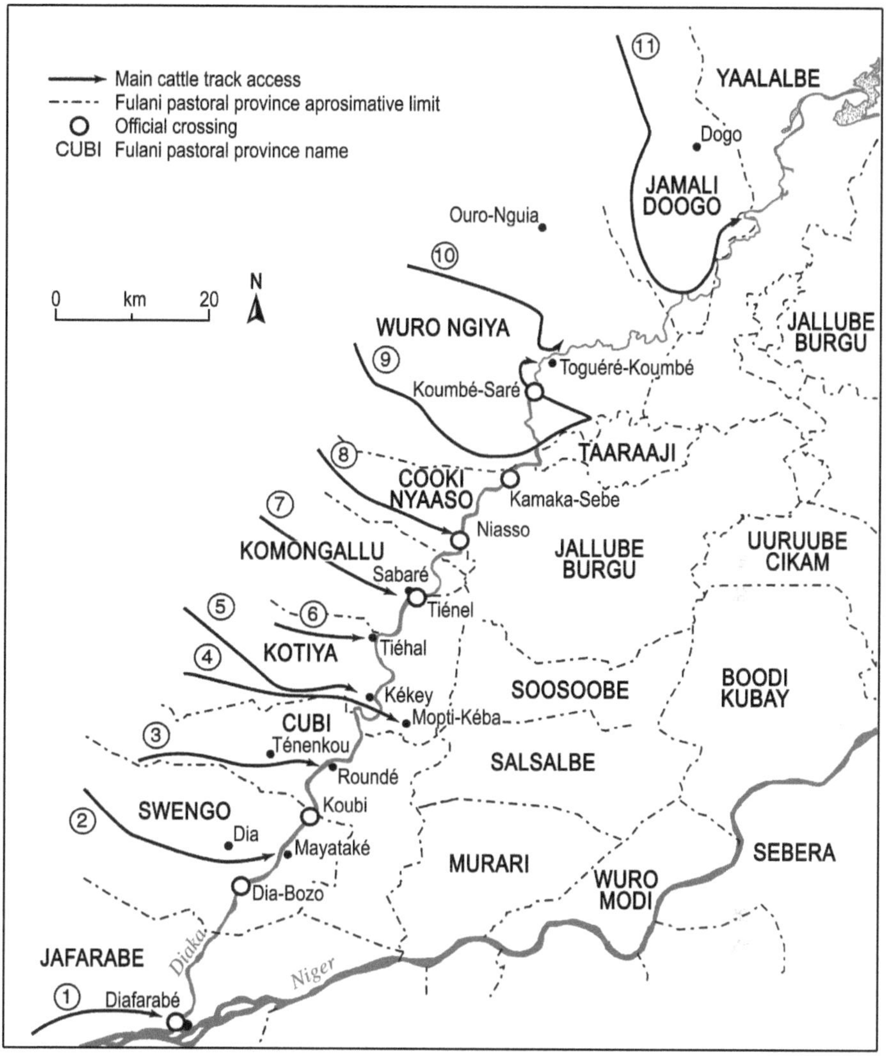

Figure 13.1 Main historic livestock access routes and leydis.

Source: Based on Legrosse (1997), redrawn by F. Ferrarese. Courtesy: Karthala.

State justice in action

In Mali, state judges are well skilled in balancing their power to enforce positive law with a seemingly conciliatory attitude aimed at convincing the social milieu of the parties that a decision is mutually beneficial to them, representing at least an acceptable solution. Should the judges fail to adopt such a pragmatic attitude, their judgements are likely to remain a dead letter (at best) or to constitute the premise

Figure 13.2 Headquarters of the juge de paix à compétence étendue, January 2017.
Source: © Ibrahima Poudiougou.

for the resurgence of the conflict over time or even for its enlargement to wider communities (lineage groups, villages, etc.).

In an interview held in January 2017 at its headquarters in Djenné (see Fig. 13.2), the judge of first instance (*juge de paix à compétence étendue*) emphasised in these terms the hybrid nature of the judicial decision-making process typically occurring in land-related matters:

> In land disputes, there are situations where the parties ask to swear on the Koran or to go and drink water from a sacred pond. If you swear on these types of things, it is assumed that the liar dies. It does not matter whether the water from the pond kills or not. That is not the important point. But psychologically, the person who is wrong will not agree to drink the water from the sacred pond. Custom therefore allows the truth to be revealed (…). The law recognises the custom of the parties. So it is the custom of the parties that I am enforcing by adopting this practice.

On one hand, judges acknowledge the existence of customary mechanisms of discovery overlapping with state-sponsored rules of judicial procedure and the need to take these into account to reach an effective decision. On the other hand, judges adopt a pragmatic attitude towards customary authorities exerting their influence on the dispute. They know that if they want a decision to be endowed with some effectiveness, they will have to face traditional leaders and notables. Thus, judges will normally convene these authorities as witnesses, knowing the customs of the parties, with the aim to find a concerted decision. In this respect, the judicial resolution of land disputes requires negotiations between different and at times conflicting legal orders, such as national state law, the vestiges of Islamic law and local community rules, which in legal jargon are rendered as the "custom of the parties".

Mediation processes are crucial in preliminary phases of the trial. When judges are called on to settle a dispute, the first action they will take is to refer the parties to an out-of-court authority that may help the parties find a negotiated solution. While in other parts of Mali, such negotiations normally take place in the vestibule of the village chief,[18] in Djenné, people are invited to turn to a body called the High Islamic Council. Presenting itself as the local branch of a national association of Islamic scholars bearing the same name (*Haut Conseil Islamique* [HCI]), the composition of this body, at least when it takes on this function of mediation of disputes, is actually the same as one would find in the vestibule of the village chief, integrated with some local Islamic scholars.

While this attitude may be misinterpreted as the judge's attempt to get rid of the case, it actually constitutes the first effort to ground an eventual judicial settlement of the dispute into the societal milieu of the parties by acknowledging that any non-negotiated decision made solely on the basis of state law would break up community ties. This caution finds a special justification in the field of land relations, where disputes touch on material and symbolic resources crucial for defining the relationship between groups. The underlying idea of justice is one aiming at compensating and reconciling the parties to restore the dignity of each of them while maintaining a common sense of legitimate political order.

If the parties find a negotiated solution in front of the HCI, the latter will draft written minutes of the agreement and send the document to the judge in a somewhat more bureaucratised fashion compared to the customary form of mediation taking place in the vestibule of the village chief. Then, the judge will convene a public hearing in order to validate the agreement and make it publicly accepted. In fact, the hearing will be attended by a number of witnesses, including the customary chiefs of land, transhumance and water, depending on both the subject of the dispute and the communities involved.

In case of an unsuccessful outcome, the judge is expected to directly investigate the case by hearing testimonies and taking other means of evidence. Even in this case, some external Islamic scholars are likely to be involved in the decision. These characters are called *alkali* in the vernacular, a possible distortion of the Arabic term *cadi*.[19] They are literate men, normally members of the local HCI, in charge of ascertaining the validity of some documents called *tarikhs*.

Literally meaning *narrative* or *history* in Arabic,[20] a *tarikh* is a written document issued by an Islamic scholar (*cadi*) expressing a judicial decision or attesting to the legality of a transaction (see Fig. 13.3).[21] In the Inner Delta, a *tarikh* is frequently employed as some sort of land certificate, becoming a central piece of evidence in a land dispute when it is necessary to document the chain of passages of certain rights over time.

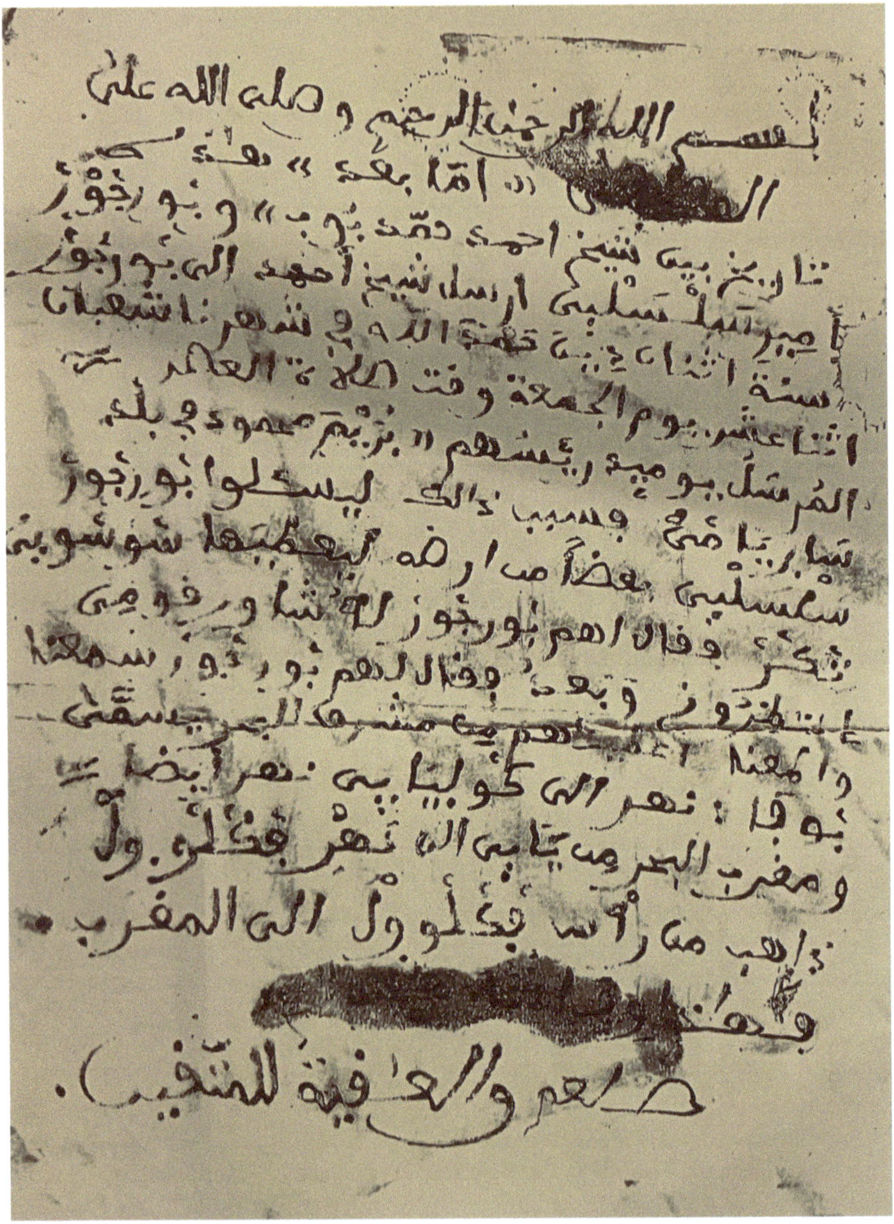

Figure 13.3 An example of a tarikh reproduced here for illustrative purposes (concerning another dispute than the one discussed in the text).

Source: © Ibrahima Poudiougou.

In practice, a *tarikh* is something less than a land title, requiring a huge administrative apparatus mapping the land and keeping a land registry, but it is something more than a private document between the parties.[22] The fact that a *tarikh* is issued by the trusted authority of a *cadi* provides the parties with a public-like certification of the overall legality of the transaction, for example, the parties did lawfully hold the transacted rights, the agreed obligations were fulfilled and so on.

Not only are these documents written (or transcribed) in Arabic – a language that state judges schooled in French do not understand – but the interpretation of their content and, most importantly, the ascertainment of the reputation of their drafters require the technical knowledge of Islamic legal scholars.[23] The more a *tarikh* is ascertained as reliable by the *alkali*, the higher will be its consideration by the judge, to the point of often being the conclusive piece of evidence.

Coupled with the strategic use of such documents by the parties of a legal dispute (Derrider, 2021, p. 35), the involvement of *alkali* reveals the hybrid character of the administration of justice in the locality and the vestiges of an Islamic legal order that can be traced back to the Macina Empire. However, it is worth observing that in official state regulations, the *alkali* figures as a "technical assistant" of the judge (*assistant technique*), as if his job were that of a translator rather than an expert consultant assessing the reliability of a decisive piece of evidence, according to a set of interpretative techniques and of both scholarly and local knowledge that the judge does not possess.

Such an understatement of the decisive role of the *alkali* is well reflected in the words of the judge:

> In Djenné, there are tarikhs (…). These are writings that the parties bring to the trial. These are texts in Arabic that are brought to the judge as proof of ownership of a given plot of land. We don't have the skills to read these documents. I don't read Arabic and I don't understand it. So I ask a marabout to help me. It's this person who will interpret the documents for the judge.[24]

Adding further complexity to the intensely mediated judicial settlement of land disputes in the Inner Delta, the Agricultural Land Act of 2017 (*Loi agricole*) introduced further innovation.[25] The act provided for the establishment of rural land commissions called *Commissions foncières locales* (hereafter, *CoFos*), including customary chiefs.[26] These bodies have two main tasks, both relevant to our discussion: the issuance of "certificates of customary land ownership" (*certificats de detention coutumière*) and the mediation of land disputes.

On the first ground, the practical advantage of the certificates issued by *CoFos* is their probative force. Like the *tarikhs*, they constitute evidence in judicial disputes because they derive from a public registry kept by the local municipality, called *registre de possession foncière des terres agricoles*. As agencies of mediation of land disputes, *CoFos* may be addressed by the judge in the pretrial phase, similar to the HCI. Overall, these land commissions represent a negotiated return of state institutions to the locality and an attempt by the public administration to capture and secularise the customary mediation of land disputes.

Insecurity

Following the insurgency of armed groups in the north of Mali in 2011,[27] the country entered a state of insecurity, rapidly spreading throughout the Sahelian region, instigating foreign military intervention and significant political instability, as testified by numerous coups.[28] Between 2015 and 2019, the Inner Delta became one of the hotspots of the security crisis in the Sahel, with predictable upheavals and disruption within the judicial system.

As a consequence of military operations in the north, since 2013, fighters of the Movement for Unity and Jihad in West Africa (MUJAO) have been displaced to the central regions of Mopti and Ségou (International Crisis Group, 2016). Reorganising their forces on new grounds, these fighters gave birth to a constellation of groups under various denominations, bringing armed insurgency to the Inner Delta. An organisation called Katiba Macina, a name significantly evoking the glory of the Macina Empire, was founded under the leadership of Fulani preacher Amadou Kouffa (Cissé, 2018).

Initially based on the idea of "holy war" and on the establishment of a new political order inspired by *sharia*, the propaganda of jihadist groups such as MUJAO readjusted gradually to the local context, targeting the inequalities between groups in accessing agricultural and pastoral lands and denouncing the privileges of the elite, particularly of the Fulani aristocracy embodied by the *jooro'en*. In jihadist political discourse, the latter's members were depicted as corrupt accomplices of state bureaucrats and officials, while the institutions of the secular state of Mali became the foes.

This initial phase had been functional in recruiting fighters among the marginalised sections of the population, including people of slave descent (*rimaybe*), the poorest herders and farmers. Interestingly to our analysis, this move represented the first step in putting land relations at the centre of the confrontation between jihadist groups and the secular state.

The second step was made when self-defence militias were mobilised, often along ethnic lines, with the aim to protect the interests of the various groups inhabiting the Inner Delta, thus filling the vacuum left by the shrinking of state institutions. By connecting to existing tensions over land relations between groups, the conflict seemed to have taken a definitive root in the locality.

Facing the withdrawal of regular armed forces, self-defence groups were formed to protect both people and goods. Targeted assassinations of non-aligned leaders (*jooro'en*, Islamic scholars and village chiefs) became the trademark of jihadist attacks, while rustling was soon established as the most important source of income for both factions (Benjaminsen & Ba, 2019; Poudiougou & Zanoletti, 2020). For example, a group called *Dan Nan Ambassagou* was founded in 2016 to protect Dogon villages against jihadist incursions (Cissé, 2018; Poudiougou, 2024). Clashes between the factions in central Mali reached their apex between 2018 and 2019.[29]

Particularly interesting to our analysis was the propaganda that Katiba Macina had turned against the administration of justice, accusing state judges of collusion with the *jooro'en* for imposing heavy fines to cattle herders. A judge was kidnapped in Niono in November 2017, court buildings were set on fire and judicial personnel were threatened via phone calls and messages.

Between 2016 and 2019, the majority of Djenne's civil servants had retreated to safer regional centres such as Sévaré and Ségou or to the capital Bamako. During the period of our enquiry (2016–2017), the only staff member that remained in place at the local tribunal was the registrar of the court (*Greffier en Chef*). The judge had taken refuge in the capital Bamako; thereafter, judicial activity was suspended. The registrar coped with the administrative business (e.g., the issuance of certificates of nationality) and the registration of claims, which understandably could not be directed and investigated any further. The civil servants remaining in place were mostly schoolteachers and some personnel of the local district office (*préfecture*), who were natives of Djenné or did not have a place to escape. The prefect himself came to town only once a month.

Addressing this vacuum in the administration of justice provided the opportunity (and also the challenge) for Katiba Macina to change its modus operandi from threatening armed incursions to dealing with the administration of civilian affairs. Legal disputes were now referred to religious scholars, *imams* and other local dignitaries, while more important cases were referred to a *cadi* appointed by Katiba Macina among the Islamic scholars who were seemingly more compliant with jihadist presence. More generally, people were threatened to not address any claim to state institutions, such as the police, the *préfecture* and the judiciary, and were cut off from public services. For example, the villagers of Mourra and Kouakourou had been isolated by the regional centre Djenné for a four-year period between 2015 and 2019.

On more general socio-economic grounds, the promise of "free access" to grazing land for the underprivileged soon had to meet both practical and symbolic needs. On a practical basis, hundreds of thousands of heads of cattle entering the Inner Delta at the same time obviously required some administration. An "entrance fee" of about 200,000 Fcfa per herd started to be tolled upon entrance, together with the payment of the mandatory charitable contribution of *zakat* when leaving the area. At the symbolic level, the collection of taxes, similar to the administration of justice, constituted a simulacrum of legitimate authority, a public display of "who rules the place". However, on both grounds, the "administrative structure" of Katiba Macina's rule did not advance much further and had to rely on local authorities.

By 2019, Katiba Macina had reinstated most of the Fulani elite apart from too-vocal opponents, who had been killed or sidelined. This move coincided with the reintroduction of customary rules in administering land and water rights, including the payment of access fees to the *jooro'en*. However, it also led to internal discontent within Katiba Macina and to clashes between two factions of fighters: those who were native from the Inner Delta, who generally joined the insurgency at a later time; and those coming from dry regions like Seeno, having joined at an early stage (in 2015 and 2016), in response to the cry of "land and water belong to God"' and not to hereditary chiefs (Baldaro & Diall, 2020; Bouju, 2020). In short, the people who embraced the armed struggle for opportunistic reasons, seeking free access to natural resources and aspiring to a better social status, were disappointed.

From 2020 onwards, the intensification of the Malian army's military operations in central Mali led to the return of most state officials and local authorities. The return of the prefect and the judge to their offices in the administrative centre of Djenné had a noticeable effect. On the other hand, it was not fully effective in rural areas because the *gendarmes* were still unable to intervene in the event of violent confrontations within (or between) villages, nor in cases of attacks by armed groups.

Conclusion

Through ethnographic analysis, we have documented overlapping sets of rules, practices and authorities concerning land relations in the Inner Delta of Niger River, with an emphasis on the legal dimension of social relations. Special attention has been paid to socio-political changes connecting the territory of the Inner Delta to more extensive projects of state formation (yet still unfinished), from the 19th-century precolonial empire of Macina to the more recent state of unrest and insecurity brought by armed groups throughout the region over the last decade.

One characteristic that has emerged with particular vigour is the interdependence among groups with different means of livelihood (based on farming, herding and fishing) and potentially conflicting legal regimes over the natural resources that they share in the same space. A number of traditional authorities ("customary chiefs" of land, transhumance and water) are in charge of negotiating the relationships and mediating the tensions among the groups.

The authority of customary chiefs finds its legitimacy in the interweaving of precolonial (Macina), colonial and post-independence trajectories of state formation characterising Sahelian societies (Bayart et al., 2019). Traditional offices such as the *jooro*, the *sila-tigi* and the *ji-tigi* are the products of social and historical transformations, connecting the territory of the Inner Delta to broader political configurations. In fact, local authorities act as intermediaries between such projects of state formation and the variety of groups inhabiting the space of the Inner Delta. Such groups are mostly based on a "segmentary lineage" kind of social organisation (Pes, 2016). Principles of "first-comership" and original occupation of the land, coupled with discourses on the mastery of symbolic and material resources, provide the grammar of a legitimate order of land and water rights, which cannot be circumvented by the state.

Following the events that have developed from the insurgency of jihadist groups in the north of Mali since 2011, the region of the Inner Delta has been the theatre of growing problems of insecurity. In this changing scenario, the role of traditional authorities has undoubtedly been put into question, especially by jihadist groups accusing the Fulani elite of colluding with state officials (particularly judges) and enriching themselves through corruption. However, after an initial period of unrest, mostly functional in recruiting local fighters among disadvantaged segments of the population, armed groups seeking to give their rule some form of legitimacy turned to (selected) local notables to administer justice.

After a series of *coups* and the abrupt change of geopolitical alliances, the government of Mali at this time seems to be in the process of regaining control of the Inner Delta, along with other parts of the country. As demonstrated by the establishment of the *Commissions Foncières* (Part III), land relations are key to this process of negotiated return of state institutions to the locality. Succeeding past hybrid forms of mediation, such as the HCI, customary chiefs are now included in state-backed institutions.

The way ahead seems long. Beyond the security concerns brought by the availability of weapons, after a decade of armed conflicts, there is now an entire economy of war to dismantle. However, there are positive signals; if anything unites the many overlapping institutions and politico-legal processes described in this chapter, this is their hybrid, flexible and adaptable nature in governing land relations in the Inner Delta.

Notes

1 This chapter is the outcome of joint reflection and work. For any bibliometric purpose, Parts I, II and V have been drafted by Luca Pes and Parts III and IV by Ibrahima Poudiougou. Fieldwork in the region of Djenné was conducted by Poudiougou and lasted for six months between 2016 and 2017. The location was revisited in December 2023. Pes carried out an extensive two-year fieldwork in the region of Manding between 2007 and 2009 (see Pes, 2016), revisited in 2024. We thank the two anonymous referees for their precious comments, the proofreaders for the careful revision of the text and Dr. Francesco Ferrarese for drawing the map reproduced in Figure 13.1.

2 In the legal debate about property, scholars have emphasised the idea of not "objectifying" resources by suggesting alternative theories of property based on non-extractive but reproductive visions of property relations (Capra & Mattei, 2015; Ellickson et al., 2014; Hann, 2007; Purdy, 2010). In Part II, we further develop this idea in relation to the holistic power/knowledge of customary chiefs.

3 The neighbouring region of the *Office du Niger*, where farming land has been developed for intensive agriculture and registered by the state, offers an example of an environment made more intelligible with a resource-minded approach. For a discussion, see Coulibaly (2015).
4 We choose to employ this expression, rather than "traditional chiefs" or others, because this is the qualification given to such authorities by Malian law and regulations. For further information, see Part III.
5 For further analysis of this expression and its capacity to convey the basic structure of politico-legal relations in the region of Manding, see Pes (2016, pp. 132–133).
6 This office bears some recognition in Malian law and regulations referring to the *chef de village* in continuity with French colonial rule. An example in the domain of land law and regulations is provided further in Part III.
7 The Macina Empire was a theocratic state founded around 1818 by the Fulani ruler Sékou Amadou Barry (1776–1845). With Djenné and later Ḥamdallāhi as capital cities, the influence of Macina extended over the entire Inner Delta, particularly over the territories located between the present-day regions of Ségou and Timbuktu. At its height, the Macina Empire bordered part of the present-day Mauritania and Mossi country (Burkina Faso). For more details, see Nobili and Syed (2021).
8 These territorial units are based on specific socio-political relations among communities practising different forms of livelihood, both sedentary and transhumant/semi-nomad, under the political leadership of a Fulani group. Currently, the *leydi* still provides an effective standpoint for reconstructing the norms governing the control, access to and distribution of natural resources in the Inner Delta. For more details, see Hampaté Bâ (1991).
9 Further implications of the Macina political domination, particularly the role of the literate and religious Fulani aristocracy, are explored in Part III.
10 For the most part, they belong to the ethnic (and socio-professional) groups called Bozos and Somonos. For further information, see Ramazzotti (1996).
11 The mystical dimension of the *ji-tigi*'s authority refers to the maintenance and cyclical renewal of the sacred pact between the ancestors of the community and the water spirits, as well as the intergenerational transmission of ritual practices. For further details, see Fay (1989) and Kassibo (1994).
12 The hybrid character of the legal order is further discussed in Part III in relation to the administration of justice.
13 A possible exception should be made for the chief of transhumance, the *sila-tigi*. Being the headman of a semi-nomad community, his authority appears less centred on the control of a bounded territory and more focused on the relationship with sedentary groups. In particular, the *sila-tigi* acts as an intermediary between cattle herding groups and *jooro'en*. A mutually respectful relationship proves particularly important at bottle-neck periods, when many herds converge at once in the same area. The *sila-tigui* also plays a key role in mediating herders' internal conflicts about pastures and herder–farmer conflicts (Bassett, 1988; Fay, 1989; Turner, 2004).
14 Similar to what has been observed in relation to the developmentalist jargon of "natural resources" (Part I), we hold that the qualification of such payments as "access fees" overlooks their substance and function in the relationships among groups in the Inner Delta.
15 On average, one taurus or dairy cow per herd is paid. The size of herds ranges from a few dozen to several hundred heads of cattle.
16 Suspects and accusations of the alleged complicity of *jooro'en* with state agents (notably officers of the *Service agricole* and the *Service pastorale et de la pêche* and judges at the local tribunal) are discussed in Parts IV and V.
17 The *ji-tigi* will receive payments in money and in kind, representing up to one-third of the catches of the group of guests. The fish will be sun-dried, stocked in warehouses

along the river and sold by the community during the year, providing an important source of income.

18 This is a shaded space, normally located within the courtyard of the village chief, devoted to conversations between elder men. For a detailed discussion of such spaces mirroring the organisation of socio-political relations, see Pes (2016, pp. 104–107, 214–224).

19 For an endorsement of this interpretation, see Hampaté Bâ and Daget (1955, p. 63).

20 *Tarikhs* are also called the historical chronicles of Muslim travellers and diplomats constituting the main written sources of the precolonial history of West Africa since the Middle Ages' history, for example, the *Tarikh-al-fattash* (the second half of the 17th century) or the *Tarikh-al-Sudan* (around 1655). See Nobili (2020).

21 The term *cadi* is frequently translated as "Islamic judge". However, as illustrated in the text, its function may also recall that of a notary, that is, a trusted authority certifying the overall legality of a transaction.

22 In other parts of Mali, such as the Mande region studied by Pes (2016, pp. 130–132), land transactions between private individuals are often accompanied by written documents called *attestations de vente*, bearing the stamp of the municipality's mayor. Although people refer to these documents as "legalised" by a state official (*legalisation*) and resort to them as pieces of evidence in land disputes, the mechanism of public certification provided by *tarikhs* proves much more effective in land dispute settlements, revealing the vestiges of an administrative bureaucratic culture transmitted by Islam.

23 The activity of interpreting *tarikhs* parallels that of Islamic legal scholars interpreting the *hadiths*, the attributed reports about the Prophet's life.

24 Interview with the *juge de paix à compétence étendue*, Djenné, January 2017.

25 The act is *Loi n° 2017-001 du 11 avril 2017 portant sur le foncier agricole*. Further regulations concerning agricultural land are provided by the *Loi d'orientation agricole* (Loi n° 06-045, 14 décembre 2005). Both regulations apply only to agricultural lands, thus rural areas. The general framework of Malian legislation on land law is provided by the *Code domanial et foncier* (Ordonnance n°00-027/P-RM du 22 mars 2000).

26 Other members include the village chief, the representative of the village council (*conseil de village*, the assembly assisting the village chief), the representative of professional associations (*corporations professionnelles*), one representative of young people (*répresentant des jeunes*) and one representative of women (*répresentante des femmes*). See Art. 2 *Loi n° 2017-001 du 11 avril 2017*.

27 The insurgency had two main drivers in a complex constellation of armed groups and changing alliances. On one side, there were "separatist" Tuareg groups, returning heavily armed from Libya after an international military intervention headed by France and Britain put a violent and disastrous end to the regime of Muammar Gaddafi (1969–2011). On the other side, there were "jihadist" groups coming mainly from the Algerian border but having proliferated in the north of Mali for about a decade, roughly coinciding with the double presidential mandate of Amadou Toumani Touré (2002–2012). Beyond engaging in illicit businesses, including kidnapping of tourists and drug trafficking, with the complicity of the Malian political élite up to presidential circles and state services (customs agency, police, etc.), these groups had developed both business and family ties in the north of Mali, which had been the main causes of the conflict's embeddedness in the locality. For an analysis of the historical dimension of such ties, see Scheele (2012).

28 Only in Mali have there been three *coups d'etat* since 2011: in 2012, 2020 and 2022. The main military interventions have been the French *Operation Sérval* (2012–2014) and *Opération Barkhane* (2014–2023) and the Multidimensional Integrated Stabilisation Mission in Mali (MINUSMA) by the United Nations (2013–2023).

29 Some of the deadliest confrontations occurred in Koumaga (more than 50 killed in December 2018), Koulongo (37 killed in January 2019), Ogossagou (more than 160 killed in March 2019) and Sobane-da (more than 100 killed in June 2019). See International Crisis Group (2020).

References

Baldaro, E., & Diall, Y.-S. (2020). The end of the Sahelian exception: Al-Qaeda and Islamic State clash in Central Mali. *The International Spectator*, *55*(4), 69–83. https://doi.org/10.1080/03932729.2020.1833566

Barrière, O., & Barrière, C. (2002). *Un droit à inventer: Foncier et environnement dans le Delta intérieur du Niger*. IRD Éditions.

Bassett, T. J. (1988). The political ecology of peasant-herder conflicts in the northern Ivory Coast. *Annals of the Association of American Geographers*, *78*(3), 453–472.

Bayart, J.-F., Poudiougou, I., & Zanoletti, G. (2019). *L'État de distorsion en Afrique de l'Ouest. Des empires à la nation*. Karthala.

Benjaminsen, T., & Ba, B. (2019). Why do pastoralists in Mali join jihadist groups? A political ecological explanation. *The Journal of Peasant Studies*, *46*(1), 1–20.

Bouju, J. (2020). La rébellion peule et la 'guerre pour la terre': Le gouvernement par la violence des ressources agropastorales (Centre-Mali, Nord-Burkina Faso). *Revue internationale des études du développement*, *243*(3), 67–88. https://doi.org/10.3917/ried.243.0067

Capra, F., & Mattei, U. (2015). *The ecology of law: Toward a legal system in tune with nature and community*. Berrett-Koheler Publishers.

Cissé, M. G. (2018). Hamadoun Koufa, fer de lance du radicalisme dans le Mali central. In M. De Bruijn (Ed.), *Biographies de la radicalisation: Des messages cachés du changement social* (pp. 181–202). Langaa Publishers.

Coulibaly, C. (2015). *Politiques agricoles et stratégies paysannes au Mali de 1910 à 2010: Mythes et réalités à l'Office du Niger*. Editions L'Harmattan Mali.

Derrider, M. (2021). *Élites, élections et transformations du politique au Mali, «ceux qui cherchent le pouvoir sont parmi nous»*. Karthala.

Ellickson, R. E., Rose, C. M., & Smith, H. E. (2014). *Property relations* (4th ed.). Wolter Kluwer.

Fay, C. (1989). Sacrifices, prix du sang, 'eau du maître': Fondation des territoires de pêche dans le Delta central du Niger (Mali). *Cahier des Sciences Humaines*, *25*(1–2), 159–176.

Gallais, J. (1984). *Hommes du Sahel*. Flammarion.

Godelier, M. (1984). *L'Idéel et le matériel. Pensée, économies, sociétés*. Fayard.

Hampaté Bâ, A. (1991). *Amkoullel, l'enfant peul: Mémoires*. Actes Sud.

Hampaté Bâ, A., & Daget, J. (1955). *L'empire peul du Macina (1818–1853)*. IFAN.

Hann, C. M. (Ed.). (2007). *Property relations*. Cambridge University Press.

International Crisis Group. (2016, July). *Central Mali: An uprising in the making?* https://www.crisisgroup.org/africa/west-africa/mali/central-mali-uprising-making

International Crisis Group. (2020, November). *Reversing Central Mali's descent into communal violence*. https://www.crisisgroup.org/africa/sahel/mali/293-enrayer-la-communautarisation-de-la-violence-au-centre-du-mali

Jourde, C., Brossier, M., & Cissé, M.-G. (2019). Prédation et violence au Mali: Élites statutaires peules et logiques de domination dans la région de Mopti. *Canadian Journal of African Studies/Revue canadienne des études africaines*, *53*(3), 431–445.

Kassibo, B. (1994). Histoire du peuplement humain. In J. Quensière (Ed.), *La pêche dans le Delta central du Niger: Approche pluridisciplinaire d'un système de production halieutique* (pp. 81–98). Karthala.

Legrosse, P. (1997). Perception de redevances de pâturage et transhumance des Peuls au Maasina (Mali). In R. Botte, & J. Boutrais (Eds.), *Figures peules* (pp. 239–266). Karthala.

Mundy, M., & Pottage, A. (2009). *Law, anthropology, and the constitution of the social: Making persons and things*. Cambridge University Press.

Nobili, M. (2020). Reinterpreting the role of Muslims in the West African Middle Ages. *The Journal of African History*, *61*(3), 327–340.

Nobili, M., & Syed, A. (2021). Introduction. The Caliphate of Hamdallahi: A history from within. *Afriques*, 12. https://doi.org/10.4000/AFRIQUES.3203.

Pes, L. (2016). *Building political relations. Cooperation, segmentation and government in Bancoumana (Mali)*. Tangram.

Poudiougou, I. (2024). Dan Nan Ambassagou, un mouvement d'autodéfense armé en pays dogon. *Anthropos*, *118*(2), 395–418.

Poudiougou, I., & Zanoletti, G. (2020). Fabriquer l'identité à la pointe de la kalache: Violence et question foncière au Mali. *Revue internationale des études de développement*, *243*, 39–65.

Purdy, J. (2010). *The meaning of property: Freedom, community, and the legal imagination*. Yale University Press.

Ramazzotti, M. (1996). *Readings in African customary water law/Anthologie du droit coutumier de l'eau en Afrique*. FAO Legislative Studies no. 58. Food and Agriculture Organization.

Scheele, J. (2012). *Smugglers and saints of the Sahara: Regional connectivity in the twentieth century*. Cambridge University Press.

Turner, M. (2004). Political ecology and the moral dimensions of 'resource conflicts': The case of farmer–herder conflicts in the Sahel. *Political Geography*, *23*, 863–889. https://doi.org/10.1016/j.polgeo.2004.05.009

14 Ségou(s), the Niger and the other side of the river

How water tells the city in the Sahel

Elvira Pietrobon

Ségou's inland side, river side

Ségou, the regional capital of Mali in West Africa, faces the Niger River. It is separated from the other bank by 1.5 km of water. There are no bridges, and the inhabitants use pirogues or the ferry from the city centre to cross the river. During the day, islands inhabited seasonally by fishers can be seen from the city; at night, only a few dim solar-powered lights are visible.

Ségou is located at the southern gateway to the Sahel. It historically represents a place of encounter and transition between the climatic and cultural influences of the Sudano-Guinean and the Sahelian zones (Fig. 14.1). It was the capital of the kingdom of Ségou, the singularity of which lay in the decentralised distribution of institutions (Sow, 2021). At that time, Ségou was a 'multiple capital' (Sow, 2021)

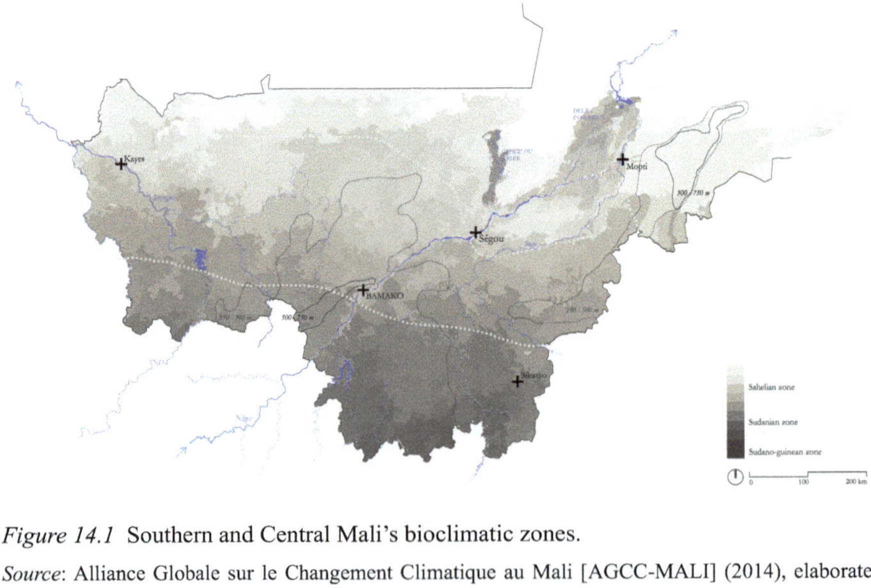

Figure 14.1 Southern and Central Mali's bioclimatic zones.

Source: Alliance Globale sur le Changement Climatique au Mali [AGCC-MALI] (2014), elaborated by the author.

DOI: 10.4324/9781003429111-20

composed of four villages, the four Ségou (*Segu nani*): Ségoukoro (*-kɔrɔ* [old]), Sé(gou)bougou (*-bugu* [hamlet]), Ségoukoura (*-kura* [new]) and Ségoushikoro (*-si kɔrɔ* [the old shea tree]) (Fig. 14.2). With the arrival of French settlers in 1890, a town centre was established at the village of Ségoushikoro, and the settlement lost its historical-multipolar character. Since then, the town began to expand radially, leaning on the axis of the Niger and the national road that borders it.

The old Ségoushikoro was destroyed during the colonial period and no longer exists. In its place, there is the commercial centre of the city, with its colonial earth-brick buildings, the river port and the light installations of the large weekly market (*dɔgɔ*). On the banks of the Niger, it is possible to recognise the 18th-century settings of the other three Ségou villages. Ségoukoura has become a district of the city. Sébougou is part of the independent rural municipality of the same name but contiguous to the city. Ségoukoro, part of the municipality of Sébougou, retains the characteristics of the traditional village surrounded by the fields of extensive family-run rain-fed agriculture (*kungo*).

Ségou is also the closest city to Bamako, the capital of Mali, about 200 km to the northeast, following the RN6 national road connecting the south and the north of

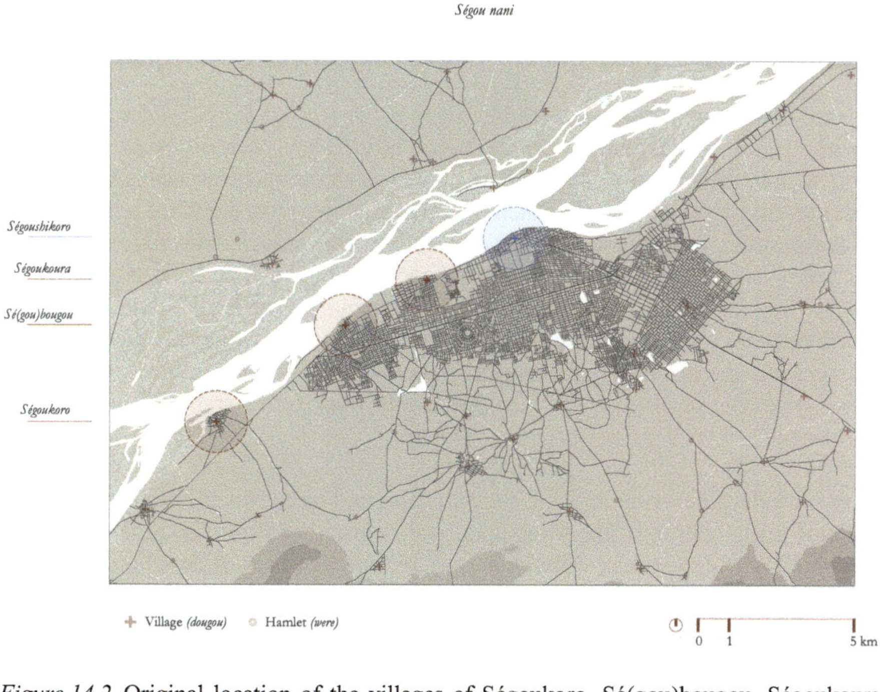

Figure 14.2 Original location of the villages of Ségoukoro, Sé(gou)bougou, Ségoukoura and Ségoushikoro.

Source: Philippe (2013), elaborated by the author.

the country. Ségou has witnessed its population grow exponentially over the last 30 years. The latest official figure available counts 135,000 inhabitants (Institut National de la Statistique du Mali (INSTAT), 2013) compared to 90,000 at the end of the 1980s (Direction Nationale de la Statistique et de l'Informatique (DNSI), 1990). Estimates of 145,000 (Philippe, 2013) and 200,000 inhabitants (De la Croix et al., 2011) are reported for 2010, while the most recent data analysis by Africapolis suggests a population of 223,008 (2020). This rapid population growth coincided with the land ownership liberalisation of the 1990s.[1] Thus began the spontaneous and dispersed expansion of the city, a 'piecemeal urbanisation' (Sawyer, 2014) at the expense of the collective land cultivated by families (*kungo*) from the three old Ségou and other villages surrounding the city.

Currently, due to the city's progressive expansion, families from the three former Ségou (Ségoukoro, Sébougou and Ségoukoura) have established different relationships with the *kungo*, the rain-fed cultivated space. Only Ségoukoro, on the outskirts of the city, has maintained proximity to and dependency on the fields for the purpose of subsistence. In Sébougou, part of the *kungo* has already been built up; the rest retains its traditional configuration, dotted with shea trees (*si*) and acacia albida trees (*balanzan*), although scattered with buildings for housing. Here, agriculture in already-sold lands is practised during the rainy season by the village families. Rain-fed agriculture is also practised around Ségoukoura, but because there is no more space for fields, the *kungo* has its place at the roadside and on unbuilt plots (Fig. 14.3).

In turn, the relationship between the Niger River (called *Badji* [water is here] or *Djoliba* [great blood]) and the three Ségou is different because of its continuity. From the periphery to the city, the Niger holds its shape and presence, and the same activities occur along its banks. Each Ségou has its own river bank transept managed by the families residing in the villages. Yet the continuous and moving space of the water is a 'world apart'. The Bozo and Somono fishers hold a preferential relationship with the river, on which they depend for their livelihood and with which they have a special intimacy. At present, the river also serves as a borderline. It represents the distance, guaranteed by the fishing community, between the area under state control, Ségou, and the other bank, where terrorist groups have ease of movement.

In this article, I describe in detail the activities and spaces of the Niger River. The scale of investigation of the three historical Ségou that still exist in their original morphology (Ségoukoura, Sébougou and Ségoukoro) is considered a case study. Local water-related practices are thus compared with the ongoing process of urbanisation and the resulting influences of the global market economy. Several surveys conducted between 2020 and 2023 made it possible to map the uses of space on the right bank of the river. Interviews with members of three families in the three villages also allowed an in-depth study of land management systems. To investigate practices in the river and exchanges with the other bank, research was

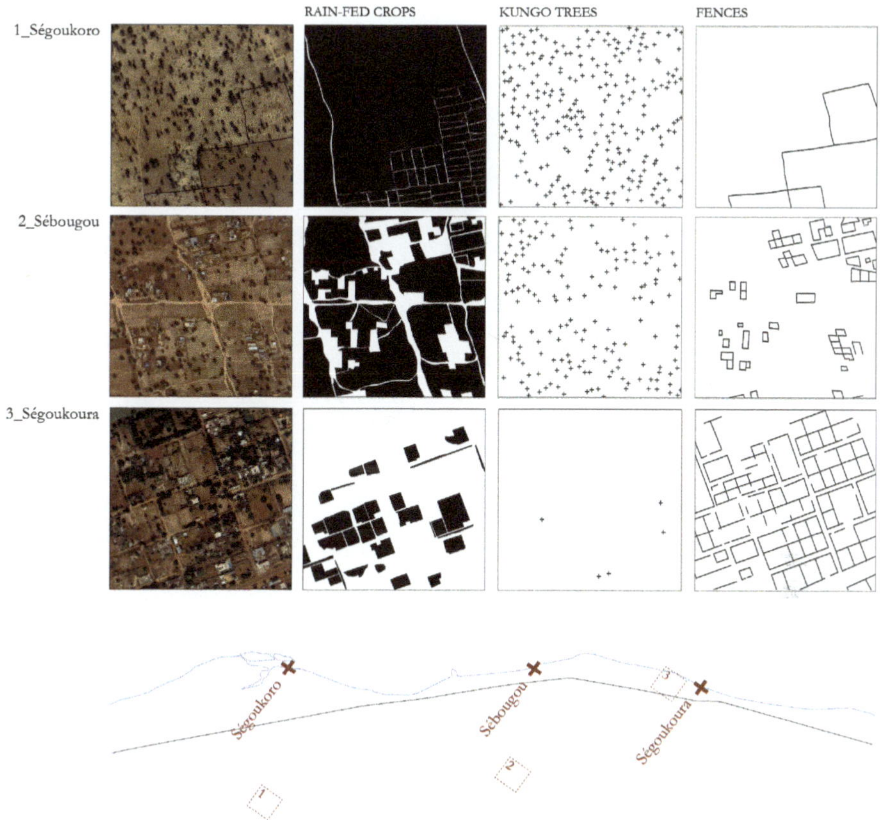

Figure 14.3 Today's *kungo* configuration in Ségoukoro, Sébougou and Ségoukoura.

Source: Elaborated by the author.

conducted with the help of Adama A. Guindo, a student at the Faculty of Social Sciences in Ségou.

Along the river: The gardens of the villages

In the three Ségou, the banks of the Niger are places of work. They are frequently used for daily activities, such as washing clothes and pots, resting places for fishers' pirogues, ports for crossing the river and for 'sand fishers'. Above all, they are the gardens of the villages (Fig. 14.4).

The vegetable gardens occupy the banks continuously in a range of about 1.5 km around the three villages. As in the rest of the territory, the silt-rich soil near the waterways and settlements, called *bɔgɔ fin* (black soil) or *bɔgɔ passa* (clay soil), is largely devoted to horticultural production (Benjaminsen, 2002). This space corresponds to the easement thickness of the river that varies seasonally

Figure 14.4 Gardens along the Niger and laundry hung out in the morning in Sébougou.
Source: Elvira Pietrobon, 2021.

and from year to year. During the river's peak flooding period between September and October, the Niger can increase its flow sixfold; on exceptionally high peak days, it can swamp almost all cultivated areas.[2] Garden work is divided into three seasons: *samiya* (rainy season), *nɛnɛ* (cold season) and *klema* (dry season). Vegetable production is most intense between November and March (*nɛnɛ*), when the river level has fallen and temperatures drop. During *klema*, production slows down, leaving room for grazing. During *samiya*, gardens are partly used for rain-fed crops, such as maize (*kaba*), niébé (*siyɔ*), peanuts (*tiga*) and okra (*gan*), in addition to *foronto bani*, a type of chilli pepper that people like to use during the rainy season for 'keeping warm'.

This shifting space is part of the public domain. In particular, the 25 m from the river limit at peak flow are by law the property of the state. Along the lines of Mali's complex land question, the thickness of the river is at the same time subject to legal pluralism (Bertrand, 2021). Here, customary law where 'l'appropriation y est-elle affectation de l'espace à des usages complémentaires' (ownership is allocation of space to complementary uses) (Le Roy, 1991, p. 15) is *de facto* applied: the cultivated banks are unfenced spaces, accessible to all but for the special yet non-exclusive use of village families.

Vegetable gardens are part of the basic practices of Bambara peasant families' economies. The fields made available to slaves were called *jɔnforo* (*foro* means field; *jɔn* means slave). By extension, the *jɔnforow* are identified as individual fields, that is, those that can be associated with a person who has the right to use one of them (Maiga, 2001). Unlike the *foroba* (large field), where cereals are cultivated under the collective management of the family, the *jɔnforo*, or simply *foro* (field), is a space commonly entrusted to women who cultivate the condiments for the sauces that accompany cereal-based meals, particularly millet and sorghum (*nyɔ*) cultivated in the *forobaw*. While it is only in Ségoukoro where families combine horticulture with cereal production, for all three Ségou, vegetable gardening is a possibility and resource available to the village community. According to custom, families living in the village settlements are assigned an area to cultivate, which they allocate for the individual use of their members once they are married (in theory, to the wife of the new household).

The gardens are divided into *balaw*, variable units averaging 10 m wide and 15 m long. Every two to three *balaw*, paths—about 50 cm wide—lead perpendicularly towards the river. Near the settlements, the paths open up into slight bays where domestic work can be carried out. Approximately every 30 m, corresponding to two *balaw*, there are longitudinal paths to the river. *Balaw* are divided into beds, the cultivated areas varying in width between 50 and 80 cm, depending on the type of crop. The beds are bordered by a 20-cm compressed and raised section of soil where feet can stand. In fact, work is done with the legs wide apart over the beds, enabling the cultivation of a larger area and adapting the space to the body's possibilities. According to this principle, the main activities in the garden are performed using short-handled tools with a metal part for working the soil facing the user. Such is the *daba*, a handcrafted hoe traditionally made by the 'caste' of blacksmiths (*numuw*) according to a *savoir-faire* associated with local myths and cosmology (Dieterlen, 1948), or the *soli*, made by twisting iron bars, commonly imported from India.

Garden space is larger in Ségoukoro (ca. 650,000 m^2) than in Sébougou (ca. 350,000 m^2) and in Ségoukoura (ca. 125,000 m^2). The cultivated area corresponds to the number of actual inhabitants of the villages and the available space. Ségoukoro is the only one of the three Ségou in which the population growth (3,339 inhabitants surveyed in 2009 [INSTAT, 2013]) corresponds to the increase in peasant spaces and practices. In all three Ségou, horticulture is conducted for the non-exclusive purpose of sale, but the degree of autonomy of production inputs differs as well as, to some extent, the type of crops.

Ségoukoro

In Ségoukoro, the vegetable gardens are located on a land flap beyond a river inlet (Fig. 14.5). In this area, only a few *balanzan* trees and *dendenw*, soil holes up to 10 m wide, interfere with the space of the gardens. Here, water is extracted

Figure 14.5 Activities along the river at the village of Ségoukoro.

Source: Google Earth (2024), elaborated by the author.

to supply the vegetable gardens farther away from the river. As in the other two Ségou, the means of irrigation are complementary: gardeners use traditional tools, such as calabash (*file*), locally produced watering cans and recycled jerry cans, as well as motorised pumps imported from China (De la Croix et al., 2011).

Kaldi Traore, a member of the Traore family from Ségoukoro, has two *balaw* (ca. 80 m^2) 500 m from his home, which he reaches by pirogue. Kaldi usually grows onions (*djabaw*) throughout the garden. The nursery is planted in the courtyard of his home with seeds bought in jars (a 100-g jar costs an average of 6,000 Fcfa) and transplanted in the garden around December. In 2023, Kaldi sold 213 kg of onions in bulk at 200 Fcfa/kg and stored and dried the damaged ones for the preparation of collective family meals. Lassine, one of Kaldi's brothers, cultivates a vegetable garden farther away, which he reaches with his donkey-drawn cart. In his four *balaw* (ca. 150 m^2), he usually grows chillies (*forontow*) and onions that he transplants from his nursery. When the chillies are ripe, they are sold by the beds to buyers who come directly to the site to retrieve the fresh fruit.

Sébougou

Modibo Coulibaly, a member of the Coulibaly family from Sébougou, has the right to use a vegetable garden 200 m from his home in the fields on the banks of the river (Fig. 14.6). His vegetable garden (about 70 m^2) is located in the shade of a few mango trees. Modibo entrusts the vegetable gardening to his nephews, who grow celery (*celeri*), parsley (*persil*), carrots (*carotiw*) and mint (*nanaye*). In Sébogou, it is common to grow herbs in vegetable gardens configured with beds of a smaller width (50 cm) in the shade of trees. In the other plots, *ngoyo* (the local variety of aubergine), salad (*salati*), carrots, beetroot (*betravu*), onions and *woso* (the local sweet potato) are grown. As in Ségoukoura, basil (*sukɔlan*) and lemongrass (*binbulunin*) are planted at plot corners. At the edges of salad beds, it is also customary to let a salad plant bloom and to plant sticks with a piece of black plastic hanging from them to ward off lizards. Fruit crops such as *ngoyo* are associated with tubers (e.g., beetroot) or bulbs (e.g., onions) and leafy vegetables (e.g., salad or parsley). Fruit trees, such as papaya and a few banana (*namasa*) trees, are found at the borders of some gardens. When the fruits are almost ripe, they are covered with bags or fabrics to signify that the owner is checking on their growth and to protect them from birds. In Sébougou, as in Ségoukoura, cassava (*bananko*) plants and henna (*djabi*) shrubs, the fruits of which are dried and sold in powder form, can also be found at the ends of the beds.

Ségoukoura

Madou Bouaré, a member of the Bouaré family from Ségoukoura, no longer cultivates a vegetable garden. He sold the land 100 m from his house ten years ago. His land is in the immediate vicinity of the village next to the plot where a hotel was later built. Despite local customs, the new owners do not allow Madou to

Figure 14.6 Activities along the river at the village of Sébougou.

Source: Google Earth (2024), elaborated by the author.

keep working on the adjacent empty lot. Madou's case is not isolated. The person entrusted with the use and management of the land is allowed to not only cultivate but also to rent out and, in some cases, to sell. In Ségoukoura, there are now three illegal constructions on officially public and flood-prone land, and in at least three other *balaw*, the sale is marked with stakes.

In Ségoukoura, where the vegetable gardens occupy a smaller space, there are no *dendenw*, but underground metal pipes connect motor pumps in the river to the more distant vegetable gardens (Fig. 14.7). Motor pumps are used to irrigate the many potatoes and salad beds, which require a lot of water. There are also external cisterns where water is collected and used to irrigate the orchards that are located on the empty lots in the immediate proximity of the vegetable gardens. These plots are managed by the plot owners or persons chosen by them, most often not belonging to the original families of the village.

Organic and chemical inputs

The gardens of Ségoukoura are located 300 m from the main road that flanks the river towards the city centre. The stretch of road in the village is particularly crowded, and every 100 m, there are stalls with fresh fruits and vegetables for sale. Some of the produce comes from the gardens along the river, grown and sold by the members of the same family. To speed up sales, the gardeners of Ségoukoura make extensive use of chemical fertilisers. Chemical fertilisers are also used in Ségoukoro and Sébougou, but in Ségoukoura, where garden space is limited and direct sales are often carried out, the use of chemical fertilisers has become indispensable. For example, an overabundance of salad is grown for sale during *nɛnɛ*, which lasts for three months. The price of a salad bed (about 10 m long × 70 cm wide) varies between 2,000 and 2,500 Fcfa. According to Madou Coulibaly, with fertiliser (1 kg is sold at 100 Fcfa and used for five beds), it is possible to harvest every month and thus earn three times as much. Without fertilisers, it would not be possible to recover the costs of petrol for the motor pumps and working time. Round-up is also often used to weed the grass at the start of cultivation (1 l costs 7,000 Fcfa, and 0.25 l is used per *bala*).

Organic fertiliser (*nɔgɔ*) is also necessarily used in the three Ségou. During *nɛnɛ*, it is common to see 50-kg bags and piles of *nɔgɔ* at different stages of decomposition in the corners of the *balaw*. *Nɔgɔ* is distinguished between manure and compost obtained from the decomposition of household waste (*nyaman*, literally rubbish) (Blanchard, 2010). In Ségoukoura and Sébougou, it is customary to buy *nɔgɔ* from the *were* (the hamlets of animal breeders, generally Peul), located on the outskirts of the city. In contrast, Kaldi and Lassine Traore from Ségoukoro use manure from cattle, which they keep in their yard and use when ploughing the cereal *forobaw*. Compost from *nyaman*, which is particularly popular for growing potatoes, is not purchased. While in Ségoukoro, *nyaman* from domestic composting pits is used, in the other two Ségou, it is common to use compost from roadside rubbish heaps, once sieved and separated from inorganic waste.

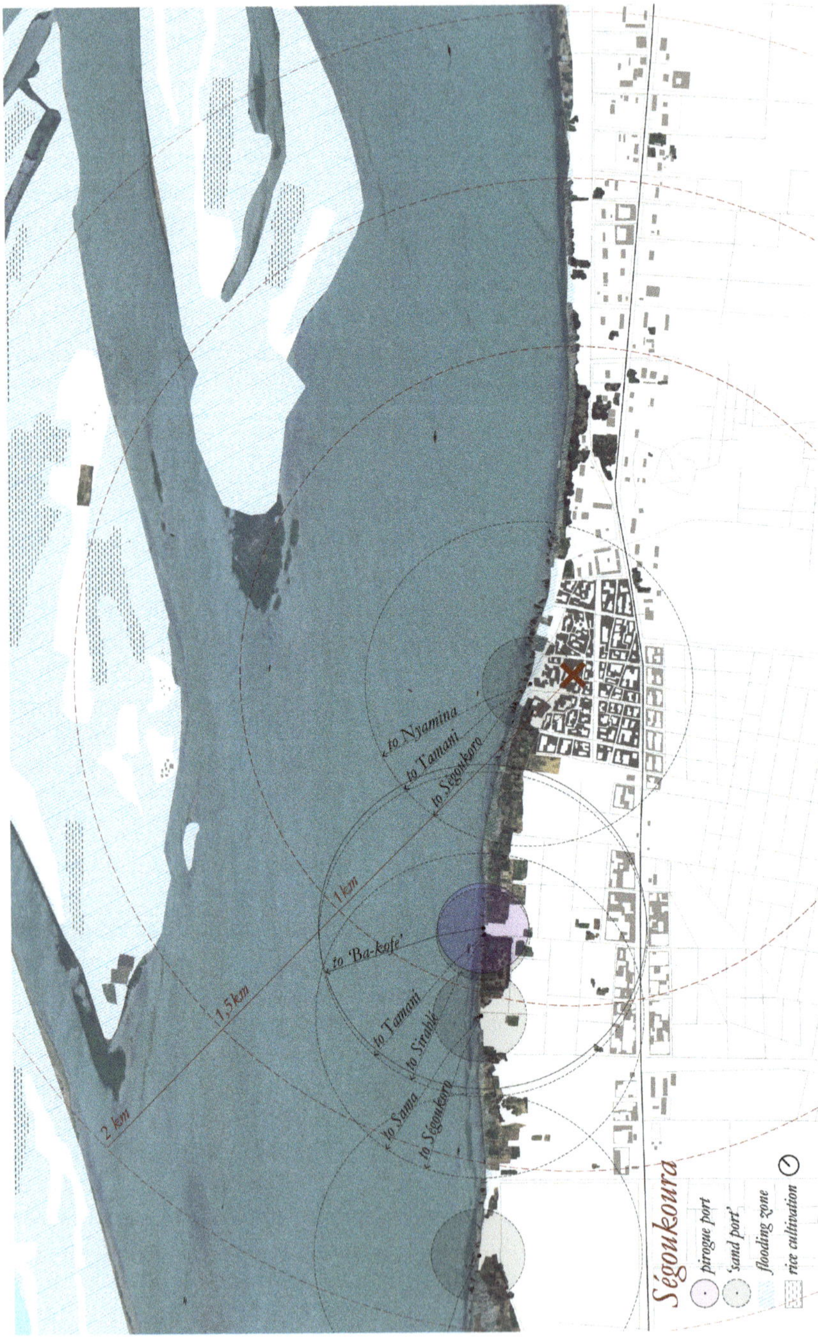

Figure 14.7 Activities along the river at the village of Ségoukoura.

Source: Google Earth (2024), elaborated by the author.

Organic input is also used to protect the soil from the sun. To maintain soil moisture, rice husks (*malo faraw*) are used, which, unlike leaves, are not lifted by the wind. Branches with neem leaves (*djirina buluw*) are only used for potatoes. In Ségoukoura and Sébougou, gardeners buy *malo faraw* (at 250 Fcfa per 50-kg bag) from the peasants of the village of Kounkoun, about 10 km away. However, in Ségoukoro, gardeners do not buy husks. *Nɔgɔ* made from cattle manure contains large quantities of *nyɔ kalaw* (millet and sorghum stalks), which help maintain soil moisture. *Nyɔ kalaw* are deposited under the shed where the cattle eat and defecate; some of the stalks are eaten, and others are triturated by the cattle.

Down the river: fishing for fish, fishing for sand

In Mali, work traditionally defines the identity of men and women, connecting them to the natural and social environment they are part of. Work is not conceived as an expression of the individual but rather one of a community social unit based on a set of horizontal external relations. From the local peasants' perspective, through work, they inhabit the territory and relate to the living (Pietrobon, 2021). This viewpoint determines a form of interdependence among different communities based on the organisation of three basic elements or resources: water, soil and grass. The settlement's spatial organisation is based on specific subsistence models:

> la combinaison de l'eau et de l'herbe détermine ainsi l'objet des activités pastorales, celle de la terre et de l'eau, celui des pratiques culturales: l'eau, combinée à un de ces éléments, en fonction de son importance devient un milieu favorable pour la pêche. D'où l'importance souvent clamée des divisions écologiques dans les rapports entretenus par les hommes dans l'occupation de l'espace.[3]
>
> *(Cissé, 1996, as cited in Coulibaly, 2015, p. 26)*

The three Ségou are known as Bambara villages. In each of the three villages, only members of certain families who define themselves as indigenous and Bambara are eligible for the role of *dugutigui* (village chief). However, in the *longue durée,* this has not excluded the Bozo and Somono fishers from being part of the village communities. Unlike the past, today there is no strict spatial demarcation separating the fishermen from the rest of the village. In addition, as a consequence of urbanisation processes, a progressive diversification of professions within the respective communities is taking place. Nonetheless, there still remains a rationale for a certain form of living that unites the fishers to the proximity of the river: for example, the Bozo and Somono fishing families are the ones occupying dwellings near the river's edge, interrupting the continuity of the gardens.

In contrast to the Somono, who are distinguished from the Bambara by their fishing occupation, the Bozo have their own language and cultural identity. They were the first to inhabit these banks of the Niger, but due to the nomadic nature of their living, they are not identified as village founders (Philippe, 2009). Today, in the three Ségou, several Bozo families have settled alongside Somono families. However, the light-thatched dwellings of semi-nomadic fishers can still be found. Sébougou has a modest-sized Bozo camp of fishers from the Mopti region. When water levels drop in the inland Niger Delta, they migrate south towards Ségou in search of deeper waters where they can find better value fish. In Ségoukoro, fishers from a nearby village (Banankoro) move around the time of some fishing campaigns. Other light dwellings are visible on the islands (*gunw*) in front of the three Ségou, particularly during *nɛnɛ* and *klema*.

Each fisher's canoe is tied to an iron pole planted near each one's house. There are 7-m long pirogues (*kurunw*) driven with the help of a long stick pushed into the river bed, as well as longer pirogues (10 m) with small motor pumps handcrafted to propellers at the back of the pirogues (De la Croix et al., 2013). The pirogues are generally made of wood from the region's *caïlcédrat*; they are decorated with paintings depicting village identification ideograms and symbols similar to those Griaule and Dieterlen (1949) associated with Bozo myth of the creation of the world. Where more boats are present, there are open-air workshops to repair and build pirogues and to manufacture fishing nets from imported nylon threads (Marie et al., 2007). These activities are carried out by the fishers themselves.

There are several fishing methods, but the most profitable in Ségou are the large nets (10 m on average) cast in the morning with floats and retrieved in the late afternoon. During all seasons, different types of fish (*djɛgɛ*) can be found, these include *tɛbɛn* (Niger carp), *manogo* (torpedoes), *bɔlo* (catfish), *naana* (elephant fish), *dodo* (puffer fish), *saalen* (African perch) and *tinɛni* (small fish of the Alestidae family). In the three Ségou, fish can be processed (dried or smoked) or sold fresh. The smaller fish are set aside for self-consumption, and the rest are sold by the women of the fishing families at the market or along the road near the villages.

There are also four floating fish-farming cages at the transect of banks under the jurisdiction of Ségoukoura, one at Sébougou and several in Ségoukoro. *Tɛbɛn* and *manogo* are generally bred in cages. Fish farming is not exclusive to Bozo and Somono fishers; entrepreneurs of all origins practise it. To start the activity, it is necessary to apply for permission from the *Direction National de la Peche*. However, the interviewed Ségoukoura entrepreneurs are also used to concluding agreements with Bozo fishers who, for a fee, check the state of the cages in the event of heavy rains or strong winds. The investment in the business consists of the manufacture of the cage and the purchase of the fry and daily feed sacks.

Pirogues are not only used by the Bozo and Somono for fishing. Wider spindle canoes (ca. 2.5 m) made of wood or iron are used by the sand fishers. Between Ségoukoura and Ségoukoro, there are seven 'ports' of sand fishers at the roads perpendicular to the river and that connect the banks to the *gudron*, the paved road parallel to the Niger (Fig. 14.8). These activities are run by private entrepreneurs, who hire labourers daily to retrieve sand (*cɛncɛn*) and gravel (*bɛlɛ*) from the bottom

Figure 14.8 ‘Sand port’ between Ségoukoro and Sébougou.

Source: Elvira Pietrobon, 2022.

of the Niger, drop them in the ports and load them into the trucks that transport the materials to the construction sites.

At the village of Ségoukoura, the Bah Construction Company works with more than ten pirogues. The sand fishers recover sand at Ségoukoro, while the ‘gravel fishers’ take a two-day trip, approximately 70 km south, to collect materials at Tamani and Nyamina. Another contractor who owns five pirogues works 700 m from Bah towards Sébougou. He retrieves sand at Sirablé, about 8 km south of Ségoukoro, and gravel at Tamani. Doucoure runs a business with four pirogues, 400 m further south. His workers also fish for sand at Ségoukoro and for gravel at Sama some 25 km south. Between Ségoukoura and Sébougou, a larger company works with fifteen 10-m pirogues. Their workers retrieve sand in Ségoukoro, while for gravel, they travel as far as Kamini, approximately 100 km up the waters of the Niger. Five pirogues, each with five fishers, set off on a three-day journey. They carry 12 kg of rice and five cans of petrol (they use 100 l during a high tide and 70–80 l during a low tide). The company based in Sébougou practises sand fishing in front of the village during receding waters and towards Banankoroni, 5 km south of Ségoukoro, during floods. A company with a dozen pirogues works in the ‘sand port’ between the banks of Sébougou and Ségoukoro. For theirs as well, the fishing point for sand is close to Sirablé; for gravel, it is towards Nyamina. The company based in Ségoukoro fishes sand in the proximity of the village, while for gravel, it moves in front of the banks of Fansongo some 30 km up the Niger (Fig. 14.9).

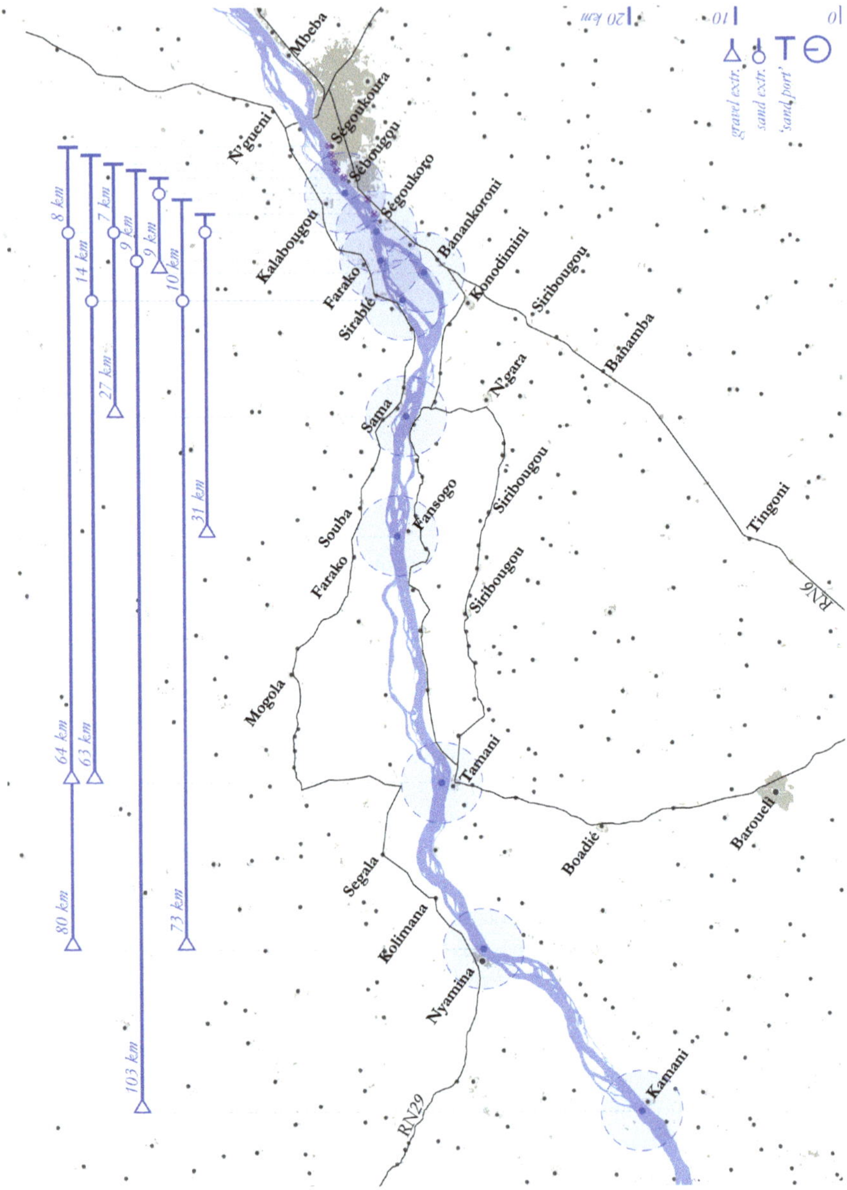

Figure 14.9 Sand and gravel fishers' supply points.

Source: Elaborated by the author.

It is not necessary to apply for permission to extract sand and gravel. However, entrepreneurs usually pay a lump sum to the traditional authorities or the municipalities of the villages where the gravel is extracted. According to a report (Collerie, 2008), entrepreneurs dealing with sand and gravel extraction in Ségou are members of the *Association des vendeurs de sable et graviers de la région de Ségou* and pay annual fees for each pirogue in service and for occupying the space at the sand port.

Crossing the river: what a pirogue can do

Every Monday, Wednesday and Friday, Peul women (*fula musow*) from breeders' villages on the left bank of the Niger cross the river to sell fresh milk. They are highly visible because they carry the milk in a *file* (calabash) over their heads and because of their braided hair smoothed with animal fat. These *fula musow* come from the Peul village of Zangou, about 5 km from the river bank. The river has three crossing points: from the Ségou shopping centre to Kalabougou by pirogue or by the daily ferry boat; by pirogue from Ségoukoura to the opposite bank on Mondays, Wednesdays and Fridays; from Ségoukoro to Farako daily by pirogue or ferry. Peul women land from the pirogues and sell milk from morning until late afternoon, when they return to the other side of the river (*ba kɔfɛ*).

By crossing the river, different communities living around its waters meet and supplement one another in terms of basic needs. In the three Ségou, until the 1980s, milk was exchanged, thanks to the fishers, owners of pirogues, for the millet cultivated by peasant families. Today, there are still forms of exchange other than currency, but these are determined by considering the selling price of a *manani* (a bowl) of grain at the market. In turn, pirogues remain a means that fishers use to supplement the needs of the families living in the villages on the right bank of the Niger. In addition to fishing and shuttling on the river, pirogues are used to reach certain resources such as straw and adobe for building bricks.

Fodder straw is a key resource for the city, given the large number of animals, particularly sheep.[4] At intersections, busy roads and near *grabal* (the livestock markets), fodder for animals is sold. The availability of fodder is linked to the cycle of the seasons and related peasant practices. In the three Ségou, the supply of fodder from the river area is of particular importance. It is categorised into straw from wild herbaceous species (called *bin* or *bulugun*) and leaves from crops such as niébé (*siyɔ*) or the local sweet potato (*woso*). *Bin* is generally accessible to all, regardless of its location. During the rainy season, *bin* may be taken from the roadside and uncultivated gardens along the river for self-consumption. The rest of the year, it is sold dry, and it is necessary to travel by pirogue to find it. What is paid for is not the resource but the means by which it can be accessed: the pirogue. Pirogue holders (*kuruntiguiw*), Somono or Bozo, are therefore necessary to the fodder supply cycle.

When the river level starts to fall, around November to December, the *kuruntiguiw* reach *ba kɔfɛ* and the river islands (*gunw*) to cut the grass (Fig. 14.10). They let it

Figure 14.10 Straw recovered from the islands and the other side of the river at Ségoukoro.
Source: Elvira Pietrobon, 2019.

dry for two days before transporting it to the right bank of the river, where it is then traded and sold at 100 Fcfa per bundle. With the beginning of the dry season, when the flocks graze along the river, niébé leaves (*siyɔ kalaw*) and sweet potato leaves (*woso buluw*) are sold as fodder at 100 Fcfa per bundle. They also sell *kaarata*, long-stemmed wild grass also used to make mats, cut from *ba kɔfɛ* during the cold season and sold during the dry season at 500 Fcfa per bundle. When the rains break out, fodder is fetched by means of pirogues from the unsubmerged pieces of land along the left bank of the river. These are wild herbs such as *kɔmalo*, similar to rice, or *badadjaraini*, leaves of a watermelon-like wild plant, sold at 100 Fcfa per bundle.

In Ségoukoro, the fodder trade is less widespread as peasant families use the by-products of cereal cultivation in the *forobaw* to feed their animals. Millet and sorghum stalks (*nyɔ kalaw*) and *siyɔ kalaw* are stored in the courtyards of the houses above the animals' hangars. *Nyɔ faraw* (millet and sorghum peels) are also used; when mixed with water, these are highly nutritious food for the sheep.

Adobe is another resource with no accessibility restrictions. Each village normally has a quarry where the inhabitants take the material to make the bricks from wooden moulds. The time to renovate the mud-brick houses is before the rainy season; around March, piles of bricks can be found in the three Ségou. In addition to the quarries on the right bank of the river, Somono fishers are used to cross the river and extract adobe to sell the bricks in the villages. The three Ségou no longer have any quarries in the immediate vicinity. In Ségoukoro, for safety reasons, the quarry is now located on the southern edge of the garden area, far from the houses. In Sébougou and Ségoukoura, quarries are located on the outskirts of the city in the area of the then *kungo*, now under urbanisation.

Another activity that the Somono people in particular carry out by means of the pirogue is the traditional cultivation of rice on river islands. The 'floating rice' is sown earlier in the rainy season in lands that will then be submerged as the waters rise. This cultivation does not require any kind of hydraulic infrastructure. In the three Ségou, this activity is practised for self-consumption by *kuruntiguiw* families.

How water tells the city

At the beginning of the rainy season, Kaldi Traore from Ségoukoro makes a sacrifice to the river. Kaldi pours fresh milk (*nɔnɔ kɛnɛ*) and ten kola seeds (*worow*), five red and five white, into the waters of the Niger as a sign of respect to Faro, a figure in the cosmogony of the Mandé culture to which the Bambara belong, associated with water, particularly the Niger River (Dieterlen, 1955). Kaldi is not the only one to make such a sacrifice; in Ségou, the Niger is currently an important space because of its width and the river bank inhabitants' deep relationships with it. It is inevitable to reflect on its symbolic significance when considering the contemporary dynamics of living. Examining the materiality of water-related practices and spaces somehow allows one to preserve the 'right to opacity' of local cultural expressions (Glissant, 1990) and, at the same time, to measure their presence in the reproduction of dwelling. The 'right to opacity', in the words of the poet and philosopher Edouard Glissant, is not only the 'right of difference' but also the right to subsist 'dans une singularité non réductible' (in a non-reducible singularity) (1990, p. 204). For Glissant, this implies shifting people's attention to the frame that singularities compose rather than to the nature (and hierarchy) of each component of that same frame.

According to Germaine Dieterlen (1955), the Niger was the geographical axis of a religious, social and political system linked to a cosmogonic myth that was found, in similar forms, among the different cultural groups that inhabited the region. Today, in Mali, the Niger River still represents the spine around which an 'opaque frame' made of 'symbolisms and praxis' is built. Indeed, the river brings along with it a system of inter-community relations and resource management that connects the settlements it encounters on its path. This guarantees a degree of autonomy and establishes a horizontal level of interactions among the inhabitants. Thanks to this link between ecological divisions and relations between men and women (Cissé, 1996, as cited in Coulibaly, 2015), the river defines a space of production as well as protection against the country's current economic crisis and political instability.

The city is not cut out of this frame. The scale of the three Ségou shows how urbanisation processes and market economy influences modify techniques and trading methods of water-related practices, on one hand, and how these are intertwined with references to the rural context and customary practices, on the other hand. This co-presence generates ambiguities at different levels that can result in social conflicts and environmental problems that I have not attempted to explore in this chapter. Nonetheless, it seems interesting to highlight what water-related activities reveal about the city of Ségou: rurality understood as the spatial and symbolic dimension that binds inhabitants to a subsistence relationship with the land (Pietrobon, 2023) is a constitutive and integral part of contemporary living practices. Ségou is the first city of the Sahel that Niger encounters on its path. It can therefore represent an entry point for reading the settlement dynamics underway in the region, starting from the spaces on which they inevitably and historically depend: the spaces of water.

Notes

1 It was only in 1986 when the first legislative text on land management in the independent state of Mali appeared: the *Code Domanial et Foncier* (CDF). The equivocal legislation left customary practices and rights in a state of informality. What actually changed and marked the beginning of a new phase in the country's history was the progressive opening to land ownership rights held by a natural or legal person.
2 The Niger's flow rate rises, on average, from less than 100 m^3/s in March to over 5,000 m^3/s in September (Autorité du Bassin du Niger (ABN), 2023).
3 The combination of water and grass thus determines the object of pastoral activities, that of land and water, the object of cultivation practices: water, combined with one of these elements, depending on its importance, becomes a suitable environment for fishing. Hence, the often-claimed importance of ecological divisions in the relationships maintained by men in the occupation of space.
4 In 2015, 10,325 cattle and 56,750 sheep were sold in Ségou (Commune Urbaine de Ségou, 2016).

References

Africapolis. (2020). *Africapolis: Maps and analysis. Countries. Mali.* https://africapolis.org/en/explore?country=Mali&poprange=1,2,3,4,5,6&year=2020
Alliance Globale sur le Changement Climatique au Mali (AGCC-MALI). (2014). *Rapport final Inventaire Mali Tome 1.* https://sifor-mali.net/v2/download/inventaires_2014/Rapport-final-Inventaire-Mali-Tome-1-AGCC-2014-VF.pdf
Autorité du Bassin du Niger (ABN). (2023). *Bulletin Mensuel de mars 2023.* http://nigerhycos.abn.ne/portal/IMG/pdf/bulletin_hydro_mars_2023_vf.pdf
Benjaminsen, T. A. (2002). Enclosing the land: Cotton, population growth and tenure in Mali. *Norsk Geograésk Tidsskrift – Norwegian Journal of Geography*, *56*(1), 1–9. https://doi.org/10.1080/002919502317325722
Bertrand, M. (Ed.). (2021). *Une Afrique des convoitises foncières.* Presses universitaires du Midi.
Blanchard, M. (2010). *Gestion de la fertilité des sols et rôle du troupeau dans les systèmes cotoncéréales-élevage au Mali-Sud: Savoirs techniques locaux et pratiques d'intégration agriculture élevage.* [Doctoral dissertation, Université Paris-Est, Creteil Val de Marne/CIRAD]. DP ASAP. https://www.dp-asap.org/productions/theses/these-de-m.-blanchard

Collerie, M. (2008). *Etude et cartographie des sites d'extraction et de dépôt de sable et graviers sur le Niger entre Ségou et Kona et sur le Bani entre Douna et Mopti*. UNESCO, Commission Européenne (CE), Institut de recherche pour le développement (IRD), Unité mixte de recherche Gestion de l'Eau, Acteurs, Usages (UMR G-EAU), Université François Rabelais (Tours), Institut sur les milieux aquatiques et les corridors fluviaux (IMACOF).

Commune Urbaine de Ségou. (2016). *Programme de développement économique social et culturel. PDESDC 2016-2020 de la commune urbaine de Ségou.*

Coulibaly, C. (2015). *Problématique foncière et gestion des conflits en Afrique noire, Tome I. Des indépendances à la faillite des dictatures, 1960 – 1990*. L'Harmattan.

De la Croix, K., Ferry, L., Landy, F., Traoré, B., Muther, N., Tangara, B., & Martin, D. (2013). Quelle «place» pour des pêcheurs urbains? Le cas de Bamako (Mali). *Cybergeo: European Journal of Geography, Espace, Société, Territoire, document 648*. http://cybergeo.revues.org/25977

De la Croix, K., Marie, J., Ferry, L., & Landy, F. (2011). De nouvelles dynamiques de mécanisation agricole: commerce, usages et spatialisation au sein de la région de Ségou (Mali). *Annales de Géographie*, *678*(2), 174–192. https://doi.org/10.3917/ag.678.0174

Dieterlen, G. (1948). L'arme et l'outil chez les anciens Bambara. *Africa: Journal of the International African Institute*, *18*(2), 105–111. https://www.jstor.org/stable/3180547

Dieterlen, G. (1955). Mythe et organisation sociale au Soudan français. *Journal de la Société des Africanistes*, *25*, 39–76. https://doi.org/10.3406/jafr.1955.1873

Glissant, E. (1990). *Poétique de la relation*. Gallimard.

Griaule, M., & Dieterlen, G. (1949). L'agriculture rituelle des Bozo. *Journal de la Société des Africanistes*, *19*(2), 209–222. https://doi.org/10.3406/jafr.1949.2601

Direction Nationale de la Statistique et de l'Informatique (DNSI). (1990). *Recensement général de la population et de l'habitat: résultats définitifs. Volume 0: ensemble du pays, tome 2: répertoire de village*. https://www.instat-mali.org/laravel-filemanager/files/shares/rgph/repvil87_rgph.pdf

Institut National de la Statistique du Mali (INSTAT). (2013). *4e recensement général de la population et de l'habitat (RGPH-2009). Analyse des résultats définitifs. Thème: ménages et caractéristiques de l'habitat au Mali*. https://www.instat-mali.org/laravel-filemanager/files/shares/rgph/rahab09_rgph.pdf

Le Roy, E. (1991). L'appropriation et les systèmes de production. In E. Le Bris, E. Le Roy, & P. Mathieu (Eds.), *L'appropriation de la terra en Afrique Noire. Manuel d'analyse, de décision et de gestion foncières* (pp. 27–35). Karthala.

Maiga, I. (2001). *Parlons Bambara, Langue et culture Bambara*. L'Harmattan.

Marie, J., Morand, P., & N'Djim, H. (Eds.). (2007). *Avenir du fleuve Niger*. IRD Éditions.

Philippe, S. (2009). *Une histoire de Bamako*. Grandvaux.

Philippe, S. (2013). *Ségou, Une région d'histoire*. Memoria.

Pietrobon, E. (2021). Territori contadini nel sud del Mali: La ruralità come dimensione spaziale, politico e simbolica della resistenza Bambara. *Contesti, città territori progetti*, *2*(2), 39–58. https://doi.org/10.13128/contest-13036

Pietrobon, E. (2023). Ruralization in urbanization in the peasant territories of Southern Mali: Everyday life in Ségoukoura, Ségou. In T. Delefortrie, C. Cavalieri, M. Gheysen, B. Grosjean, R. Lassalle, & L. Lerchs (Eds.), *Beyond metropolization. Exploring new hybrids. Proceedings of the 10th international U&U PhD seminar* (pp. 44–57). Louvain Research Institute for Landscape, Architecture, Built Environment, UCLouvain. https://www.lille.archi.fr/wp-content/uploads/2023/06/UU_Proceedings_FINAL_compressed.pdf

Sawyer, L. (2014). Piecemeal urbanisation at the peripheries of Lagos. *African Studies*, *73*(2), 271–289. https://doi.org/10.1080/00020184.2014.925207

Sow, M. (2021). *L'Etat de Ségou et ses chefferies aux XVIIIe et XIXe siècles – Côté cour, côté jardin*. PU Bordeaux.

15 Changing patterns of resource conflicts in the Logone floodplains (Cameroon/Chad)

Between actors' practices and development narratives

Aboukar Mahamat, Dobah Marsala and Francesco Staro

Introduction

The Waza-Logone floodplain is the largest wetland in the Far North of Cameroon, in the middle of the Sahelian zone. Covering an area of around 8000 km^2, this ecosystem is transformed throughout the year by the hydrological rhythms of Logone River and its tributaries. Historically, it has represented a strategic area for the lives of farming, fishing and pastoralist communities (nomadic, sedentary and transhumant) in the region and elsewhere, whose activities are based on the availability of surface water for seasonal flooding of farmland, regeneration of pastures and reproduction of various species of fish.

In recent decades, the populations of the Waza-Logone floodplain have been faced with a growing crisis, marked by increasingly frequent violent clashes. The emergence of new practices of natural resource management has called into question the interdependence among local communities, thus exacerbating their competitions. The communities most affected by these conflicts are located in the Zina district, where the dynamics of access to resources (agricultural land, water and grazing areas) have been heavily affected by the development of irrigated schemes since the 1970s. These interventions have reshaped the geography of the floodplain and consequently the organisation of different production systems and the links between them.

The historical matrix of the episodes of conflicts in the area has long been dominated by internal tensions and socioeconomic asymmetries within groups, as well as the related mediation and conflict resolution strategies. In contrast, the conflicts that have occurred since 2020 seem to confirm the dynamics of interethnic violence and growing tensions among different groups. In this chapter, we attempt to account for the changes in the scale and patterns of violence in the area under study. Thus, our aim is not to present an exhaustive account of the conflicts that have taken place in recent years but to propose a few keys to understanding the transversal character of these conflicts and their interethnic and intraethnic dimensions.

DOI: 10.4324/9781003429111-21

In this chapter, we present the results of the field research carried out by Aboukar Mahamat in Cameroon and Dobah Marsala in Chad within the framework of the AtlaSahel project. Their contributions were not limited to gathering detailed information on the most recent conflicts but were also crucial in situating local tensions within the broader framework of the local history of natural resource management and access. This information was later analysed and organised by Francesco Staro, an anthropologist and researcher at the University of Padua.

Beyond project failure: From project logics to local practices

Since June 2022, there has been a series of violent episodes between Arab groups from the village of Abangoula and Kotoko groups from the village of Ngamé in the Makary area, following a series of episodes of cattle thefts, encroachment on or interference with grazing areas and cattle corridors, and the destruction of cultivated fields and fishing channels. Clashes between these groups had been occurring since November 2020 over the management of the Moulaga waterhole; in fact, this site has been under dispute since the 2000s.

Previous research in this region (Moritz, 2006; Turner, 2002) has shown the diversity of interests at stake in the clashes between farmers and herders, as well as the impossibility of understanding in a dichotomous manner the tensions between these populations considered as homogeneous groups. Other research has called into question the idealised and stereotypical portrayal that describes the relationship between sedentary farmers and pastoralists in terms of a shift from a historical situation of symbiotic coexistence during the precolonial era to an increasingly violent competition during the postcolonial era, often attributed to the scarcity of natural resources. From different perspectives, these studies account for the need for researchers and policymakers to adopt a more nuanced view of local conflicts, distancing themselves from a consensual and simplistic discourse on interethnic clashes, particularly between farmers and pastoralists, starting with a more detailed analysis of the conflict dynamics within different groups.

Table 15.1 shows a short selection of the episodes of conflicts that took place in the Logone plain from 2000 to 2022. This information is presented here to show the diversity of the dynamics at stake that are analysed in this chapter.

Other destabilising factors are related to the transborder dimension of these conflicts. This point is best illustrated by the clashes between the Musgum and Choa Arab communities in August 2021. At first, two villages were involved, Hinduk and Machichka, inhabited by the Choa Arab and Musgum communities, respectively. The origins of the tensions are related to both communities' use of a natural pond for fishing activities. Some Musgum men started to dig deeper holes to catch more fish, which became traps for cattle that risked falling into them and drowning. Following the first armed clashes for the control of this pond, men from

Table 15.1 Selection of conflicts in the study area.

Date	*Location(s)*	*Major events*
November 2000	Ngodeni, Arrondissement of Zina	Clash between the Kotoko and Musgum communities over management of the Moulaga waterhole.
December 2006 and January 2007	Waza, Tchédé, Garlé, Moroma and Zinave National Park Arrondissement of Zina	Conflict between the Kotoko and Musgum communities over theft of fish inside Parc National de Waza (PNW).
April 2007	Zina, Arrondissement of Zina	During a reconciliation ceremony between the Kotoko and Musgum communities, the governor announced that from now on, "if the MP is Musgum, then the mayor is Kotoko and vice versa".
2 November 2014	Bourgouma, Borough of Logone-Birni	The Musgum communities contested the agricultural development of an area by an elite Kanuri ethnic group for the benefit of their own people.
2 November 2014	Kousseri	The Kotoko, Musgum and Massa communities were at loggerheads over land management at a former site of SEMRY III (Société pour le développement et la modernisation de la riziculture de Yagoua) project.
June 2020	Waza, Borough of Waza	Choa Arabs and the Kanuri people clashed over the management of a borehole.
December 2020	Ngodeni, Arrondissement of Zina	Clash over management of the Moulaga waterhole between the Kotoko and Musgum communities.
July 2020	Koumboula, Arrondissement of Zina	The Kotoko communities of Zina and the Musgum communities of Misdé clashed over the management of a piece of land; a Musgum farmer started work on a canal running through a field belonging to a Kotoko farmer.
10–15 August 2021	Canton of Madiako and EL-birké, Borough of Logone-Birni	The tensions of 8th August turned into war between the Arab and Muslim communities.
7 January 2022	Maroua	The UNDP organised a meeting to present its conflict resolution plan to the parties involved in the clashes, following the peace agreement signed by the Musgum and Choa Arab communities.
June 2022	Villages of Abangoula and Ngamé, Arrondissement of Makary	Clashes between Arab and Kotoko groups due to disputed resources.

Source: A. Mahamat's fieldwork.

the Choa Arab community living in neighbouring villages crossed the river and sought refuge in the Guilmey camp in Chad, where Dobah Marsala was able to carry out fieldwork in 2022.

This case study is interesting for two reasons: firstly, it allows us to introduce the issue of deregulation of fishing practices that is discussed throughout this chapter. Secondly, it shows that the local roots of conflict over natural resources intersect with cross-border mobility practices. Indeed, the strategic use of borders by the communities inhabiting the Logone floodplains is a crucial element, not only in understanding ongoing competitions but also in analysing how flows of goods and people contribute to reshaping local practices.

Furthermore, the violent clashes constitute part of a regional insecurity that results in obstacles to the mobility of people and goods in cross-border areas. In Cameroon, local communities accuse the Chadian services of controlling all motorised pirogues that navigate the Logone River, demanding Chadian permits in order to impose taxes on all transported products. Often, the refusal to dock at a checkpoint set up along the river is considered a crime, and offenders are shot at. These tensions occur more frequently between October and December, when fishers set up nets and traps in the beds of Logomatya and Loromé Mazera Rivers, obstructing the movement of motorised pirogues. This practice gives rise to disputes, often leading to fights between fishers and traders.

As we shall show, the resurgence of tensions and the emergence of new forms of conflicts are embedded in the adoption of new production practices and strategies by local populations. The sustainability and viability of these practices have a major detrimental impact on peaceful coexistence among local populations.

Our analytical approach to these dynamics refers to institutional economic anthropology and to the actor-oriented approach. Firstly, we analyse how social institutions for the management of common resources are transformed in such a way as to privilege the interests of those groups with stronger decision-making and bargaining powers. This situation is the result of structural changes brought about by the development of trade networks, connections and available infrastructures. This point is interesting because it does not reduce institutional change to ecological dynamics alone (e.g., the decrease in flooding following the construction of dams and hydraulic infrastructures) or to related social dynamics (agricultural migrations and increased pressure on available resources) or to the supposed impacts of agricultural development schemes. The prices of the produced goods (especially fish, as we shall show) increase because of higher demand in the interconnected regional markets. This process of commercialisation (in terms of quantities sold and monetary value) then consolidates the decision-making power of the interest groups that benefit the most and are thus able to change social institutions and rules of access to common resources to their advantage.

From this perspective, socioeconomic change does not exclusively reflect local or contextual dynamics but is part of a broader process that includes interconnections among different scales. At the same time, this approach also questions the theories of management of common pool resources, based on "formal variables, abstract models, deductive typologies, one-size-fits-all interpretative frameworks" (Olivier de Sardan, 2013, p. 289), by closely examining the role of local authorities and the issue of power asymmetries and inequalities of resources between actors who are able to shape new institutional settings (Ensminger, 1992; Haller, 2010).

We thus pay attention to the practices and strategies of local actors and refer to the concept of agency, which is useful in studying the actors' own actions and perceptions in the context of institutional change. The actor-oriented approach focuses on individuals' experiences and actions that emerge at the interfaces among different lifeworlds (e.g., among development workers, government agencies and local groups or across ethnic or intercultural boundaries). Our study concentrates on networks of social relations by investigating the discrepancies in values, interests and resources that can be bridged and accommodated or contested and renegotiated through new social practices. This strategy also allows us to understand how knowledge and power domains are co-produced, challenged or transformed (Long, 2001). By focusing on actors' practices, we avoid equating informal institutions with cultural traditions considered as a static and unchanging set of norms; rather we examine how formal and informal institutions are modified, adapted or even reinvented over time (Helmke & Levistsky, 2004).

The need to mobilise appropriate tools to analyse the dynamics at work in the Logone floodplain seems to us even more urgent in light of the importance given to the narrative of the failure of agricultural development programmes, which has become a dominant element of the development discourse. There is the risk of assimilating the conflict dynamics in the Logone floodplains into a simple consequence of a 'wrong' development model. This step limits the attention to the internal logics of projects, favouring a self-referential discourse that relegates the analysis of local practices and logics to the background.

Regardless of the success or failure of development interventions, we will rather focus on the analysis of what a development intervention produces from a material and symbolic standpoint, as well as the performative and normative character of the development discourse. This approach makes it possible to enrich the debate on project evaluation. It is not just a question of pointing out 'good' or 'bad' practices but of understanding what the social life of the project is, and what the different actors' (often conflicting) perceptions of a project's success or failure are.

Hydraulic and social engineering in the Logone floodplain

Since the 1970s, the ecosystem of the Waza-Logone plain has been the focus of major hydraulic and agricultural development interventions. Driven by the need to ensure Cameroon's food security and economic development in response to the series of droughts that had hit the Sahel countries, these projects aimed to improve the productivity of the floodplain by building dykes and hydraulic structures to control the flow of water and thus support rice cultivation.

These initiatives started in 1979 with the construction of the Maga dam and a 21-km dyke on the left bank of Logone River, designed to support the development of irrigated rice fields as part of the activities of the Société pour le développement et la modernisation de la riziculture de Yagoua (SEMRY II). This was followed in 2015 by the construction of a 7-km dyke on the right bank of the Logomatya River as part of the Projet d'urgence de lutte contre les inondations (PULCI Emergency Flood Control Project), which also contributed to the rehabilitation of the rice-growing facilities of SEMRY II and the setting up of an early warning system. These projects resulted in a significant change in the hydrological regime of the Waza plain, making rain-fed rice cultivation extremely vulnerable to flooding in the far south of the plain, particularly in the part located in the commune of Zina.

These projects embody a modernist vision of agriculture, which was initiated with the installation of infrastructure to control the flow of water and modify the ecological rhythms of the Waza-Logone floodplain. In this area, water and agricultural development projects should have made it possible to organise and optimise production processes and protect local populations from flooding, while meeting the political objective of securing the border with Chad.

Hydraulic works to support irrigated agriculture in the Logone floodplains follow other examples of large-scale projects to ensure development through large infrastructures on a regional scale. These projects significantly represent the policymakers' fascination for large water-based developments in Africa. They include the hydraulic engineering interventions endorsed by the climate change narrative to avert the disappearance of Lake Chad, while enabling navigation and transport, fishing on the lake and major irrigation in the basin, as well as supplying hydroelectricity (Commission du Bassin du Lac Tchad/CBLT, 2012). This case is interesting because it shows how discursive constructions of development are based on an oversimplification of ecological dynamics and variability; it also engendered the myth of the disappearing lake that was used to raise new funds and legitimise new projects (Magrin, 2016; see also Chapter 2).

The construction of irrigation perimeters in the Logone floodplains contributed to transforming the demographic balance and social composition in the area. Starting in the 1970s, new allottees arrived from all over the region because the project needed the labour force that was not available locally. A good example is the village of Simatou, so called because it is inhabited by the Siratà (a local name for the Kanuri, who come from Borno), Massa (whose area of choice is along the river in the Yagoua area) and Toupouri, who came from the Kaélé area, in the lowlands at the foot of the Mount Mandara (Bertoncin & Pase, 2015). These long-standing processes have led to a reconfiguration of the relations among different user groups and called into question the local institutions' effectiveness in regulating access to the floodplain and its resources.

In the early 1980s, other populations (Kotoko, Musgum and Choa Arabs) came from the north forced to abandon their territories because of the drought. Between 1976 and 1987, the population of the Maga district rose from 4680 to 8734, and its density increased from 24.1 to 45.1 inhabitants per square kilometre. Conflicts soon broke out as the Musgum regarded the newcomers as profiteers. However, more than half of them abandoned the perimeters as soon as production conditions in their homelands improved. As a result, grazing lands to the north and west of Mayo deteriorated. The traditional territorial patterns of transhumance were disrupted, and sedentary livestock farming, which was already unstable, declined in the areas to the northwest of the dam.

In addition to these issues, the SEMRY project also achieved poor results due to problems with the maintenance of the hydraulic infrastructure, the lack of agricultural machinery and the non-observance of irrigation calendars. Due to these limited outcomes, further measures were taken to renaturalise the area from the end of the 1980s onwards. Under the Waza-Logone project funded by the Dutch development cooperation, the hydrological rehabilitation of the plain by reopening a canal blocked by the embankment along the Logone River would make it possible to release a significant quantity of water from Lake Maga, thus restoring the flexibility of the wetland system (de Noray, 2002).

We point out two critical aspects of these interventions. Firstly, the reflooding projects were not adapted to support the local population, as happened during the exceptional rainfall in 2001, which caused abnormal flooding and had a major disastrous impact on the villages along the Petit Goroma. These events confirm the difficulty of controlling the unpredictable and highly variable ecological dynamics of wetlands. At the same time, at the local level, these events make it difficult to attribute responsibility and identify who is accountable for dealing with these environmental disasters. This situation produces a condition of uncertainty, which (as we shall show) further exacerbates the fragmentation of common resource management systems.

Secondly, the denaturalisation of the floodplains' dynamics that existed before the construction of the dam and the dykes was based on the assumption that the project would restore traditional resource uses and livelihoods. The study of common resource management systems is relevant here to highlight the limitations and inconsistencies of this approach. In the particular ecosystem of floodplains in African drylands, the management of common pool resources is shaped by ecological processes, as well as by changes due to human-made hydraulic works. This case shows the anthropogenic character of social institutions, which is not limited to ecological features (e.g., rainfall, floods, quality of pasture) but also to political and institutional interactions among user groups on different scales. The process of renaturalisation and restoration of the ecological rhythms of the plain therefore shows its limits if it is unable to integrate these socioecological dynamics. The transition from flood control to rehabilitation of the ecological rhythms of the plain therefore refers to a 'natural' condition prior to a hydraulic intervention that in fact did not exist. This approach echoes the colonial discourse that falsely idealised the relationship between local societies and their ecological environment as 'natural', referring to a mythical time that failed to account for the social factors at play in organising access to resources (Haller et al., 2013).

Following the reflooding projects, new practices have emerged, particularly the opening up of new fishing channels downstream of the Petit Goroma from Logone and Logomatya Rivers. These spread to the point where the problem did not lie in opening breaches to reintroduce flooding, but eventually in closing them and managing the huge increase in fishing activities.

Deregulation of fishing activities

In the Logone floodplain, fishing is practised by members of the Kotoko, Musgum and Massa communities, as well as by transhumant or foreign fishers who arrive each year from the shores of Lake Chad and Nigeria. Several fishing techniques exist, and they are practised within the framework of different management systems. Management of the canals in the Logone floodplain is essentially individual, unlike the ponds or reserves where management is communal. Family management of fishing canals mainly concerns those acquired by inheritance, which therefore represent genuine family properties (Belal, 2003). In localities such as Lahai, Sifna, Guidiba and Kazéré, which have several community ponds, community fishing is much more widespread, and fishers organise themselves into committees for collective fishing in March each year.

Fishing canals have been dug or rehabilitated since the end of the flooding. This is a very old technique in the plain. Historically, fishing canals were few and used mainly by Kotoko groups. They were dug and then closed using racks interspersed with traps to hold back the water and the fish when the floodwaters receded in November and December. The social institutions of the Kotoko chiefs played an important role in supervising these practices. Among Kotoko groups, fishing practices are traditionally organised around rituals to celebrate the opening of the fishing season, and restrictions are put in place to preserve fish during the reproductive period (Landolt, 2010).

Shifting patterns of resource use and the temporal nature of rights agreements are key features of these norms regulating fishing practices. Because the availability of fish changes due to the variability of water levels, there is open access during high floods. When water recedes, a local group or village can consider fish as common property and decide to restrict access or grant permissions to other groups, depending on their existing relationships (Haller, 2002).

From the 20th century onwards, in conjunction with major migratory movements in the region, other groups, particularly the Musgum communities, began to dig and use the canals. Since the 1970s, following the construction of the Maga dam, this practice has become fairly widespread and has made a significant contribution to transforming the hydraulic profile of the plain, amplifying the effects of the hydraulic infrastructures and irrigation schemes implemented in the region (Fig. 15.1). The increase in the number of canals has caused the 50 or so large permanent ponds to dry up before February, whereas they used to hold water throughout the dry season. The only exception is the Tchouvn'no pond, located 7 km east of Mazera in the Zina district. It is only connected to the Logone River by a single fishing canal that is more than 100 years old, and the prohibition on increasing its dimensions (depth, width and length) is still respected by the family that exploits it.

These canals have irreversibly disrupted the normal flow of water, reduced pastoral areas and are at the root of more than 70% of local conflicts among families, particularly among fishers, but also between fishers and herders and between fishers and farmers (CBLT & Association Camerounaise pour l'Education Environnemental/ACEEN, 2018). In 2017, the local organisation ACEEN estimated that there were 7000 fishing channels in the Cameroonian part of the floodplain and more than 2000 in the Chadian part, but it is difficult to provide an accurate estimation of their total number. Laborde et al. (2016) report that the dimensions of the canals are also impressive: 70–20,000 m long, 2–15 m wide and 1–2.5 m deep. The ban on this practice since 2006 by an order of the governor of the Far North Region has not been enough to stop the digging of new canals and the renovation of old ones, and the number of fishing canals continues to grow.

Figure 15.1 Fishing canals.

Source: Aboukar Mahamat, 2024.

Today, despite the significant investment required for digging and maintenance work, as well as the uncertainty over the quantity of fish caught, the use of canals is considered one of the most profitable fishing techniques (Laborde et al., 2016), particularly by wealthier individuals who can mobilise financial resources (Ziébé, 2015). This activity, which can generate significant profits, is a major source of economic inequalities within fishing communities. This problem is due to a major difference in the productivity of fishing techniques (between an average catch of 6011 kg for canal fishing and 2785 kg for other techniques) (Labara et al., 2020).

Changing patterns of resource conflicts

The construction of new fishing canals has increased tensions on different scales: inequalities and reduced availability of fish caught upstream and inequalities and new competitions within fishing communities. At the same time, there is a significant divergence between the episodes of conflicts (that mainly involved fishing groups) and more recent tensions. In particular, livestock farmers are increasingly faced with a reduction in grazing areas due to canals and the development of irrigated fields, resulting in frequent conflicts.

Fishers who own canals are frequently opposed to one another, more so than canal heirs and owners opposed to non-owners (Khari, 2011). The latter blame the pressure exerted on resources by canals, which are said to monopolise the bulk of catches (Loth, 2004), leaving a minimal share for other traditional techniques. However, surveys of conflict resolutions in relation to fishing canals show that 79.6% of the conflicts have been resolved, compared with 20.4% that have remained (Labara et al., 2020). While conflicts are frequent around canals, strategies for managing them are nevertheless being undertaken through several channels and at a number of levels (community, administrative, security and judicial). Local Kotoko institutions have ensured that tensions are managed through customary conflict resolution mechanisms, but the emergence of new fishing practices, as well as the involvement of new outside groups attracted by the rentability of the canals, creates rivalries that cannot be absorbed and resolved by existing social institutions.

Other examples coming from the fieldwork carried out by A. Mahamat show the local institutions' importance in regulating fishing practices and the key role played by customary authorities in resolving conflicts. For instance, the ban on the use of all fishing techniques harmful to the resources is respected in the waters of Lake Maga. The Dam Management Committee, which was set up more than 10 years ago, is still implementing these measures. Any unauthorised fisher is taken to the Sultan's Court and required to pay a fine. The majority of fishers have embraced the initiative and are actively participating in the fight against the use of beach seines, locally known as *taro*. The biological resting period of the fish in the dam is respected, so fishing is suspended during that season.

New patterns of resource conflicts are also related to the emergence of other fishing practices and techniques. These include individual fishing reserves, clumps of grass that fishers plant in a portion of the riverbed and harvest three or four times a month, encircling them with beach seines made from extremely fine-meshed nets. These practices started on the Logomatya plain in the early 1990s, before spreading to Logone River and its tributary, Loromé Mazera. They contribute to the silting up of riverbeds and the destruction of fish habitats.

Since 1994, near the Maga dam, fishers have been using clay mixed with rice bran to catch fish using hawks. While some of the rice bran is ingested by the caught fish, the clay has encouraged the growth of algae over more than half the surface of the water body.

It is also important to consider drying practices and fish storage. To date, chilled storage is practised exclusively in Maga, where there is a cold storage facility and two ice factories. The most widespread traditional methods for preserving fish products so that they can be marketed are smoking by burning wood or its by-products or ox dung or drying on racks on the ground, on raised racks or on straw spread out on the ground. Fishers may also use highly toxic chemicals, which have caused numerous cases of poisoning (ACEEN, 2002).

These fish processing practices contribute to the deforestation process, which has worsened tensions at the local level. A case in point is the Al-Hamdoulillah community forest at Madiako in the Logone Birni Commune, which was created to protect around 400 ha of woodland (notably palm and date trees). By 2022, this reserve became in danger of disappearing, while incidents of violence over the control of this area have been recorded in the vicinity of Kidam since 2008. Some communities have mobilised, as in the case of the village of Hinalé, where a committee has been set up to combat the destruction of the bush.

Cultural change as a dimension of new fishing practices

Our research shows how institutional change related to fishing activities is part of broader cultural dynamics that need to be accounted for. Whereas Kotoko fishers have been converted to Islam, their religious life retains many animistic aspects, including rituals for spirits of the water. Animism legitimised autochthony and notions of ownership, while Islam provided norms and procedures for handling rights of access to water and pastures. The institution of religion allowed the establishment of monitoring and sanctioning devices to organise access to natural resources. These regulations were "embedded in the religious system, creating a strong legitimacy for the position of authority" (Haller et al., 2013, p. 43).

The power of the traditional authorities who are responsible for applying these rules at the community level is now being called into question since the digging of fishing channels is increasingly perceived as an individual activity and an economic investment that is not subject to community control. Sometimes, during interviews, farmers explained the emergence of these practices by referring to religion; for example, local traditions such as the worship of certain reptiles as sacred animals are said to be disappearing in many localities and are considered to be beliefs incompatible with Islam. These practices show the local actors' capacity to use their social belonging and religious beliefs to renegotiate the balance of power among different interest groups.

As mentioned, these cultural dynamics are also intertwined with demographic trends on a regional scale. In the mid-1900s, the Musgum community started to question the authority of the Kotoko leaders, making reference to the principles of democracy with regard to resource management, after the independence of Cameroon and Chad, and again in 1990. Since the 1990s, the decentralisation reform in Cameroon has further removed the legitimacy of the Kotoko people as resource owners, as the majority rule for political decision-making was applied. As some authors observe, state institutions put the nomad groups and the Musgum community in a dominant position (e.g., Haller et al., 2013).

In this context, religious or ethnic categories fail to acknowledge the historical character of identity dynamics. In fact, the process of Islamisation is linked to the contextual emergence of new forms of belonging to the territory and its resources. It is from this perspective that researchers should consider the new divisions and tensions based on religious beliefs. Therefore, we need to distinguish the cultural dimension of fishing activities, which is rooted in the history of local social institutions, from current claims that reflect and are influenced by economic interests and the logic of speculation and enrichment of new fishers who decide to invest in fishing channels as sources of profit.

How old and new restrictions in accessing grazing land are taken into account in development programmes

If fishing practices are emblematic for understanding the conflicts in the Waza-Logone floodplain, there are significant differences in the competitions related to other natural resources, especially grazing areas. As we shall show, these differences concern not only the causes of tensions but also their representation in the development discourse and the projects that this discourse legitimises.

Each year, after the onset of flooding (mid-October), transhumant and nomadic herders from Cameroon, Nigeria, Chad and Niger enter the plain in search of pasture until the annual floods begin (mid-June). The annual presence of livestock was previously estimated at around 200,000 animals (Seignobos & Iyébi-Mandjek, 2000), but the number of herders and herds accessing the plain's pastures each year is now on the increase due (among other reasons) to the fragile security situation in the Lake Chad basin.

The social institutions that have historically regulated the organisation of access to pasture land are characterised by their ability to integrate multiple uses and social logics. Previous research showed how Kotoko groups entered into contracts with nomadic pastoralists, who came to use the local resources and were granted access rights through a social institution named *ngalway*, responsible for coordinating timing and access to pastures (Fokou & Haller, 2008; Moritz et al., 2002). These institutions also regulated the coordination of other economic activities, such as trading fish and milk and herding of Kotoko livestock by nomadic groups.

The multiplication of fishing channels has resulted in a fragmentation of the territory and made it more difficult for herds to access grazing areas during the dry season. Other factors of competition must be considered, starting with the land tenure reform which in 1972 formalised land ownership by the state and transformed common property pasture into state property, which has thus become less accessible to nomads. At the same time, agricultural land also became less accessible, following the dam construction for irrigated rice cultivation and the expansion of the protected area of the Waza-Logone national park. The interpretation of an ethnic-based conflict between farmers and herders flattens and simplifies a long-standing process in which various factors are intertwined, which have changed the way of gaining access to natural resources and amplified the dynamics of rivalry at the local level (Pase et al., 2022).

In this context, grazing land has become scarce due to new restrictions and environmental regulations. The introduction of taxes by the state of Cameroon was a determining factor that led herders' groups from neighbouring regions to stop recognising the social institutions of Kotoko communities and respecting agreements to distribute access to resources (Fokou & Haller, 2008; Haller, 2020). As Haller et al. (2013, p. 8) puts it, "the nomads viewed this tax as a paid contract for free access to the pasture and were no longer interested in collaborating with coordination of land and fisheries use administered by the Kotoko".

These revenues are often used to support the livestock production chain. For example, the Mazera livestock market in the Zina Arrondissement and the Hounangaré market in the Logone-Birni Arrondissement owe their existence to the pastures of the plain. Statistics from the livestock services of the two communes show that since 2014, over 10,000 head of oxen and more than 15,000 head of small ruminants are sold annually at these two markets, before being transported to the Bogo livestock market in the Department of Diamaré, or directly to Nigeria.

Today, this situation is even more fragmented as a result of the decentralisation process. Access to grazing land on the plains is an important source of revenue for local administrations, and its use is subject to the payment of an annual tax levied by each of the communes of Bogo, Maga, Petté, Zina, Logone Birni and Waza. The amount of this tax varies from one commune to another. For example, the commune of Zina collects 10,000 CFA francs per herd of 50 heads of oxen. In this context, the conflicts that can arise from the lack of access to key resources during the season enable local administrators to receive bribes and other types of payments from all sides (fishers, agro-pastoralists and nomads), who may be somehow interested in maintaining the conflicts (Haller et al., 2013), following the logic of revenue generation and rent-seeking (Moritz, 2017).

In this framework, it is fairly common to observe that the initiatives of development actors are focused on other priorities, such as the constitution of professional organisations of herders or the strengthening of herders' cooperatives. For example, local organisations complain about the lack of coordination among the several hundred Common Initiative Groups that exist in the Logone floodplain or the current absence of any professional organisation exclusively dedicated to livestock farmers in the region. The issues of corruption to secure access to grazing areas and the breakdown of social institutions that have historically ensured the preservation of complementary links among different communities, interest groups and productive specialisations remain in the background. Instead, improving the management of production processes is prioritised to increase the value and rentability of economic activities, thus addressing the risk of impoverishment of local populations.

This approach is often based on a representation of local realities as lacking in evidence. In fact, these problems often refer to the lack of reliable statistics on pastoral livestock farming because the local livestock services and communes in the area do not have detailed data and information. This is a good example of the normative dimension of the development discourse, which can shape subjectivities centred on the divide between 'developed' and 'underdeveloped' subjects (Escobar, 1995). This discourse is also reinforced by the reference to climate change, portrayed as caused by catastrophic, unstoppable processes that are dissociated from current social dynamics and local practices. The priority given to the strengthening of productive processes is also influenced by the ideology of the sustainability of development interventions. It reflects the notion of environmentality or the making of environmentally sound subjects, who prioritise economic evaluation of natural resources, thus incorporating the state-driven view of sustainability (Agrawal, 2005).

Farmers' insecurity in irrigated areas

Here, we focus on the relations between conflicts and new agricultural practices that have been adopted by farmers who cultivate within the irrigated perimeters in the Logone floodplain. Although rain-fed rice is still grown (from June to December), it has declined considerably due to the change in the hydrological regime of the plain since some parts of the banks of Logone River have been dyked. Off-season millet, commonly known as *mouskwari*, is grown in the flooded areas on the periphery of the plain. Farmers have also developed the cultivation of irrigated rice from streams or boreholes using motor-driven pumps.

In this framework, manual weeding is mostly practised, and the use of animal or motorised traction is limited. At the same time, the use of herbicides and chemical fertilisers for the majority of crops is increasing and often done inappropriately

(CBLT & ACEEN, 2018). This issue is raised in an ecological context where soil fertility can be very important, thanks to the regular input of organic matter during floods. This has been observed elsewhere in the region, for example, in the Mamdi polder on the shores of Lake Chad, where plots can be cultivated for several years without the need for fertiliser, provided that drainage conditions are maintained.[1] This asset has been compromised by the development of irrigated areas, which have altered the natural drainage conditions of the soil.

The fluctuating price of fertilisers is a factor that can significantly weaken the purchasing power of the farmers of the plain. In January 2022, a 50-kg bag of fertiliser costs between 42,000 and 46,000 CFA francs, whereas the price for the same month in 2021 was 16,500 CFA francs. Additionally, the use of herbicides and chemicals to clear rice fields has a major harmful impact on grazing land and is a source of tension between farmers and livestock breeders.

Other factors have further undermined the farmers' livelihood and increased the sources of tensions with other groups. Firstly, we mention the institutional weakness of the public bodies that manage the irrigated areas, which are currently unable to maintain the hydraulic equipment or provide adequate support to local farmers. This problem is reflected in the working conditions of SEMRY employees, whose salaries are paid very late, sometimes in kind (e.g., bags of husked rice). In 2022, the company had only five machines of all types for its production work, particularly the ploughing of around 11,000 ha. Under these conditions, farmers ploughed their fields by hand using pickaxes and were obliged to pay a royalty of 51,000 CFA francs for a 0.5-ha plot.

The second issue concerns land grabbing in the dry plains of Diamaré by national and international actors. This involves the arrival of international operators to exploit oil resources on the plains. In 2009, a partnership was signed between the Government of Cameroon and Yan Chang Logone Development Company SA. Its area of operation covers around 85% of the plain's surface. Exploration work began in 2011 and confirmed the existence of a series of reservoirs of hydrocarbon 18 m high. It was suspended in 2015 when Chinese workers who were hired to rehabilitate National Highway No. 1 were taken hostage by armed groups in Waza. These events come on top of the episodes of land insecurity we mentioned earlier and contribute to increasing the feeling of dispossession and injustice among the people of the plain.

The poor management of irrigated areas also results in land tenure insecurity, which encourages something similar to land grabbing by local political elites. This dynamic has been observed at the regional level in other lowland areas. For example, in Chad, plots of arable land within irrigation schemes set up by SODELAC must be allocated first to the inhabitants of one of the villages benefiting from the projects,[2] and then to the most vulnerable families. These rules have not prevented the concentration of land in the hands of wealthier owners, who can hold up to 20–35 ha per person.

It is difficult to assimilate these practices into land grabbing. In fact, the informal allocation of plots of agricultural land occurs within the irrigated perimeters, without the involvement of foreign actors or investors. Nevertheless, these practices result in the impoverishment of farmers and an increase in socioeconomic inequalities. The dynamics at play in the Logone plain seem to correspond to the phenomenon of 'slow grabbing' observed in Central Asia from the 1990s onwards, where the agrarian reforms that followed the dissolution of the USSR favoured land appropriation by local elites and former civil servants, who were in a position to mobilise their networks and political influence (Pase et al., 2022). This shows the diversity of land grabbers and provides a better understanding of the land-grabbing practices in the Extreme North Region from the 1990s onwards, when the Cameroonian state's decentralisation reforms have given rise to land appropriation at the local level.

Local arrangements for the use of irrigated land have also been marginalised in favour of private operators. In Cameroon, the Ministry of Agriculture and Rural Development awarded three national operators for the use of 6521 of the 9521 ha of land in the Zina perimeter in a decision dated 26 March 2020.[3] Furthermore, the organisation of public initiatives, such as a lottery to allocate plots to young people in the Zina district, was perceived as an attempt to conceal the ongoing expropriation process.[4] Farmers who had been dispossessed took individual or collective action to demand compensation for lost land, which sometimes led to episodes of violence, as happened in Mazera (ACEEN 2020). National authorities described these actions as acts of rebellion and threatened farmers with imprisonment.

These clashes are the most recent episodes in a series of conflicts that should be taken into account. We might recall here the uprising of farmers on the plain in 2008, when the CEO of SEMRY set the surface area of a plot at 0.25 ha and raised the fee to 51,000 CFA francs. The situation degenerated into a riot that some authorities described as an insurrection. Clashes between the protesters and the army resulted in casualties and extensive damage, including the destruction of the private residence of the CEO, Maga and other SEMRY executives.

The alarming situation of hydraulic infrastructures is also evident in SEMRY I of Bongor, where the cultivated area was reduced due to the lack of water and the breakdown of electrical equipment for pumping water due to power cuts. These trends are clearly shown in the SEMRY maps presented in Chapter 8.

In fact, the pumping stations run on electricity, and in 2022, the energy deficit resulted in the drying up of at least 500 ha of plots that had already been replanted. In this situation, conflicts among farmers linked to episodes of water theft are becoming increasingly frequent.

These practices were already observed among local and immigrant farmers within the perimeter of failed agricultural schemes, such as SEMRY II (Bertoncin & Pase,

2015). Farmers reclaimed the use of wetland areas by using improvised pumps to irrigate a few hundred hectares, and in the lower zones where rainwater accumulates or inside the channels of the project, they returned to cultivating sorghum (Pase et al., 2022).

In this way, local farmers reaffirm the logics and social relationships that historically regulated access to shared resources, thus calling into question the technical and managerial know-how governing agricultural work under large-scheme development programmes. These practices emerged in the context of economic development and *mise en valeur* of local resources that marginalised subsistence-oriented production systems. These practices can be traced back to the local history of the Logone floodplain, when local resources were not considered exclusively as market goods but were at the centre of socioeconomic networks of mutual aid and reciprocity based on common interest.

Conclusions

In the Waza-Logone floodplains, conflicts over natural resources between farmers and herders are the most recent and visible phenomena in a broader process of growing competition over natural resources. We have shown how the breakdown of customary fishing practices led to the proliferation of canals and the individual appropriation of canals and fishing reserves. These dynamics undermined the complementarity among the various production systems (agriculture, fishing and livestock farming) that were historically established in the area. At the same time, changes in the organisation of access to other resources (agricultural land and grazing areas) reflect other dynamics and call into play other destabilising factors, such as land insecurity and corruption that amplify local inequalities.

Today, the causes of conflict and the measures to be taken have also become sensitive issues in the local political debate. This situation threatens to widen the divide between local communities, all the more so if we consider the migratory phenomena in the Logone floodplains and the impact of these dynamics on opposing groups.

In this chapter, we have also attempted to show how the analysis of conflict dynamics in the Logone floodplains can be strengthened by considering development as a discursive construction. The different narratives at play, such as the tragedy of the commons theory, only partially reflect local dynamics and risk simplifying a more complex situation. It is better to focus on the practices and strategies of local actors (digging of new fishing channels, disproportionate use of chemical fertilisers, new techniques for fish preservation, corruption and water theft) to account for the transformation of the social institutions regulating access to shared resources.

Our study's findings can contribute to the debate on the evaluation of development interventions. Beyond indicators that quantify their performance, it is also important to question the keys to understanding the strategies and practices of local actors, which often stand in contrast to the projects' objectives, and to introduce new tools to integrate these considerations into project planning.

Notes

1 Interview with the head of the research and development division for irrigated perimeters at the Société de développement du lac (SODELAC), conducted by D. Marsala in Bol on 24 May 2022.

2 For the Djadaya rural training and development centre (otherwise known as the Taiwanese area), these villages are Blabli, Zaraf, Seyaba and Bourmaneyé et Jardin présidentiel.

3 These operators are S.C.P SARL, AFRIFOOD and AGRI-INVEST. The remaining 2415 ha are allocated to communities comprising 8344 residents, representing 3120 households in 10 villages along the perimeter.

4 The press release for the event organised by the Ministry of Youth and Civic Education on 29 July 2021 states that "530 young people will be selected (...) to settle in the second-generation rice-growing village of Zina (...). The young people selected will receive free rice-growing plots, agricultural inputs and technical support to help them maintain their plots (...)" (source: Press Release). This initiative puts young people in the Logone plain at a disadvantage, as many of them ignore the existence of the National Youth Observatory.

References

ACEEN (Association Camerounaise pour l'Education Environnementale). (2002). *Enquête sur l'usage des produits chimiques pour la conservation des poissons séchés ou fumés dans la Plaine de Waza-Logone*. [Unpublished report].

ACEEN. (2020). *Périmètre hydro-agricole de Zina, un investissement à "fort regret" pour les communautés et l'écosystème la Plaine de Waza-Logone*. [Unpublished report].

Agrawal, A. (2005) Environmentality: Community, intimate government, and the making of environmental subjects in Kumaon, India. *Current anthropology*, *46*(2), 161–190.

Belal, E. (2003). *Les pêches amplifiées dans la plaine d'inondation de Waza-Logone (Yaérés) province de l'Extrême-Nord, Cameroun. Rapport général de l'atelier international sur l'amplification des ressources halieutiques à travers la gestion participative: leçons et perspectives*. Ministère de l'Agriculture, de l'Hydraulique et des ressources Halieutiques (Burkina Faso).

Bertoncin, M., & Pase, A. (2015). Progetti di sviluppo e trasformazioni territoriali: L'esempio di una zona umida saheliana (Camerun). *Rivista Geografica Italiana*, *122*(1), 15–32.

CBLT (Commission du Bassin du Lac Tchad). (2012, February 6–8). *Rapport de l'atelier de validation du projet de rapport final de l'étude de faisablité du projet de transfert d'eau de l'Oubangui au lac Tchad*. CBLT.

CBLT, & ACEEN (2018). *Plan de gestion de la plaine d'inondation de Waza-Logone* (1ère version révisée). CBLT.

de Noray, M.-L. (2002). *Waza Logone*. IUCN and University of Cambridge.

Ensminger, J. (1992). *Making a market: The institutional transformation of an African society*. Cambridge University Press.

Escobar, A. (1995). *Encountering development*. Princeton University Press.

Fokou, G., & Haller, T. (2008). Are local stakeholders conservationists? Livelihood insecurity and participatory management of Waza National Park, North Cameroon. In M. Galvin, & T. Haller (Eds.), *People, protected areas and global change: Participatory conservation in Latin America, Africa, Asia and Europe* (Vol. *3*, pp. 325–362). Perspectives of the Swiss National Centre of Competence in Research (NCCR) North-South, University of Bern.

Haller, T. (2002). Common property resource management, institutional change and conflicts in African floodplain wetlands. *African Anthropologist*, *9*(1), 25–35.

Haller, T. (Ed.) (2010). *Disputing the floodplains. Institutional change and the politics of resource management in African wetlands*. Brill.
Haller, T. (2020). Institution shopping and resilience grabbing: Changing scapes and grabbing pastoral commons in African foodplain wetlands.. *Conservation and Society*, *18*(3), 252–267.
Haller, T., Fokou, G., Mbeyale, G., & Meroka, P. (2013). How fit turns into misfit and back: Institutional transformations of pastoral commons in African floodplains. *Ecology and Society*, *18*(1), 1–16.
Helmke, G., & Levistsky, S. (2004). Informal institutions and comparative politics: A research agenda. *Perspective on Politics*, *2*(4), 725–740. .
Khari, B. (2011). *Impact socioéconomique des conflits liés aux ressources halieutiques dans la plaine d'inondation de Waza-Logone (Extrême-Nord Cameroun)*. Mémoire d'Ingénieur de conception en Agronomie, Economie et sociologie rurale, Faculté d'Agronomie et des Sciences Agricoles, Université de Dschang, Cameroun.
Labara, K., Djaowe, A., Désiré, T., Chouto, S., Chendjou, F. X., & Tagne, R. (2020). La pêche artisanale par canaux dans la plaine du Logone: Une rentabilité source de conflits entre pêcheurs. *Bulletin de la Société Géographique de Liège*, *75*(2), 83–100. https://doi.org/10.25518/0770-7576.6109
Laborde, S., Fernández, A., Phang, S. C., Hamilton, I. M., Henry, N., Jung, H. C., Mahamat, A., Ahmadou, M., Labara, B. K., Kari, S., & Duran, M. (2016). Social-ecological feedbacks lead to unsustainable lock-in in an inland fishery. *Global Environmental Change*, *41*, 13–25. https://doi.org/10.1016/j.gloenvcha.2016.08.004
Landolt, G. (2010). Lost control, legal pluralism and damming the flood: Changing institutions among the Musgum and Kotoko of the village Lahaï in the Waza-Logone floodplain (Cameroon). In T. Haller (Ed.), *Disputing the floodplains. Institutional change and the politics of resource management in African wetlands* (pp. 171–194). Brill.
Long, N. (2001). *Development sociology: Actor perspectives*. Routledge.
Long, N., & Long, A. (Eds.) (1992). *Battlefields of knowledge: The interlocking of theory and practice in social research and development*. Routledge.
Loth, P. (2004). *The return of the water: Restoring the Waza Logone floodplain in Cameroon*. IUCN (International Union for Conservation of Nature).
Magrin, G. (2016). The disappearance of Lake Chad: History of a myth. *Journal of Political Ecology*, *23*(1), 204–222. https://doi.org/10.2458/v23i1.20191
Moritz, M. (2006). The politics of permanent conflict: Farmer-herder conflicts in Northern Cameroon. *Canadian Journal of African Studies*, *40*(1), 101–126. https://doi.org/10.1080/00083968.2006.10751337
Moritz, M. (2017). Misreading a pastoral property regime in the Logone floodplain, Cameroon. *Ecology and Society*, *22*(1), 13–21.
Moritz, M., Scholte, P., & Kari, S. (2002). The demise of the nomadic contract: Arrangements and rangelands under pressure in the Far North of Cameroon. *Nomadic Peoples*, *6*(1), 124–143.
Olivier de Sardan, J.-P.. (2013). Embeddedness and informal norms: Institutionalisms and anthropology. *Critique of Anthropology*, *33*(3), 280–299. https://doi.org/10.1177/0308275X13490307
Pase, A., Hofman, I., García, A. K., Haller, T., Cirillo, D., Giger, M., & Bertoncin, M. (2022). Large-scale agricultural investments in drylands: Facing some blind spots in the grabbing debate. In A. K. García, T. Haller, H. Van Dijk, C. Samimi, & J. Warner (Eds.), *Drylands facing change* (pp. 73–92). Routledge.
Seignobos, C., & Iyébi-Mandjek, O. (2000). *Atlas de l'Extreme-Nord du Cameroun*. IRD Éditions.
Turner, M. D. (2002). *Moral dimensions of 'resource struggles' and their treatment in political ecology*. Association of American Geographers.
Ziébé, R. (2015). *Fisheries in the Waza Logone floodplain: An analysis of the status of the fisheries sector and mitigation of conflicts within the sector in North Cameroon*. Leiden University.

Part V

Towards a Fluid Research Agenda

16 COVID-19, polycrisis and metamorphosis

A reflection on doing collaborative research in, on and with the Sahel

Angela Kronenburg García, Mariasole Pepa, Andrea Pase, Dobah Marsala, Moumouni Ouande, Aboukar Mahamat, Ahmed Abdelwahid Abdelrahim and Francesco Staro

Introduction

On 11 March 2020, the World Health Organization (WHO) declared COVID-19, the coronavirus disease that was first detected in China in 2019, a global pandemic.[1] Soon after and following in the footsteps of China and Italy, more and more countries started to close their borders, shut down schools and introduce curfews, lockdowns and other measures to contain the spread of the coronavirus. The COVID-19 pandemic and the introduction of containment measures across the world coincided with the start of the Contested Wetlands in Sahelian Drylands project, a three-year research project funded by the Italian Ministry of University and Research that aimed to study the politics of development in Sahelian wetlands and large-scale irrigation projects.[2] The research was designed around fieldwork to be carried out by members of the research team, which initially comprised three University of Padua researchers (Marina Bertoncin, Andrea Pase and Angela Kronenburg García). However, as the COVID-19 pandemic dragged on and movement restrictions persisted, it soon became clear that the project needed to be redesigned. International travel restrictions in particular made it difficult for the Padua researchers to plan and conduct fieldwork in the Sahel.

Inspired by an emerging model of collaboration in another project that also involved the Padua researchers (see Roque de Pinho et al., 2023),[3] the Padua project team quickly adopted the other project's research methodology and started to recruit collaborative researchers ("co-researchers") from and based in the countries of research interest. The co-researchers included a young professional working in a company (Ahmed Abdelwahid Abdelrahim), an environmentalist leading an NGO (Aboukar Mahamat) and two students, Dobah Marsala and Moumouni Ouande, pursuing their PhD degrees at the University of Yaoundé II and the University of Joseph Ki-Zerbo, respectively. Ahmed was based in Khartoum, the capital of Sudan, and Moumouni resided in Ouagadougou, Burkina Faso. Both Aboukar and Dobah lived in the Lake Chad region, the former on the western side of the Logone River

DOI: 10.4324/9781003429111-23

(a tributary of Chari River) in northern Cameroon and the latter on the eastern side of the river in Chad. The co-researchers were tasked with collecting data on the current state of specific irrigation projects and wetland areas, as well as the impacts of the COVID-19 pandemic on their countries. Their research on wetlands and irrigation projects is featured in Chapters 11, 12 and 15 of this volume. In this chapter, we report their work on the COVID-19 impacts *and* reflect on the experience of doing collaborative research in times of crisis. In doing so, we use the concept of "metamorphosis" (Beck, 2016) to explore how the COVID-19 crisis opened an opportunity to critically rethink our ways of doing research.

The research material on the COVID-19 impacts that is used for writing this chapter comes from the reports, images and videos produced by the co-researchers. The reports are based on fieldwork (interviews and observations) and document/archival research, following a list of guiding questions, initially developed by the Padua researchers but redefined collectively over time. The co-researchers sent the reports to the Padua researcher with whom they were paired. Ahmed (Sudan) was partnered with Angela. Moumouni (Burkina Faso), Aboukar (Cameroon) and Dobah (Chad) first worked with Andrea, later with Mariasole Pepa, who joined the Padua team in 2021, and in 2024 during the last (writing) phase, also with Francesco Staro. Ahmed sent three reports on the COVID-19 situation in Khartoum, covering December 2020, January 2021 and February 2021. Moumouni sent two detailed reports of a survey conducted between 30 January and 21 February 2021 in Sourou Valley and between 6 May and 13 June 2022 in the town of Tougan. Both Aboukar and Dobah included a section on the COVID situation in their update reports on the wetlands and irrigation projects rather than sending separate COVID reports. Aboukar's regular reports on Cameroon covered September, October and December 2020; January–May, June–August and September–December 2021; and January–February, March–April and June–August 2022. Dobah sent reports on the Chad situation in September and December 2020 and March 2021.

The reflections on collaborative research in this chapter include the experiences and perspectives of co-researchers Moumouni, Aboukar and Dobah, as well as Padua researchers Angela (based in Mozambique), Andrea and Mariasole (based in Italy) and Francesco (based in France).[4] They also comprise thoughts on the research process as the design, approach and team composition changed; on the writing process; and on the impacts of the COVID-19 pandemic and other crises on our work and lives. The reflections from the co-researchers come from their responses to a short and open questionnaire geared towards understanding their challenges, ideas and experiences. The reflections of the Padua researchers were added directly to the text during the writing of this chapter.

The Sahel, particularly the wetland areas (as crucial livelihood resources for dryland people; see Kronenburg García et al., 2023), is currently facing numerous escalating crises, including jihadist insurgencies (Chapter 4), wars (see Chapter 10 for the war in Sudan), widespread food insecurity, climate change (Chapter 3) and many more (see Olivier de Sardan, 2023). The term "polycrisis" seems fit for capturing much of what is happening in the Sahel since it refers to "the compounding nature of the interactions between multiple intersecting crises" (Jakupec et al., 2024). Although the concept of polycrisis was originally coined to describe the situation in the world (climate change, pending nuclear war, a pandemic, a global economic downturn, etc.), we argue that in the Sahel, it is more acute, so much so that it is restricting scholars' ability to do research, producing a research crisis as well. Henig and Knight (2023) advance an anthropological take on the concept that helps make sense of the complexity of polycrisis in particular situations. They propose a lens of "knotted eventedness" (a set of knotted events) to trace the connections and relations within and across spatial and temporal scales in order to understand "life-in-(poly)crisis" (pp. 4, 6).[5] In this chapter, we examine the particular "knot" in time (roughly March 2020–August 2022) when COVID-19 was added to the Sahelian crisis mix. Early reports on COVID-19 impacts in dryland areas (Roque de Pinho et al., 2023) suggested that it was not so much the disease (in terms of casualties and health impacts) but the harsh lockdown measures that mostly had detrimental consequences and challenged the lives and livelihoods of people in the Sahel. In this chapter, we therefore focus on the impacts of COVID-19 *containment measures*; additionally, we explore how they interacted with other crises in the region.

The chapter is structured in two parts. Part I is dedicated to the COVID-19 work of the co-researchers, showing how the COVID crisis interacted with the various ongoing crises and challenges in the different study sites in the Sahel. Part II is devoted to our reflections on doing collaborative research in times of COVID-19 and polycrisis, as well as our ruminations on how this experience might be expressive of a broader metamorphosis of the way we do research. In line with the latter, we share our final thoughts in an unconventional manner.

Part I: COVID-19 lockdown impacts in the Sahel

Khartoum City, Sudan

Khartoum, a large metropolitan city located at the confluence of the White and Blue Nile Rivers, was under strict lockdown for about three months (between April and July 2020). During this period, schools and universities were shut down, group gatherings and prayers in mosques suspended, government services halted, local markets and restaurants closed and a 15-hour curfew imposed, while small grocery stores and pharmacies were allowed to remain open for limited hours.

The COVID-19 pandemic occurred when Sudan was undergoing a political transition and a worsening economic crisis. Sudan had been ruled for nearly 30 years by Omar al-Bashir until a 2019 coup d'état put an end to his dictatorship. The coup was preceded by four months of popular protests (called the December revolution) that had their precedent in mass demonstrations against austerity measures, shortages of essential imported products (food, fuel and medicines) and rising prices since 2011 when the secession of oil-rich South Sudan threw the country into an economic crisis (Kostelyanets, 2022). Given that Sudan imported most of its food, food prices were severely affected by the economic crisis.

One of the groups active in the mass protests were the so-called resistance committees, which would play a role in mitigating the negative impacts of the COVID-19 pandemic. Resistance committees are youth-led groups organised at the neighbourhood level and known for setting up non-violent gatherings ("sit-ins") that can last for weeks (Bishai, 2023). They emerged in 2013 in Khartoum, following a deadly crackdown of a protest, and have since then evolved into a decentralised network with branches in most of Sudan's towns organising civil disobedience campaigns (Kostelyanets, 2022). However, resistance committees are more than that. Bishai describes how resistance committees "have offered – at the level of the neighbourhoods – many of the services that the state stopped offering years ago. They clean up and take out the trash, they aid needy students, and have organized community-based health check-ups" (2023, p. 1481). During the COVID-19 lockdown in Khartoum, they once more took up the role of public service providers, assisting the transitional government in the fight against the coronavirus.

Ahmed reported that resistance committees in Khartoum spread awareness of the coronavirus. They delivered bread and cooking gas to households in their neighbourhoods. They sterilised vehicles and public spaces; furthermore, in cooperation with pharmacists, they produced local sanitisers, sanitising soap and face masks according to WHO standards and distributed them to the public. One resistance committee (from the Amarat neighbourhood in central Khartoum) worked to provide health workers and patients with meals in hospitals. When people were again allowed to attend services at the mosque, the committee placed markers on the floor to ensure social distancing during prayers, which was immediately adopted by resistance committees in other locations.

Though resistance committees brought some relief during the pandemic, the combination of an ongoing economic crisis and COVID-19 restrictions aggravated people's living conditions. Mobility restrictions hindered trade and contributed to increasing prices and food insecurity.[6] People were more concerned about rising prices, lost incomes and the few hours per day when they could buy food and other necessities than about coronavirus infection. The lockdown coincided with Ramadan; some considered it a blessing because the fasting automatically cancelled two meals a day.

Lake Chad region (Chad and Cameroon)

The Lake Chad region is found at the intersection of four countries: Niger, Chad, Nigeria and Cameroon. Long before the COVID-19 pandemic, this region was already facing multiple challenges. It is considered one of the most insecure places in the world, mostly due to attacks by Boko Haram and the counter-insurgency, and has been plagued by extreme poverty (Iocchi, 2022; Owonikoko, 2021).

Starting in March 2020, the COVID-19 lockdown measures introduced in the region included border closures, curfews, travel restrictions, bans on large gatherings and the closure of markets, shops, restaurants, schools and mosques (Owonikoko, 2021). Although some governments (e.g., Chad) put in place a number of initiatives to mitigate the impacts of the COVID-19 containment measures, the latter made things worse, especially food insecurity, in the region. Co-researcher Aboukar reported that in the Cameroonian portion of the Lake Chad region (the Far North Region), people perceived the measures as burdens and preferred to die of COVID than from hunger.

The economy of the Lake Chad region depends on cross-border trade, and border closures reduced the flow of people, livestock and goods, with rising prices as a consequence. For example, Chad depends heavily on basic foodstuffs and medicines coming from Nigeria, Cameroon, Sudan and Benin, and with the imposition of COVID-19 measures, the prices of these commodities soared. Livestock markets and (cross-border) transhumance had already been disrupted by insecurity, but market closures during the lockdown meant that pastoralists now also struggled to sell their animals and therefore had less cash to buy agricultural produce, which was rotting on the stalls along the main roads due to the lack of buyers (Association pour la Promotion de l'Elevage au Sahel et en Savane (APESS), 2020).

The Far North Region of Cameroon was already declared an "economic disaster zone" in 2019, that is, an area where economic activity is disrupted by insecurity and natural disasters (e.g., floods).[7] COVID-19 cross-border restrictions intensified the economic downturn. Price spikes of products imported from Chad and Nigeria forced many Cameroonian households to change their eating habits. Some reduced their meals from three to two per day; others, who used to eat rice and other imported products, began consuming mainly local produce. To make things worse, Cameroonian rice (particularly the one produced by SEMRY, a large state-owned irrigated rice project; see Chapters 11 and 15) was fraudulently exported to Nigeria, which was no longer able to import rice from Asia. Never in the history of the Far North Region had husked rice cost so much as during the pandemic (55,000 francs per 100-kg bag). Rice was bought in the fields in advance (i.e., before being harvested). Local authorities were complicit in the fraudulent export of food products from Cameroon to Nigeria. The Governor of the Far North Region, who was also the Chairman of the Board of Directors of SEMRY, had acquired over 200 tonnes of rice, which he sold to Nigeria.

According to Owonikoko (2021), COVID-19 lockdowns and containment measures in the Lake Chad region not only exacerbated poverty and hunger but also worsened the security situation. With their renewed attacks, Boko Haram and other violent groups seemed to have taken advantage of the military re-deployment to the COVID-19 emergency. It was also feared that these groups might exploit the situation of widespread poverty aggravated by the pandemic to recruit new members.

Sourou Valley, Burkina Faso

Sourou Valley in northwestern Burkina Faso, along the border with Mali, was already under a state of emergency and under curfew due to multiple terrorist attacks when the first COVID-19 cases appeared in the country in March 2020.[8] The curfew was extended to the rest of the country, and additional measures were introduced to curb the spread of the virus, including the closure of schools, borders and places of worship (mosques and churches), a ban on urban and interurban public transport and the quarantine of infected towns and people. However, unlike the markets of the large cities (Ouagadougou and Bobo Dioulasso), which were closed, those of Sourou Valley (Di, Gouran and Lanfiera) remained open.

Border closures, travel bans and the quarantine of major cities significantly affected trade in Sourou Valley. Traders and transporters coming from Ouagadougou and Bobo Dioulasso and neighbouring countries (Mali, Ghana and Ivory Coast) could not supply the valley anymore with finished products and buy local produce in turn. The prices of fresh produce (tomato, cabbage, cucumber, etc.) thus fell, reducing the incomes of local farmers and, conversely, increasing the costs of imported products (oil, salt, soap, flour, fuel, beverages, mechanical equipment and medicines). However, the immediate impact on food security did not seem to have been as severe as in the Lake Chad region because most farmers do not rely on the purchase of basic foodstuffs for their food security and had already stored most of the crops (corn, millet and sorghum) that they usually consume. However, the stoppage of transport meant that agricultural inputs (fertilisers and seeds) arrived late. It was also not possible to bring in the technicians and mechanical parts needed to repair machinery that had broken down in the areas where irrigation agriculture schemes had been implemented (see also Chapter 12). Together, all of these factors have had adverse effects on crop yields and thus potentially on household food stocks in the future.

The COVID-19 pandemic's impacts on agricultural production and food security must be contextualised in the longer history of the worsening local economy and deteriorating security in the region. The latter is related to long-term tensions and clashes, often between farmers and herders over access to land, that have more recently been capitalised on by terrorist groups (see Chapter 12). In fact, it was not COVID-19 but the terrorist attacks since 2018 that were locally considered the main causes of food insecurity in Sourou Valley because such violence prevented farmers from working on their fields. As a representative of the Village Development Committee of Di commented:

> In our commune, several inhabitants of villages bordering Mali were found dead in their fields, killed by the terrorists; some farmers received threats.

> Since then, many farmers have abandoned their fields. Until now, they cannot return there to carry out their work.

Terrorist attacks thus hinder farmers' ability to meet the food needs of their families.

COVID-19 most certainly made things worse, but the insecurity that coincided with the pandemic overshadowed the latter's impacts and seemed to dominate the local discourse and crisis experience.

Zooming in on the COVID-19-insecurity "knot" in Sourou Valley

The report by Burkina Faso co-researcher Moumouni on COVID-19's socioeconomic impacts on Sourou Valley makes several references to the local insecurity situation. This gives hints on how the pandemic became entangled in various ways with the insecurity situation. Here, we focus on some of these hints.

The security crisis featured in some of the local explanations regarding the coronavirus disease. Some people interpreted COVID-19 as a divine punishment because of the killings. For them, the coronavirus was a curse, a warning, a manifestation of divine wrath to punish humans (the unbelievers) for their various transgressions. Only prayer was thought to help remedy this situation. This reasoning circulated next to other explanations, such as the idea that the coronavirus disease was a reaction to the loss of customary values, particularly the reduced use of sacrifices to prevent disease and the belief that the disease only affected the rich and westerners (whites) and would therefore have no adverse effects on the poor and Africans in general, whose bodies were also more resistant to the virus. The latter fed the general reluctance among the youth to be vaccinated. As a transport operator asserted, "I don't even believe that COVID-19 exists in Burkina; it's a white man's disease. Rumour has it that vaccines have side effects, making men sterile and reducing life expectancy".

The combination of COVID-19 containment measures and terrorist attacks also interrupted education and environmental conservation. During the lockdown, public and private primary and secondary schools were closed. However, well before the advent of the COVID-19 crisis, some schools in Sourou Valley were already closed due to terrorist attacks and threats. When containment measures were relaxed in Burkina Faso and schools were allowed to reopen, schools in two municipalities of Sourou Valley (Di and Lanfiera) remained closed due to the insecurity situation. Containment measures have led to the state's reallocation of financial resources to prioritise the fight against COVID-19. It resulted in reduced funding for environmental organisations, which were thus unable to carry out their conservation-related activities properly. However, it was recognised that the main factor restricting conservation was the prevalence of terrorist attacks. As the chief forester of the commune of Di put it, "These days, all our colleagues have left their posts in Sourou Valley because they have been threatened by terrorists. As a result, they are unable to patrol the protected areas". In fact, with the absence of forest guards, protected areas (*forêts classées* [classified forests]) have now become hiding places for terrorist groups. This means that no one dares to venture there anymore.

The insecurity situation is a context that cannot be ignored. Between 2018 and 2022, several incidents and terrorist attacks occurred in Sourou Valley, including destruction of assets and buildings, roadblocks, kidnappings and murders. The attacks seemed to be directed at people, infrastructures and institutions associated with the state, such as the police and national security forces, as well as hydraulic infrastructures, agricultural development projects and agribusinesses. The fear of violent attacks is palpable from the Burkina Faso report, and the hints mentioned above show how such anxiety affected various dimensions of life and work in Sourou Valley. This stress was added to the public dread and panic concerning the coronavirus infection, especially at the initial stages of the pandemic.

COVID-19 pandemic as a minor crisis in the Sahelian polycrisis

From the preceding discussions, we can tentatively conclude that the COVID-19 pandemic was considered a minor crisis in the Sahel. Its impact as a disease that infects and kills people was markedly lower than in other parts of Africa (Al-Saidi et al., 2023) and the world,[9] feeding beliefs of it being a white person's (and Chinese) disease and even scepticism about its existence. It is therefore no surprise that COVID-19 was not experienced as a "real" crisis in the Sahel, negligible in the face of other more severe and acute crises in the region (e.g., insecurity in Sourou Valley and in the Lake Chad region). However, the measures introduced to contain the disease, particularly the lockdowns, did make life more difficult for people in the Sahel by either adding new challenges or deepening existing ones in an already challenging situation. It may well be said that COVID-19 further thickened the knots of the Sahelian polycrisis.

Part II: Reflections

On doing research in times of (poly)crisis

We, both researchers and co-researchers, found ourselves trying to conduct research in times of crisis. The COVID-19 pandemic prevented the Padua researchers from doing fieldwork in the Sahel. Although for the co-researchers, fieldwork was arguably easier because they were already on-site – or at least close(r) to the study sites – the presence of other crises in the region added another layer of difficulty, if not impossibility, in doing research. This issue points to another crisis in the Sahel, a research crisis, especially research based on the collection of primary data through fieldwork (see also Deb & Baudais, 2022).

In 2022, the Lake Chad region experienced the heaviest rainfall recorded over the past 30 years, resulting in extreme flooding in the area (see also Chapter 11).

Aboukar from Cameroon reported how the flooding complicated access to the Logone plain (his study site), needing at least an outboard pirogue to reach the place. In Sourou Valley, it was primarily the security crisis that restricted the work of the Burkinabe co-researcher. As Moumouni explained:

> The majority of respondents refused to be filmed, citing their concern about the security situation in the area. In addition, despite the presentation of the document signed by your university, the survey in urban areas was faced with refusals to collaborate. We preferred to stop collecting data in urban areas.

The widespread fear and mistrust related to the insecurity situation also had an emotional impact on the co-researcher. Reflecting on the challenges of doing research in such circumstances, Moumouni explained:

> On an emotional level, there was the psychosis caused by the presence of terrorists in the study area. Terrorist groups can be recognised by the way they dress, but their informers, who slip in and out of the population, cannot be identified. In addition, those who have attended school [like Moumouni] are [perceived as] systematically assimilated into the west and treated as such. Mistrust of anyone you meet in the field is obvious because you don't know who you're dealing with. Finally, there is the fear of being kidnapped or ambushed by these terrorist groups.

In 2021, the area even became completely inaccessible due to the insecurity, and further fieldwork plans there were cancelled. The research crisis in the Sahel is not only about access to study sites but also concerns the safety of researchers and study participants, thus creating clear ethical issues.

The concept of polycrisis, as advanced by Henig and Knight (2023), draws attention to situatedness, that is, how people are situated at the intersection of multiple crises. This situatedness also applies to us as researchers since we might find ourselves in (crisis) situations that constrain our ability to do research. During our project, these situations included the COVID-19 pandemic, floods, political turmoil, lack of security and precarious access to electricity and the internet. Some co-researchers "disappeared" for periods of time, in the fog of unconnectedness, leaving the others worried about their wellbeing and fate. Prolonged sickness was one of the reasons, as well as the outbreak of war (e.g., contact with Ahmed, who fled the violence in Sudan after the coup of 2021, was only restored in 2024). Thus, researchers' situatedness in (poly)crisis involves more than just context affecting research because it concerns our very own bodies and how emotions shape our bodies (Ahmed, 2004), wellbeing and lives, as well as how we navigate this vulnerability (Behar, 2022). No one is safe from infection, as we learned so well in Europe with the COVID-19 pandemic, which, we could say, was for many people the first experience of a real-life crisis at home, not witnessed in Europe since World War II.

The research crisis in the Sahel has been shaped by the evolving polycrisis in the region, particularly the mounting insecurity and political instability, and affects both local and foreign researchers. Padua researchers Marina and Andrea, who have more than 30 years of research experience in the Sahel, experienced the development of the Sahelian research crisis up close. In the 1990s, they carried out fieldwork in Mauritania, Senegal, Mali, Burkina Faso, Nigeria, Niger, Cameroon, Chad and Sudan without much problem. Any difficulties were related to logistics – it was, after all, before the era of mobile phones, to mention just one practical aspect. At the turn of the millennium, things started to change, and the space for doing research started to close. The Italian Ministry of Foreign Affairs began to mark some areas in red, advising its citizens against travelling to those locations because they were too dangerous and eventually forbidding them to do so. The red mark started to invade the entire Sahel, leaving only a few scattered places untouched. Hostility against Andrea and Marina started to emerge in many forms, and they increasingly felt as if they were crossing moorlands swept by an "angry wind" (Tayler, 2005). For example, in Mauritania, they felt threatened by gestures of knives slashing throats directed at them. They wondered what images, emotions and discourses were evoked by their presence in the field. At the same time, there was always the desire for exchange and dialogue, mutual curiosity and openness in interactions with Sahelian colleagues, as well as with the staff of large-scale irrigation projects, farmers, herders and people in the villages and along the roads. These moments of friendly interaction kept them going for as long as possible.

Even before the COVID-19 outbreak, it was already decided that the fieldwork by Padua researchers (Angela and Andrea) would be limited to Sudan, given the insecurity in the other study sites (the Lake Chad region, Sourou Valley in Burkina Faso and the Office du Niger in Mali). At that time, Sudan was full of hope for a better future, following the December 2019 revolution and the formation of the transitional government, and was considered safe for fieldwork purposes, but when the pandemic broke out, that country also became off-limits. Angela remained in Maputo, Mozambique, where she was living. Initially, she experienced a period of panic when many foreigners were evacuated or left the country, fearing that the already fragile Mozambican health system would collapse as infections would increase. However, against all expectations, registered COVID-19 cases and deaths stayed relatively low, especially in comparison to the countries where foreigners and expatriates had gone. The lockdown in Mozambique was tough and strictly enforced in the city, but Angela was not greatly affected, privileged as she was in being able to stay and work from home, unlike many (urban) Mozambicans whose livelihoods depended on being out there on the streets and working in informal markets. Andrea's experience of the pandemic was different. Italy was the first European country to be affected by the coronavirus and, with almost a thousand deaths a day at its worst, was hit very hard. Moreover, the place where Andrea lived, a small town in the Veneto region, was one of the first red zones in Italy, with red – in this case – indicating the danger of viral infection and the need for strict lockdown (Pase et al., 2021). COVID-19 had a strong impact on Andrea's family; his wife, Morena, was working as a doctor in the hospital that became the first designated COVID hospital in Italy. Days were filled with worries of infection (especially of Morena) and feelings of danger. Everyday life was severely affected in Italy; in the worst period, people were not allowed to move beyond a 200-m radius around their homes. Streets were deserted; only the lights and sirens of the police and ambulances broke the silent desolation of the cities and villages.

Amid the COVID-19 pandemic, Angela and Andrea, together with other researchers working in dryland areas, started to meet online and initiated a working group to follow the impacts of and responses to COVID-19 restrictions in dryland areas in Africa, Asia and North America. They reconnected with friends, colleagues/students and former research assistants living in the drylands and invited them to the working group as co-researchers and, with the support of several grants, they began documenting life under lockdown in the drylands. The working group grew and was joined, among others, by the other Padua team members (Marina, Mariasole and the co-researchers). The working group, with its weekly meetings held over a two-year period (approximately), offered some relief and an opportunity to connect (online and remotely) at such a time of stay-at-home and social distancing policies. It provided a platform for sharing personal experiences and concerns, especially at the beginning of the pandemic when much uncertainty and anxiety existed, as well as for developing a research project based on collaboration with co-researchers. Researchers and co-researchers communicated through phone calls, WhatsApp messages and email. This collaborative project culminated with a workshop held in Kenya in 2022 (when COVID-19 restrictions were lifted), which brought together most of the researchers and co-researchers of the working group.

Faced with the pandemic, the Padua researchers, particularly Andrea and Angela, had to find ways of still making the research project happen. Inspired by the evolving collaborative methodology of the COVID-19 in the drylands working group, they reached out and started organising the work with co-researchers.

On doing collaborative research

Our collaboration has been an ad hoc (it was unplanned) and experimental (trial-and-error) process, as well as a new experience for the researchers and co-researchers involved. It has been an adventure, a valuable co-learning journey and an eye-opener. For the Padua researchers, it has involved letting go (understanding that not everything can be controlled and predicted), adapting and being sensitive to unexpected events and the living circumstances of the co-researchers (see also Roque de Pinho et al., 2023). The co-researchers appreciated the trust placed in them to carry out the research, the autonomy to plan the various activities and the flexibility to adapt the study to the context of each country and integrate new research topics. However, the process has encountered challenges. There is much to be learned and unlearned from this adventure, step by step and together.

In their written experiences that reflect on the collaboration, the co-researchers express their gratitude for the opportunity to participate in the research project and the experience of working with people from different cultures and disciplinary backgrounds. However, in the same breath, they mention that the research could have been more fruitful if, for example, they had been involved in the project design from the beginning. In this case, the co-researchers flag a common feature of North–South academic collaborations, whereby researchers in the North are often in decision-making positions, setting research agenda objectives, while those in the South are "merely" involved as data collectors and considered passive researchers rather than key actors of the research process (Tilley & Kalina, 2021). Power imbalances such as these, which are partly shaped by funding structures, are difficult to dislodge, and this issue was a recurring preoccupation of the Padua researchers. As Mariasole explained:

> It was the first time for me to "coordinate" co-researchers. I understood it as a horizontal process, in which I was the medium between co-researchers and the research. Yet, even though my intention was good, that [did] not mean that I [was able] to mitigate power imbalances throughout the process.

Additionally, for Mariasole, as a young, white woman working with male co-researchers, usually older than her, it took time for her to feel comfortable in her position as a coordinator.

Francesco, who joined the team towards the end of the project, during the writing of outputs, expressed similar feelings:

> It was difficult to review the material sent by the co-researchers and evaluate which parts were to be included in the final publication. Personally, I felt the risk of reaffirming hierarchies between us, and this led me to search for an equilibrium in the process of writing between their field-based knowledge and my external point of view.

Communication was regarded as a key challenge of the collaboration. Communication was unavoidably remote and online (via Zoom, email and WhatsApp), saddled by difficulties in accessing the internet and electricity on the side of the co-researchers. Zoom meetings needed a translator because of the language barrier, which complicated mutual understanding. In Dobah's case, communication started indirectly, through an intermediary: "In the beginning, it was quite difficult because we didn't know exactly what the Padua researchers wanted .... But as time went on, things improved [with direct contact], and we were able to make the most of the collaboration". In fact, over time, communication between Dobah and Mariasole became more frequent, which led to an inspiring and mutually enriching journey. As they got to know each other better – personally and, for example, regarding each other's stance about "development" – they discovered a shared interest in postcolonial and decolonial studies and decided to explore this further, beyond the Padua project (see Pepa & Marsala, 2024). As Mariasole explained:

> With Dobah, we decided to dive into a process of collective autoethnography to reflect on our ways of thinking about "development". I learnt a lot in the process, [not only] about the materiality of attempting to weave alternative academic partnerships, but also about other possible ways of thinking, doing, feeling in the research process.

The Padua side also encountered administrative challenges that needed to be overcome, particularly by Andrea as the project leader. The proposed way of working with co-researchers and remunerating them for their work was unprecedented at the Department of Historical and Geographic Sciences and the Ancient World, as well as quite unusual and atypical for the University of Padua at large. Italy has very strict audit and transparency policies, further complicating the matter. The question was how to justify the awarding of contracts (even if these covered small sums of money) in a direct manner (i.e., without institutional mediation), without a competitive selection and at such a great distance, measured not only in kilometres but also in terms of language and administrative and accounting procedures. Andrea and the technical research managers (Roberto Galati and Valentina Casari) spent hours thinking about and constructing new procedures to be able to mandate, monitor and remunerate the co-researchers. All of these happened in the midst of the pandemic (and thus while working remotely) and would not have been possible without the willingness and inventiveness of Roberto and Valentina.

For Angela, who had no research experience in the study sites and thus lacked a good understanding of local realities, it was challenging to contextualise the reports sent by Ahmed – and subsequently, to analyse the reports prepared by the other co-researchers and draft or contribute to chapters without having "been there". Mariasole and Francesco were both familiar with Sudan but struggled with the same issues in the other regions. For their part, the co-researchers also worried that their reports were not enough for the Padua researchers to really understand what was occurring on the ground.

Finally, as mentioned before, it was a long-distance collaboration. The co-researchers indicated that they missed face-to-face interactions and the opportunity to get to know their colleagues in person because the formality of this remote collaboration was not easily overcome. Nevertheless, even if carried out at a distance and mediated by screens, research collaborations still entail human interactions ("We are not robots", Mariasole said); as such, we have feelings. Mariasole was so happy to see a picture of Dobah and Miriam's wedding. When Aboukar and Ahmed reconnected with the Padua team after a long period, there was a collective sigh of relief. Andrea and Mariasole, who were in Sudan shortly before the war broke out, passed through periods of apprehension and emotional stress, fearing to receive bad news about their friends and colleagues back in Sudan. As Mariasole said, "It is not just about research, it is about real life, it is about weaving relationships that go beyond fieldwork, it is about keeping in contact and caring about others' [lives]".

On things that happen during metamorphosis

The world is "metamorphosing", according to sociologist Ulrich Beck (2016). He argues that out of the turmoil of our world, "something quite new is emerging" (p. 3). The world is not simply changing in the face of significant global events, such as climate change, but transforming radically, challenging and destabilising our certainties and understandings of the world. Metamorphosis is about "the positive side effects of bads" that give way to "entirely different ways of conceptualizing the world" (p. 6). In this sense, immersion in a destabilising context filled with "bads" – the (poly)crisis – may open up fissures from which new ways of knowing this metamorphosed world can sprout. We would like to think along these lines and consider the COVID-19 pandemic as an influencer of metamorphosis in the space of knowledge production and research.

The COVID-19 experience presented an opportunity to rethink the research approach to which we (particularly the Padua researchers) were accustomed, which involved travelling to the places in Africa where we conducted research, collecting data "individually" (which often still entailed working with a research assistant, guide or translator) and returning to our countries of residence for data analysis and writing. The pandemic forced us to shift from this well-established research practice, which is at risk of being extractive, to a more collaborative way of doing research. It made visible the value of collaborating with researchers who lived and worked in (the countries of) the study sites, especially in places such as the Sahel where a research crisis exists as well. After all, the empirical basis of the Padua project would not have been possible without the fieldwork carried out by the co-researchers. Although our collaboration was born out of necessity (it was a pragmatic solution to a problem), in hindsight, we believe that we took steps towards new epistemic paths that are valuable for research contexts of (poly)crisis. As we have shown in the previous sections, these baby steps were tentative, confusing, emotional, discomforting, riddled with ethical concerns and posed problems. They destabilised our certainties of doing research. Here, we would like to share some thoughts and an analysis of this "epistemic storm" in which we found ourselves.

First of all, it is worth reiterating that the pandemic contributed to or even accelerated the processes that had already begun under the surface, namely the metamorphosis of research within "the metamorphosis of the world" (Beck, 2016, p. xi) and, more specifically, the thickening of the polycrisis in the Sahel. At the heart of Beck's theory of a metamorphosing world is the sharing of a global risk capable of being fatal to humanity and even to life on the planet (an existential risk). Global events such as climate change or the reappearance of a nuclear threat envelop the entire planet in a haze of uncertainty. Awareness of sharing this level of danger can lead to "goods" obtained from "bads", that is, the formation of "communities of risk". What happened to the research team during the COVID-19 pandemic precisely involved their common feeling of being exposed to great danger and the relative uncertainty of the future. While for those living in the Sahel, this feeling was primarily rooted in polycrisis (there, the COVID-19 outbreak was only a minor crisis, as we have observed), for the group members from the "global North", this experience was violently triggered by the arrival of the coronavirus, challenging mainstream ideas that only the "global South" was an unsafe place. In this sense, the little virus momentarily made the world a little more homogeneous, levelling what were previously absolutely non-comparable risk contexts. The sharing of a common experience of risk paradoxically brought the researchers closer, while keeping them at a distance. Geographical distance became insurmountable, confining everyone to a tiny handkerchief of the world, rendering even the best-performing passports (those that easily open borders) useless. It was not as if inequalities disappeared with the pandemic; after the pandemic peak, the vaccination certificate became essential for travel, but who had first access to the vaccine? Other discriminations began to (re-)appear as well. Simultaneously, however, other distances (social, cognitive and organisational; Boschma, 2005) decreased, precisely because all researchers belonged to the same risk community.

One outcome of participating in the same community of risk is a shift in the axis of relations among researchers towards greater horizontality. The quest towards more equal relationships in North–South academic collaborations is obviously not new (see, e.g., Coetzee, 2019 and other articles on "ethical collaboration" published in the same issue of the *Journal for African Cultural Studies*). What is unprecedented is that during the COVID-19 pandemic, this orientation became a necessity and everyone had to make the immediate transition simultaneously. This adaptation entailed a sudden redefinition of power relations, which was evident, for example, in the Padua researchers' loss of control over data collection (see also Roque de Pinho et al., 2023) and the co-researchers' increased power regarding decision-making on the spot (what data to actually collect, how and when), as those doing the fieldwork. Some Padua researchers experienced a sense of disorientation about being unable to "feel" the data. At the same time, the co-researchers wondered whether their Padua counterparts really grasped the local situation. The uncertainty of the metamorphosing world thus entered the team's research and collaboration arena with gusto, forcing a transition (not easy) towards accepting (and perhaps we should embrace it) a greater degree of uncertainty, indeterminacy, creative ambiguity and multiplicity of viewpoints.

Another outcome relates to the complexity of working and co-writing at a distance, which was historically made possible by the emergence and development of digital (communication) technologies and which brought about the socio-technical process that Serres (2009) describes as a shift from relationships to connections. In this context, the COVID-19 pandemic obliged us to test new practices of remote research, which turned out to be an opportunity and a constraint at the same time. After all, it is not evident to work together at a distance, in very different situations, in contexts fragmented by polycrisis, with difficulties relating to both concrete aspects, such as the availability of electricity and the internet, and more general conditions, such as the security situation in Sahelian countries. Such remote research entails a marked differentiation, in time and space, of the types of data that are available for collection and causes disparate processing times (i.e., different rhythms for processing data) that cannot be easily coordinated. In many respects, these uncalibrated connections are part of the epistemic storm of a metamorphosing world, where uncertainty extends from the difficulty of interpreting the processes being analysed to the provisionality and instability of the data that describe them, especially for those regions most affected by the condition of polycrisis. In the storm, one must reduce the surface of the sails and exercise all possible caution in charting the course on the rough seas, knowing that sailing by sight is part and parcel of being able to navigate.

Reducing the surface of the sails means slowing down. There is a need for care and solidarity in doing collaborative research in times of polycrisis, but it takes time to build caring relationships and accommodate researchers' (changing) needs along the process (including unforeseen circumstances such as illness and power cuts). However, the time needed to care and work in periods of uncertainty clashes with the temporality of neoliberal academia and its demands for responding quickly to emails, meeting tight deadlines and publishing at ever-increasing rates. Moreover, how can researchers work in solidarity within a system that is extractive in nature? Our journey points to the importance of having space to write about the difficulties of doing collaborative research; sharing strategies, fears and doubts; and constructing mechanisms collectively, which resonate with our ethical understanding of doing research that goes far beyond doing so for its own sake. The "publish or perish" culture of our academic system closes this space while demanding that our articles and books meet the required standards in terms of both quantity and quality, with "first authorship" scoring the highest. For (co-)researchers who do not operate in such an (academic) environment, are not other research outcomes more important to them, perhaps more creative ones that can be widely disseminated in the areas of study? Calls for slow science or slow scholarship and the ethics of care and solidarity in research had already been made before the pandemic (Caretta & Faria, 2020; Mountz et al., 2015), but COVID-19 has made it particularly clear that it "is not just the world but our ways of producing knowledge that are in crisis" (Gonda et al., 2021, p. 2). How can we avoid resuming business as usual now that the pandemic is over? How can we create more space for co-learning about non-extractive research in an academic system that is based on the commodification of our time and emotions? How can we hold on to the ethics of time and care in working with others when we feel the pressure to publish in order to advance or secure our own careers and futures? How can we "become better researchers in our connectedness with differently situated others?" (Gonda et al., 2021, p. 1). How can authorship be attributed to out-of-the-box products and collaborative work when each author is equally indispensable to its elaboration? Will this "widespread authorship" then be understood and appreciated by those who decide on academics' careers and research funding allocations in Europe and elsewhere?

A chorus of final thoughts

Angela: This chapter has been the hardest to (co-)write in my entire career. When you read this chapter from beginning to end, you might notice a change. It starts very "academic" (in the sense of it being canonical) with the observation and analysis of local realities written in the third-person point of view – the way that many of us have been trained to do and are (or rather, were) comfortable with, certainly myself. As the chapter progresses, it gradually moves to more personal considerations with reflections and self-reflections about our collaboration and our positionality as researchers therein, with evidently more use of "I", "we" and "our". There is a tension in this chapter that reflects both a struggle and a co-(un)learning experience about the way we do and write research. I think it nicely reflects the metamorphosis of research and knowledge production that we have discussed in the previous section; therefore, rather than smoothing out this tension, as I would usually do for the sake of academic coherence, we let it be. In the spirit of letting go of conventional academia, we experiment even further and would like to end this chapter in an unconventional way. In the following quotations, you will hear the voices of most of the researchers and co-researchers featured in this chapter, with their final thoughts in random order.

Aboukar: When I agreed to participate in this collective research, I had no idea of what it would be. Overall, it has been an inspiring experience. I feel I learnt a lot about my own capacities to work side-by-side with a group of international researchers, although the conditions of this collaboration were very unusual. If everything has to be done again, I would say yes without hesitating. One day, when I was interrogating people about their experience of COVID in Northern Cameroon, a woman said, "Abou, do you want this pandemic to become even more serious? The more you speak about it, the more it kills. Stop worrying about this disease". I realised that in these regions, the real problem is related to the labour conditions in the rice-growing perimeters of the SEMRY project, where people work the land manually. This is very sad. I hope that this research will help find solutions to these concerns.

Andrea: If I try to take stock of what remains of these years of research together, deeply marked first by the pandemic emergency (although today there are phenomena of removals of what happened, which are all the stronger the harsher the experience was, at least in Europe) and then by the polycrisis in the Sahel, these legacies come to mind: the gift of thinking the unexpected (what was previously only a remote possibility: all of humanity affected by the same problems/fears at the same time); the awareness of being part of the same "community of risk", exposed at all times to the destabilising processes of metamorphosis; the ability to hold together what is distant: in space, in cultural backgrounds, in academic and non-academic roles; the courage to try new paths and the honesty to face the difficulties of these new paths; the recognition of the materiality of problems: money, visas, the physicality of our bodies in their ageing, in their being different; and the unequal distribution of this materiality (not all ages have the same problems, not all visas have the same value…).

Francesco: I believe that COVID has made it clear, in the Sahel as elsewhere, that the crisis also concerns our interpretive models for thinking about environmental change in its social dimensions. The eruption of the virus on a micro-scale left us speechless because we did not imagine that ecological processes and environmental dynamics could be so "close". I realised that in order to do research in this context of multiple crises (not limited to health issues), it was necessary to scale down the impact of the pandemic. In the Sahel, this not only concerns the difficulties in monitoring the number of contagions due to the absence of reliable data. Doing research at the time of the COVID [outbreak] also reminded us of the importance of not neglecting, and leaving in the background, local inequalities and injustices that pre-existed the pandemic. Another dimension that struck me about these crises is their generative, innovative dimension. I think [of] the immobility that was imposed on us during the lockdown period and how this condition helped to build new research relationships over the long term.

Moumouni: This research collaboration in the context of multiple crises in the Sahel would not have been able to achieve any tangible or significant results without the co-learning approach, based on the independence of the co-researchers. In fact, unexpected events made it difficult to carry out the research activities as planned. The initial focus of irrigation projects and Sahel's wetlands was disrupted by security issues and the COVID-19 crisis. It was difficult to carry out fieldwork in the Sourou Valley due to increased risks, and I was forced to limit my movements to urban areas and to hire local agents to collect additional data via the KoboToolbox server and the KoboCollect application. This solution, although it allowed the work to continue, did not ease the emotional burden and stress of conducting research in such an unstable environment. Another challenging point was the difficulty in building a professional relationship with our partners at the University of Padua, due to the change of researchers involved in the project (Andrea at the beginning, then Mariasole and finally Francesco).

Mariasole: As the COVID-19 pandemic drew to a close, I often found myself in discussions with family and friends, wondering, "What will remain of the struggles we faced during the pandemic? Will COVID-19 have a lasting positive impact on how we will navigate the future?" My personal answer is no – everything in Italy seemed to return to the way it was before. I believe [that] research faced a similar fate. Many of us in academia were forced to reconsider how we conducted research or even to question why we did it in the midst of a global pandemic. For those of us in Italy, it was perhaps the first time we fully understood the limitations of our mobility – we could no longer travel! New collaborations emerged, yet new burdens also fell on co-researchers as new processes of knowledge production took shape. Being part of this journey was illuminating in many ways. It made me reflect, "Do we need to travel so often to the 'field'? Is it sustainable, from any perspective, to think that 'we', as Western researchers, are so essential to conducting research 'there'?" For me, this was a moment to pause and reflect – on my privilege, on the enduring coloniality in knowledge production and on the reality that I was part of the very system I sought to critique. Yet more importantly, I realised that while things might return to normal for some of us after COVID-19, they would not for everyone. There are still people, such as the colleagues with whom I have collaborated throughout this research process, and there remain margins for writing together otherwise (Acetino et al., 2024) for those who are navigating forced and profoundly different directions.

Dobah: Regarding the mode of knowledge production and research, COVID-19 was the straw that broke the camel's back, because it highlighted what had been contested and contestable for decades, namely the epistemological injustice in the mode of knowledge production and research. At the same time, it relaunched the debate around the need to explore other more inclusive and collaborative paths. In this sense, this research project is an example. Indeed, as citizens of the global South, we are well aware of discrimination and epistemological injustice. This is why this project was an opportunity for us to participate in building something inclusive, based on respect for everyone's experience. This project is all the more important since such projects remain rare today because the world of knowledge production remains largely dominated by Eurocentric logic. Participating in a project that explores other modes of knowledge production and research, more collaborative and inclusive, and that goes beyond the purely academic dimension to integrate a human dimension, can only be interesting. The outcome of this work is proof that by taking a little more risk, as the initiators of this research project did (by choosing to explore research modes different from those they are used to and by trusting the co-researchers), we can discover other equally interesting paths.

Acknowledgements

Our gratitude goes to Joana Roque de Pinho and Carla Braga for their careful reading of and thoughtful comments on an earlier version of this chapter. This work was funded by the PRIN 2017 project "Contested wetlands in Sahelian drylands: Which kind of development and for whom?" (2017B4JBWN). It was also part of the "COVID-19 in African, Asian, and North American Drylands" working group (supported by the NSF-funded SSEER Network and the CONVERGE facility at the Natural Hazards Center at the University of Colorado Boulder - Award #1841338) and the "Mobilities and Socialities: COVID-19 in the Drylands of Africa and Beyond" project (funded by DFG under Germany's Excellence Strategy - EXC 2052/1 – 390713894), and has benefited from discussions with team members from the "Mystical Natures" project (funded by FCT - PTDC/SOC ANT/2990/2020; doi.org/10.54499/PTDC/SOC-ANT/2990/2020).

Notes

1 https://www.who.int/director-general/speeches/detail/who-director-general-s-opening-remarks-at-the-media-briefing-on-COVID-19—11-march-2020 (accessed January 18, 2024).
2 The project officially started on the 1st of March 2020.
3 The COVID-19 in African, Asian and North American Drylands interdisciplinary working group consisted of researchers and co-researchers, who focused on tracking the impacts of and responses to COVID-19 containment measures in dryland areas. See https://converge.colorado.edu/working-groups/COVID-19-in-african-asian-and-north-american-drylands/ (accessed March 12, 2024).
4 The Padua team lost contact with co-researcher Ahmed for some time, and therefore he did not participate in this exercise.
5 On the idea of "knots" in the social sciences, see also Part I "Knotting" of Tim Ingold's book *The Life of Lines* (2015).
6 http://www.ispionline.it/en/publication/sudans-terrible-combination-existing-humanitarian-crisis-and-COVID-19-26193 (accessed October 5, 2023).
7 https://www.crtv.cm/2019/09/economic-disaster-zones-north-west-south-west-and-far-north-regions-given-special-status/ (accessed October 5, 2023); https://reliefweb.int/report/cameroon/cameroun-extr-me-nord-logone-et-chari-mayo-danay-inondations-situation-report-no-01 (accessed June 3, 2024).
8 Following the Burkinabe co-researcher, we opt here to use the term "terrorist" rather than the term "jihadist" which is more commonly used in the scientific literature to qualify episodes of armed violence in the Sahel. In Burkina Faso, there is local opposition to the label of "Jihad", which is perceived as an externally imposed classification that does not reflect local realities. See Chapter 12 for more information on this issue.
9 https://theconversation.com/the-impact-of-covid-19-has-been-lower-in-africa-we-explore-the-reasons-164955 (accessed August 10, 2024).

References

Acetino, F., Autiero, A., Bergesio, N., Bonato, M., Brollo, B., Colla, F., Eccher, L., Ferrante, G., Garozzo, E., Loi, M., Mingo, R., Pepa, M., Sabatini, F., & Salimbeni, A. (2024). Blank(et) spaces. Intrecciando una geografia transfemminista del precariato in accademia. *Revista Lüvo*, *11*(1), 45–50.

Ahmed, S. (2004). *The cultural politics of emotion*. Edinburgh University Press.

Al-Saidi, M., Gayoum Saad, S. A., & Elagib, N. A. (2023). From scenario to mounting risks: COVID-19's perils for development and supply security in the Sahel. *Environment, Development and Sustainability*, *25*, 6295–6318.

Association pour la Promotion de l'Elevage au Sahel et en Savane (APESS). (2020). *Note d'analyse des premiers impacts de la pandemie du COVID 19 sur les exploitations familiales agropastorales membres de l'APESS*. APESS.

Beck, U. (2016). *The metamorphosis of the world*. Polity.

Behar, R. (2022). *The vulnerable observer: Anthropology that breaks your heart*. Beacon Press.

Bishai, L. S. (2023). Resistance is life: How Sudan's resistance committees perform democratic power. *Journal of North African Studies*, *28*(6), 1473–1491.
Boschma, R. A. (2005). Proximity and innovation: A critical assessment. *Regional Studies*, *39*(1), 61–74.
Caretta, M. A., & Faria, C. V. (2020). Time and care in the "lab" and the "field": Slow mentoring and feminist research in geography. *Geographical Review*, *110*(1–2), 172–182.
Coetzee, C. (2019). Ethical?! Collaboration?! Keywords for our contradictory times. *Journal of African Cultural Studies*, *31*(3), 257–264.
Deb, S., & Baudais, V. (2022). The challenges of data collection in conflict-affected areas: A case study in the Liptako-Gourma region. *SIPRI Insights on Peace and Security*, *7*, 1–7.
Gonda, N., Leder, S., González-Hidalgo, M., Chiwona-Karltun, L., Stiernström, A., Hajdu, F., Fischer, K., Asztalos Morell, I., Kadfak, A., & Arvidsson, A. (2021). Critical reflexivity in political ecology research: How can the COVID-19 pandemic transform us into *better* researchers? *Frontiers in Human Dynamics*, *3*, 652968.
Henig, D., & Knight, D. M. (2023). Polycrisis: Prompts for an emerging worldview. *Anthropology Today*, *39*(2), 3–6.
Ingold, T. (2015). *The life of lines*. Routledge.
Iocchi, A. (2022). *Living through crisis by Lake Chad: Violence, labor and resources*. Routledge.
Jakupec, V., Kelly, M., & McKay, J. (2024). *Foreign aid in a world in crisis: Shifting geopolitics in the neoliberal era*. Routledge.
Kostelyanets, S. (2022). Sudan's December revolution and the demise of the Al Bashir regime. In L. Issaev, & A. Korotayev (Eds.), *New wave of revolutions in the MENA region: A comparative perspective* (pp. 33–55). Springer.
Kronenburg García, A., Pase, A., Haller, T., Artur, L., Nogueira Lisboa, S., Metrak, M., Bertoncin, M., & Suska-Malawska, M. (2023). Wetlands in drylands: Large-scale appropriations for agriculture, conservation and mining in Africa. In A. Kronenburg García, T. Haller, H. van Dijk, C. Samimi, & J. Warner (Eds.), *Drylands facing change: Interventions, investments and identities* (pp. 53–72). Routledge.
Mountz, A., Bonds, A., Mansfield, B., Loyd, J., Hyndman, J., Walton-Roberts, M., Basu, R., Whitson, R., Hawkins, R., Hamilton, T., & Curran, W. (2015). For slow scholarship: A feminist politics of resistance through collective action in the neoliberal university. *ACME: An International Journal for Critical Geographies*, *14*(4), 1235–1259.
Olivier de Sardan, J.-P. (2023). *L'enchevêtrement des crises au Sahel: Niger, Mali, Burkina Faso*. Karthala.
Owonikoko, S. B. (2021). *Implications of COVID-19 for peace and security in the Lake Chad Basin region*. Policy Brief (15/06). IPSS, Addis Ababa University.
Pase, A., Presti, L. L., Rossetto, T., & Peterle, G. (2021). Pandemic cartographies: A conversation on mappings, imaginings and emotions. *Mobilities*, *16*(1), 134–153.
Pepa, M., & Marsala, D. (2024). Weaving an alternative partnership: A collaborative autoethnography on 'failing' to research epistemic freedom. *Third World Quarterly*, https://doi.org/10.1080/01436597.2024.2442565
Roque de Pinho, J., Kronenburg García, A., Hashimshony-Yaffe, N., Sternberg, T., Pase, A., ole Neboo, S., Seenoi, D., Mayiani, D., ole Mowuo, L., ole Mowuo, M., Wario, S., Batjav, B., Battsengel, B., & Sainbayar, E. (2023). Pastoralists under COVID-19 lockdown: Collaborative research on impacts and responses in Kenyan and Mongolian drylands. In A. Kronenburg García, T. Haller, H. van Dijk, C. Samimi, & J. Warner (Eds.), *Drylands facing change: Interventions, investments and identities* (pp. 215–235). Routledge.
Serres, M. (2009). *Temps de crises*. Pommier.
Tayler, J. (2005). *Angry wind: Through Muslim Black Africa by truck, bus, boat, and camel*. Houghton Mifflin Company.
Tilley, E., & Kalina, M. (2021). "My flight arrives at 5 am, can you pick me up?": The gatekeeping burden of the African academic. *Journal of African Cultural Studies*, *33*(4), 538–548.

17 Ways of being-knowing and doing with water and land in Africa

Carla Braga, Evelien de Hoop, Angela Kronenburg García, Esther Miedema, Andrea Pase and Mariasole Pepa

Introduction

This chapter seeks to offer an outline of an analytical approach to studying diverse and unequally connected ways of doing and modes of being-knowing, which much research on land and water tends not to capture and/or tends to ignore in the design of its research processes.

Our thoughts on this research agenda have been prompted by a series of crises, debates and events. We find ourselves in an 'era of polycrisis' (Henig & Knight, 2023), including conjoined environmental and financial crises across the globe. At the same time, the COVID-19 pandemic called into question the complex interactions between humans and other living beings, especially in the form of global fluxes of human bodies and entities such as viruses, thereby also posing challenges to the research canon (see Chapter 16). Likewise, hunger and poverty persist across the globe, and the Ukraine-Russia war is spiking energy and food costs. The war in the Middle East is fraught with complex underpinnings of intergenerational suffering, pain, fear, hopes and desires, as are other more 'invisibilised' wars, such as the war in Sudan (see Chapter 10) and the war in northern Mozambique. We consider that discourses on development, globalisation and sustainability offer little recourse (Meneses, 2020) and argue that we are facing a civilisational crisis (Escobar, 2018) or even a 'change of age', as Mbembe (2024) termed it, in which life itself is at stake.

We thought through these crises in dialogue with scholars writing from and towards several fields of study, including but not limited to Anthropology, History, Feminist scholarship as well as Science and Technology Studies, which include Arturo Escobar, Sabelo Ndlovu-Gatsheni, Marisol de la Cadena, Mario Blaser, Donna Haraway, Francis Nyamnjoh and Tania Li as well as Gloria Anzaldua and Oyeronke Oyěwùmí. Through this dialogue, we were searching for futures fostered by decolonial ways of thinking and collaborating—futures that are underpinned by relational ontologies and ethics of care, departing from the premise of our (human—other-than-human) interdependence. We were (thus) searching for futures, which gave room to a plurality of worlds and the emergence and reemergence of life

DOI: 10.4324/9781003429111-24

itself. Interweaving these ideas with some chapters of this book and research from Mozambique, where some of the authors live and work, this contribution imagines a research agenda which 'takes people seriously' and brings to the fore ways of relating amongst earth beings (humans and other-than-human beings) (cf. de la Cadena & Blaser, 2018; Kohn, 2013; Tsing, 2015), forms of knowledge and situated practices. How do these ways of relating take shape in the flowing condition of droughts and floods in the Sahel? And, how can the foregoing help us better understand possible forms of coexistence and conviviality within that context?

Two conditions make it sensible to attempt this research path. The first is the breath of the Sahel, with its movements of rains and floods (see Chapters 5 and 7) that manifest water and land as fluid, which do not rhyme with the cognitive and operational 'traps' of modernity. The second is failure (see Chapter 1). That is, the modern, Western project of political and economic territoriality has floundered in the Sahel, thereby opening the space, albeit often in dramatic ways, to the need for a radical rethinking of approaches.

Therefore, we try to unsettle the modern binaries of nature/culture (Haraway, 1989, 1991; Latour, 1993; see Chapter 14) and subject/object (Harding 1987, 1998). We focus on the inseparable dimensions of being-knowing (i.e. onto-epistemic politics) and doing (i.e. practices). As for the study of 'being-knowing', we first aim to focus on the other-than-human relations that unsettle the centrality of human beings as an analytical focus, which Donna Haraway (2016) refers to as 'human exceptionalism'. We also do not treat humans as a bounded, undifferentiated category, but instead consider diversity (in terms of gender, race, class, spatial roots and mobilities) as our starting point. Finally, we consider being and becoming as taking place in tandem with knowing—be it cognitively, embodied and/or sensory—performed by humans and other-than-human beings together. As for 'doing', we understand this as the ever-ongoing emergence of ways of being-knowing. In our engagement with 'doing', we seek to make visible the invisibilised ways of being-knowing, including their sometimes conflicting interactions with other dominant ways of enacting, performing and bringing about worlds, i.e. worlding.

When we say "different ways of being-knowing and doing", one may ask, 'different from what'? We suggest that the phrase '*diverse* ways of …' carries no implicit hierarchy amongst the 'ways of', in contrast to the phrase '*different* ways of …', which implies a standard from which to measure or compare, but which rarely explicates what standard that is. Here we borrow from Sylvia Wynter the observation regarding how the concealed and dominant standard of 'the human' is Man, which, since the Renaissance and the Enlightenment, became the standard by which to rank other men and women, including European women (Mignolo, 2015; Wynter, 2003), thus resulting in what Maldonado-Torres (2008, p. 122) referred to as "the reduction of the human to Man".

This reflection makes it possible to highlight how the process that led to the silencing of alternatives to the arrogant, rational, patriarchal, capitalist way of seeing and acting on the world had already exercised its apprenticeship in pre-modern and modern Europe with the repression of subordinate groups (e.g. women, 'heretics', believers in other religions and bearers of diverse and ancestral knowledges). Suffice it to refer to Carlo Ginzburg's studies (1980, 1991) that dug into the faint traces—paradoxically found precisely in the files of the trials of the Holy Inquisition—of cosmologies and forms of communication with the other-than-human deeply rooted in the culture that was transmitted orally in the mountains and deep countryside of Europe between the 14th and 17th centuries. This suppression of diverse ways of reading, perceiving and acting on the world soon spread to the colonised New Worlds.

Development, modernity/coloniality

Irrigation in the Sahel, as pointed out in Chapter 1, has deep roots both in colonisation and in the high modernist view of water and land (Scott, 1998). It can be argued that large-scale irrigation projects were the laboratory, or, if you prefer, the playground, of the construction of colonial and modern territory in a context considered 'alien' by most Europeans. One of the most evident aspects in the establishment of the "total institutions" (Goffman, 1961) of large-scale irrigation schemes was precisely the determination to nullify every pre-existing form of conception of relations with the environment that human groups had developed over time in the regions that would be radically transformed by agricultural innovation. Modernisation means eradicating what is perceived as backward, unproductive, anti-scientific and irrational: from this blinkered backward world, according to colonisers, comes 'underdevelopment'.

In the face of the pervasiveness of such 'underdevelopment' discourses, which are particularly evident in narratives used to legitimise and promote large-scale irrigation schemes in the Sahel, we ask ourselves what are the necessary conditions of the possibility of rethinking, re-imagining and aspiring to futures otherwise (Escobar, 2018; Kothari et al., 2019; Mpofu & Ndlovu-Gatsheni, 2019; Sarr, 2020; Tamale, 2020): that is, is it thinkable to conceive of plural futures, regardless of these dominant paradigms?

To address this question, we start by pointing out that 'development' and 'underdevelopment' as concepts, whether or not we put them within quotation marks, have a genealogy; they emerged in particular historical contexts (Escobar, 2012). Development discourses were and are 'legitimised' through a rhetoric of salvation, civilisation, rationality, scientism and liberalism, the horizon of which is universalism (Ndlovu-Gatsheni, 2020) as well as the 'pursuit of modernisation', which is mostly reductively conceived as a process of Westernisation (Ndlovu-Gatsheni, 2020; Nyamnjoh, 2006; Escobar, 2012). The power mechanisms of development work through discursive formations through which power is exercised and 'development' becomes experienced/embodied. This discursive formation was well captured by Nyamnjoh (2006, pp. 396–397), whom we will quote extensively:

> By modernization Western theorists understood the process of change towards the types of social, economic and political systems which developed in Western Europe and North America … and which subsequently spread to other regions of the world. According to this definition, modernization touches on all aspects of existence: social and political organization, family, kinship, belief systems, and economy. It possesses an original model: that of Europe and North America. This model is placed under the sign of scientific rationality, inherited from the century of the Enlightenment and perceived as universal… . It promises the only rational way, the-one-best-way, therefore the universal way of doing anything.

However, modernity—or better, Western modernity—would arguably not have been possible without coloniality as its 'underside' (Mignolo, 2018). At the risk of reiterating a well-known distinction, we will point out that coloniality is not the same as colonialism. The latter refers to "the political and economic relation in which the sovereignty of a nation … depends on the power of another nation" (Maldonado-Torres, 2007, p. 243). Coloniality, instead, refers to persistent patterns of power that emerge from colonialism and continue "to define culture, labour, inter-subjective relations and knowledge production beyond the limits of colonial administrations" (ibid). Coloniality hence outlives colonialism (Maldonado-Torres, 2007), and, as authors such as Ndlovu-Gatsheni (2020, p. 25) have remarked, the effects of coloniality are still present today along with the necessity to dismantle coloniality in African geographies (Daley & Murrey, 2022) as well as recognise the impacts of coloniality in development studies in Africa (Ndlovu-Gatsheni, 2012).

Indeed, development has also been associated with the coloniality of time intended as a linear process and with the colonial appropriation of places through mapping (Pase, 2024). Authors such as Icaza and Vázquez (2016) highlight development's foundations on Western-centric linear notions of social evolution, progress and civilisation (Icaza & Vázquez, 2016). Crucially, the space of the so-called global North has commonly been, and continues to be, depicted as the future and the space of Africa as the past. As Rao (2020) observed, colonial and postcolonial projects were justified by claiming that the 'temporally primitive' would become initially 'civilised', subsequently 'developed' and ultimately, potentially, human. Development discourse, and underpinning assumptions concerning what is modern or not modern, are thus closely related to the partition of humanity, which portrays those put in the past and rendered as pre-modern (i.e. nonmodern or sometimes 'traditional') as devoid of humanity (Escobar, 2012; Ndlovu-Gatsheni, 2020). This discourse laid the groundwork for sub-humanisation and even dehumanisation, as we will address in the next section.

Along with critiques of the sub- and dehumanisation processes associated with development, others have argued that development 'has proved inappropriate in Africa' (Ferguson, 1994; Nyamnjoh, 2006, p. 395). The latter critique of development, articulated from Africa, builds on various counts, including (a) its decontextualisation and inadequacy to address the problems in Africa, as development entails "a multiplicity of exogenously generated ideas, models and research paradigms" (Nyamnjoh, 2006, p. 393) and (b) its imposition of a singular mode of being-knowing and doing grounded in assumptions that there is only one (way to a) 'good life' (i.e. following, mimicking and replicating Europe and North America). As noted earlier, development is thus rendered synonymous with Westernisation and produces what Meneses (2020) refers to as homogenisation. Nyamnjoh also spoke of a reductionist perspective focused on 'achieving growth' and ending poverty 'within the context of neoliberalism' (Nyamnjoh, 2006, p. 393).

Recently, critiques of development have become deeper and broader on at least two counts, one being its relation to extractivist logics and the other being its enactment of binary spatial logics. With regard to the first, critiques of the reductive focus of development, and economic development in particular (Obeng-Odoom, 2020), have been strengthened by studying and questioning the extractivist logics that underpin growth-based development models in Africa. Extractivist logics and growth-based development models are arguably tied up with one another, as growth-based development depends on the continuous intensification of the extractivist logics that have emerged since colonial and imperial times and that now proceed with new forms of extractivism[1] as well as new actors, such as Brazil, Russia, India, China and South Africa (the BRICS countries). At the same time, the imperative of growth is being increasingly questioned across the globe, even within the global North, and has intensified particularly in the face of the present planetary crisis, which has led to reconsideration of the importance of exploring alternatives to 'development' within the North itself (Bendix et al., 2019).

With regard to the second count, the consolidation of the BRICS countries, particularly China's growing influence in Africa since the early 2000s as a promotor of South-South cooperation (Banik & Mawdsley, 2023), has intensified the need to rethink traditional binary divisions, such as North/South and developed/underdeveloped, which have long dominated spatial understandings in development studies. These categories now fail to adequately capture ongoing global transformations (Mohan, 2021). This shift not only challenges North/South distinctions but also raises crucial questions, such as 'where is the South?' (Gonzalez-Vicente, 2019), whilst opening new epistemological and ontological inquiries regarding the geographies of South-South cooperation (Mawdsley et al., 2019).

For example, members of the global South or North need not necessarily be defined in terms of nation-states (each with different ways of belonging in the international system) but instead may be defined in transnational social terms (Hurrell, 2013). In this vein, Gonzalez-Vicente (2019) critiqued the conventional view of the global South as territorially bounded nation-states and called for an intersectional approach that recognises global inequalities. The global South and North can also be understood to refer to sets of 'practices, attitudes and relations' and "a disavowal of institutional and cultural practices associated with colonialism and imperialism" (Grovogu, 2011, p. 177). These modes of understanding challenge simplistic and simplifying conceptions of 'global South' (or North) as referring to a coherent category, with a central command and defined scale and form (Grovogu, 2011). Waisbich et al. (2021), drawing from Chimamanda Ngozi Adichie, argued against the homogenising meta-category of the global South and advocated instead for polyphony to reflect its diverse voices. Moreover, Murrey (2018) emphasised using 'South(s)' and 'North(s)' to acknowledge pluriversality, which is intended as a practice that recognises the existence of multiple worlds without attempting to homogenise divergent ontologies, epistemologies, ways of life and political systems (de la Cadena & Blaser, 2018). This reimagination of the spatial perspective challenges mainstream development categorisation by seeking alternative, transformative frameworks (Kothari et al., 2019). The effort attempted in this volume to map the Sahel in a diverse, imaginative and metaphorical way also moves in this same direction (see Chapter 9).

To conclude, in the face of the apparent 'unthinkability' and incapacity to conceive a better life or lives beyond the developmentalist paradigm, we suggest that 'development' as an analytical or standard-setting tool is neither the only nor the most helpful concept with which to think about how to improve people's lives and well-being and to tackle hunger, poverty and inequalities. As Gustavo Esteva (quoted by Escobar, 2012, p. xix) stressed: "development failed as a socio-economic endeavor but the development discourse still contaminates social reality". Ndlovu-Gatsheni (2020, p. 32) similarly claimed how "Within the colonial context, development meant the transformation of African society according to the needs, demands and imperatives of colonial regimes". Hence, we argue that it is problematic to use 'development/underdevelopment' to refer to what could be generally understood as improvements or betterment of quality of life and to equate these desires and life projects to 'development'. Yet, as Felwine Sarr (2020) argued:

> The expression [i.e. development] does not exist in Senegal's national language… . How can people be mobilized around a notion which does not find an echo in a certain linguistic and mental universe? … This concept, as it came about, has become ideological and can be criticized for many reasons: it implies that developing countries are lagging behind developed countries, and that there is a mimicry, a path already set out.

Making visible diverse livable lives

We intend to explore visions and ways of living that imply improvement of standards of life to achieve livable lives. We seek to do so without presupposing the primacy of humans, the domination and control of 'nature', the obliteration of diversity or the perpetuation of social and economic injustice. Examples of concepts and philosophies that may inspire different ways of living include Ubuntu, Kiboita and Sumak Kawsay/Suma Qamana.

Ubuntu is a crucial epistemic and ontological category of African philosophy, particularly of the Bantu-speaking peoples (Ramose, 2009; Terblanché-Greeff & Nel, 2023), and some authors have circumscribed it geographically to avoid including North Africa and parts of West Africa (Pereira da Silva, 2023, p. 175). Most importantly, Ubuntu denotes "simultaneously a particular state of being and a becoming",[2] which implies movement and exchange (Ramose, 2009, p. 169). Within Ubuntu, "To be human is to affirm one's own humanity through the acknowledgment of the humanity of others", and it is associated with respect, dignity, sharing and mutual care (Ramose, 2009, pp. 170–171). The 'other' in Ubuntu philosophy encompasses human and other-than-humans, both of which are crucial to becoming fully human (Terblanché-Greeff & Nel, 2023).

African cosmologies variously consider all beings that populate the earth, including ancestral spirits, as a whole: as co-constituting one another or as otherwise intimately entangled (Steiner quoted by Meneses 2020, p. 30). Moreover, plants, animals, rivers and other other-than-humans are in many ways central to African ontologies and religious identities (Meneses, 2020). For example, Maasai, a non-Bantu African people, use the expression *kiboita* when they talk about their relationship with 'wildlife'; it translates as 'we stay together' or 'we are together/being together', which highlights how wild animals are part of *Enkop* (which translates as 'land' or, perhaps better, 'earth'), which is all-encompassing and includes everything on the land—human and nonhuman—all of which are in relationship with each other without hierarchies (Roque de Pinho, 2009).

Sumak kawsay and *suma qamana*, in the Quechua and Aymara languages, respectively, are difficult to translate with exactitude to other languages. They mean something approximate to 'harmony or plenitude', 'to live and convive well' and 'to live in peace' and have been translated to Spanish as *buen vivir* and in English as 'good living' (Pereira da Silva, 2023). Although it emerged mainly from indigenous social movements in the Andes, *buen vivir* is also influenced by critical thought within the West and has been envisaged as contributing to and influencing global debates, for instance, on the possible forms buen vivir might assume in diverse contexts (Escobar, 2018, p. 148). It is worthwhile to stress that *buen vivir* prioritises human dignity, social justice and ecology over economic objectives and envisages the relationality of all forms of life (Escobar, 2018, p. 148; Kothari et al., 2014).

Researching otherwise

Our intention is to depart from humans and other-than-humans from/in Africa for a reflection, analysis, theorising and creation of new concepts, practices and connections. The creative work that we strive to do is geared to fostering rich, situated and adequate (per diverse African peoples' views) understandings of lived experiences and lives, conceptions of the cosmos and worlding practices—that is, the way people make, imagine and enact their worlds. The challenge here is to think and unthink living with and alongside land and water from African experiences whilst decentring developmentalist discourse.

We, the authors of this chapter, are each in a process of unlearning/relearning to respect and recognise the knowledge and ideas of the people with whom we are working and co-thinking to give them due credit and not conceal how their ideas prompted our own thoughts in what Ramon Grosfoguel (2016) (quoting Betasamosake Simpson and Cusicanqui) referred to as epistemic extractivism and cognitive extractivism. Thus, we propose a grounded approach that is based on ethnographic detail of peoples' ways of being-knowing and doing. We call for taking people seriously because "… not to take others seriously is to silence them and in that way to dominate them", as stressed by Viveiros de Castro (as quoted by Holbraad & Pedersen, 2017, p. 193). However, should we not also take other-than-human beings seriously? If we do not, are we not privileging humans over other forms of being and actually reproducing the very same binary we want to unsettle?

We propose researching diverse ways of being-knowing and doing (Blaser, 2009, 2013; Matusse, 2019) to render visible the unequal encounters between worlds and enlarge spaces for the worlds currently in danger. It goes without saying that none of our intentions are straightforward. Our search for ways of researching otherwise may clash with, amongst other things, the material reality of doing research within academic structures (see also Chapter 16).

Concepts

The concepts we use to refer to 'land' and 'water' become grids through which we perceive our worlds and influence the very questions we ask and the way we analyse/interpret research materials. We take inspiration from Helen Verran, who regarded concepts as 'world-making tools' (quoted by de la Cadena & Blaser, 2018). As such, we suggest unsettling the very terms in which research is thought of and attempting to answer the call to create new concepts (Holbraad & Pedersen, 2017; Ndlovu-Gatsheni, 2020), new vocabularies and new maps (Davis, 2016; see also Chapters 5 and 9). Beyond creating new concepts, we argue for the need to better attend to practices and terminologies that have always been there but have been often marginalised, invisibilised, silenced and erased. In doing so, we seek to work towards epistemic freedom in Africa (Ndlovu-Gatsheni, 2018).

When attending to ways of being-knowing and doing otherwise, one faces great challenges in terms of language and conceptual translation. The words we use to express ideas and concepts are already value laden (wa Thiong'o, 1986) and provide evidence for a particular way of sensing and conceiving 'the world'.[3] Moreover, the words we use in themselves may be considered generative and enabling and disabling of particular understandings and worldings (Mavhunga, 2018). We therefore try to be mindful of the analytical language we use—in particular, how we write to take plurality seriously. That is, analytical language can trap the researcher in a binary logic and other frames that s/he/they are trying to challenge and destabilise (Holbraad & Pedersen, 2017).

The world of Sahelian waters and lands is also a world of words of many languages and with a range of meanings and references as extensive over time and space. Reference can be made, for example, to the *Atlas de la province Extrême-Nord Cameroun* (Seignobos & Iyébi-Mandjek, 2000), which contains many entries devoted precisely to the ways of saying 'land', 'water' and 'living beings' that the different human groups have elaborated over time in this region among the Logone River, Lake Chad and the Mandara Mountains. Referring to this edited volume, in Chapter 14, Elvira Pietrobon is particularly attentive to the local language and its meanings and categorisations. For instance, the name of the Niger River can be translated as *Badji* (great/mother water), which denotes the closeness and proximity between humans and the river, which are thought of, expressed and possibly experienced in the language of kinship. This example illustrates wa Thiong'o's (1986) argument regarding the decolonising of the mind. That is, it offers insight into the ways in which language shapes the processes of colonisation of beings-knowings and doings in and with the world and the potential value of reclaiming language discarded during and for the purposes of colonisation.

We think of research as an inter-subjective process which acknowledges knowledge as constructed in dialogue (Holbraad & Pedersen, 2017, p. 20), that is, as a co-creative process in which researchers and interlocutors (human and other-than-human entities) can influence ideas and ways of doing, thinking and feeling research. When studying ways of being-knowing and doing, our intention is neither to essentialise these ways (as trapped in space and time) nor romanticise them. Our intention is to contextualise emergent, ever-changing ways of being-knowing and doing, thereby bringing to the fore historical legacies as well as present-day phenomena variously named neoliberal capitalism and extractivism, amongst other names.

This type of research must grant researchers time to question their own assumptions. Such questioning is neither a purely intellectual or rational enterprise nor solely an individualised one. The reflexivity we refer to demands being open to the unexpected and to ways of being-knowing and doing as well as the ability to let go, accept failures (Harrowell et al., 2017) and rethink and/or reshape one's own taken-for-granted ideas (Kapoor, 2004; Rodriguez & Ridgway, 2023) with transformative effects. Understanding how our ideas were formed in and by particular socio-historical and political junctures and institutions is key to the process of unthinking and rethinking.

We must pay attention to a plurality of ways of knowing. We envisage knowledge not only as an academic endeavour but also in terms of practices of knowledge making and how knowledges are performed and embodied. We hope to unsettle the privilege of sight and the verbal (Oyěwùmí, 1997) in knowledge production and emphasise the multiplicity of sensorial approaches so that smell, sound and touch are sensed and 'registered' as well. In short, our intention is "to learn and unlearn in order to relearn" (Mignolo & Walsh, 2018, p. 73), which implies what Ndlovu-Gatsheni called epistemic freedom, that is, "the right to think, theorize, interpret the world, develop own methodologies and write from where one is located and unencumbered by Eurocentrism" (Ndlovu-Gatsheni, 2018, p. 3).

Attending to ways of being-knowing and doing otherwise

Through the exploration of plural ways of doing and modes of being-knowing, we intend to challenge the central and dominant position of the human or the logic that humans are superior to other beings, which they should or could dominate and control. The proposition to decentre anthropocentric rationality resonates with Pietrobon's Chapter 14, which presents the Niger River as the main entity through which the history of Segou, its inhabitants and their economic practices are narrated. Even more importantly, the author writes in a relational way, decentres human practices and gives primacy to the Niger River. It is through the river that the story is narrated rather than through people.

But, how far are we willing to go in decentring the human? If we follow through, the notion of nature as something outside humans and of which humans are not part is called into question and exposes the social construction of 'nature' (Coole & Frost, 2010; Escobar, 2017, p. 240). That is, nature as singular, pristine and untouched is a crucial part of the Western imaginary (Law & Lien, 2018, p. 149). As Coole and Frost (2010, p. 8) remarked, such a Cartesian-Newtonian understanding of matter implies the conceptual and practical domination of nature associated with the modern ethos.

Indeed, nature has often appeared as a universal category separated from culture and has been treated as a given (Blaser, 2013; Haraway, 1989; Law & Lien, 2018, p. 131)—a category that may be considered very specific to Western modernity wherein 'Nature' is envisaged as 'inert space' and the inorganic and nonhuman are treated as "objects to be extracted, eradicated or destroyed …" (Escobar, 2017, p. 247; Yusoff, 2018). Other beings, alive or not, are perceived as just 'resources' for human use and consumption (Escobar, 2017).

Borrowing from Tania Li (2014, p. 591), we suggest that the word 'nature' "carries a cultural baggage that we need to make strange for the purpose of analysis". Nature can be decentred. And we could instead study emergences of other-than-human beings-doings and knowings on their own terms. Thus, our quest is for ways of being-knowing and doing which imply relations amongst human and other-than-human beings and between several types of beings, including entities such as rivers or mountains, which are not dominance- and control-driven and do not presuppose hierarchies.

Whilst attempting to question the boundaries between humans and other-than-human beings and to decentre the human as the focus of analysis, we attend to entanglements and assemblages that connect several kinds of beings, living or not, thereby emphasising what Escobar (2012, p. xxviii) designated as relational ontologies—"those that eschew the division between nature and culture, between individual and community, and between us and them that are central to the modern ontology". As Ndlovu-Gatsheni (2020, p. 10) put it, "What is envisioned is another possible world, with a different ethics of living together".

We also intend to raise additional questions: What kinds of spaces allow all beings to thrive? How do humans relate to what is often referred to as mountains, rivers, forests, ancestors and spirits? And, how may and do humans and other-than-humans co-constitute each other?

Power relations that underlie the categorisation and hierarchisation of humans as well as the attribution of reason to some and its denial to others need to be taken seriously. Because this book is about Africa and the Sahel in particular, it is not possible to write about being-knowing and doing without considering the historical legacies of enslavement and colonialism that are associated with the categorisation and racial hierarchisation which underlie sub-humanisation and even dispensability. The coloniality of being (Maldonado-Torres, 2007) implies the denial of the humanity of colonised subjects. The degradation of those classified as 'slaves', 'natives', 'blacks' and 'females' into a sub-human category remains a key existential concern today and continues to influence intersubjective relations (Ndlovu-Gatsheni, 2020).

However, there seems to be a tension in our efforts. That is, on one hand, we call for a decentring of anthropocentrism, whilst on the other hand, we propose a reflection on the 'human', both in terms of the re-humanisation of those who have been dehumanised, animalised and bestialised in reconceiving what it means to be fully human. The challenge might be to combine in our research and writing the exposure of the racial and gendered hierarchisation of colonised peoples whilst at the same time decentring the human by weaving together several life forms.

The coloniality of being (Maldonado-Torres, 2007), that is, the denial of the humanity of colonial subjects, is associated with the denial of reason, as eloquently exposed by Chukwudi Eze (2008), and thus implies the incapacity to know and the incapacity to produce knowledge. Colonised subjects were considered not fully human beings with no reason and no knowledges, thereby revealing the entanglement of ontology and epistemology (de la Cadena & Blaser, 2018; Ndlovu-Gatsheni, 2020; Sousa Santos & Meneses, 2009), hence the onto-epistemic approach adopted in this chapter. Work related to the notion of ontological deprivation, that is, the refusal to acknowledge the other as totally human (Sousa Santos, 2018), is relevant here too.

When one refers to 'human and other-than-human' who is *this* human? Are we not bundling all human beings into the single category of 'humans'? What does this line of thinking do? Does this line of thinking not deflect attention away from inequality amongst human beings? The argument regarding ontological deprivation can be extended to other dimensions of sub-humanisation, which include the homeless, the poor, queers, the elderly, children and immigrants. Do they not, in varying ways, appear as in the lower ranks of what counts as 'fully human'?

Whilst in this chapter we wish to avoid hierarchisation of difference, we direct our attention in particular to the relationship between dehumanisation and the absence of reason and thus the capacity to know and to produce knowledge in relation to colonised peoples. The horizontal dialogue between knowledges has been termed 'ecology of knowledges' (Sousa Santos & Meneses, 2009, p. 13), which does not conceive of knowledges in the abstract: "there is no knowledge without practices and social actors", the authors underscored. To be successful, ecologies of knowledge need to articulate amongst subaltern experiences through local-global relationships (Sousa Santos, 2009, p. 50).

To what knowledges do people resort in relation to land and water? Which senses do people deploy to know their world? How are those knowledges, senses and ways of sensing and knowing transmitted? Through which practices are knowledges about land and water produced and reproduced? How are knowledges about water and land acquired? And how do they change?

Inspired by Tania Li's (2014) provocative question 'what is land?', we might ask, 'what is water in the Sahel?'. Li noted that the indigenous highlanders in Sulawesi, Indonesia do not have a word for 'land'. There are words for more specific qualities, Li writes, such as soil (*petu*), primary and secondary forest (*do'at* and *ulat*), fallowed and active garden (*abo* and *jo'ong*) and grassy patches (*gio*). Crucially, these words denote "assemblages of material substances and social relations" (Li, 2014, p. 590). A primary forest, for instance, is a forest where no one has ever taken an axe to a tree.

Returning to water, we need to ask if there is even a word for water (see Pietrobon, Chapter 14). Drawing on Li (2014), which entities, beings and relations come together in the answer to the question 'what is water?' Is water a thing? Or a home to other things? Which type or kind of entity is water? Which characteristics may several kinds of 'water' be endowed with in the Sahel? Or, similar to 'land' in the Sulawesi highlands, are different 'forms' of water assemblages of materials and social relations (Li, 2014)? Is 'water' endowed with agency? Water in the Sahel is so relevant that it has become, in many cases, the focus of cosmology, for example, amongst the Kotoko in the Far North Region of Cameroon (Lebœuf, 1973). In the Logone plain, which is composed of water or land, depending on the rhythm of the flood and the quantity and distribution of seasonal rainfall, rain is unpredictable, and in the conception of the local populations, it depends on the action of supernatural forces. The rain god was the most important of the deities worshipped. When rain did not fall, it meant that the deity was irritated and withheld it (de Garine, 2003, p. 362).

What do the use of the words water and land do? What do practices associated with land and 'water' (e.g. demarcation, title deed emissions or rerouting of rivers) do? Which worlds do these practices bring about? How can relations with land and water be captured as an exploration of indigenous geographies (Dieckmann, 2021)?

Besides looking at the commodification of land and water, we also need to pay attention to cosmologies and to ways of being-knowing and doing with and alongside land and water that constitute worlds, which are worlding practices. We should also ask, "What are the entanglements among humans and other living beings, such as animals, livestock and crops and entities, such as spirits, and 'things', like documents and hoes, in the Sahel?"

Chapter 2 lifted the veil on some of the entanglements of the ever-changing Lake Chad with its rhythms of high and low waters, receding water and advancing desert (and vice versa), as well as its constantly changing shorelines that generate a multiplicity of amphibious spaces (e.g. islands, marshes, open waters and sandy extensions) that are teeming with living beings and interacting with a plurality of changing human practices and mobilities. In parallel to Escobar's (2017, p. 246) mangrove-world, "There is a rhizome logic to these entanglements, a 'logic' that is impossible to follow in any simple way, and very difficult to map and measure, if at all; this logic reveals an altogether different way of being and becoming in territory and place". In Lake Chad, these complex entanglements and these different ways of being and becoming were ignored by mega-irrigation development projects (all of which eventually failed) as they imposed modernity and coloniality (see Chapter 1).

Doing

> Is it simply that people believe different things about reality? Or is it that there are different realities being done in different practices? If the first of these positions is right, then we're in the business of beliefs, perspectives and epistemologies. If it's the second position then we're back into issues of ontology. Here's the analytically radical nature of this second position. We are in the business of treating reals as effects of contingent and heterogeneous enactments, performances or sets of relations.
>
> *(Law, 2011)*

Practices as enactments of particular ontological configurations interact with each other "and are constituted and transformed through different socio-material practices and infrastructures" (Holbraad & Pedersen, 2017, p. 38). From this point of view, what happens during and after the severe crisis of the large-scale irrigation schemes is undoubtedly significant: it is in the ability of allottees to act in the interstices and remnants of large-scale projects that innovative practices are developed which integrate—in doing differently—well-rooted knowledge with the impulses coming from agricultural modernisation (see Chapter 1). It is precisely in focusing on doing and on the continual change of practices that it becomes evident how our approach stays away from an essentialist conception of alternative cultures.

How are ways of being and knowing materialised? Through what practices? We intend to explore the materiality of ways of being-knowing of people whose voices have been silenced, made invisible and secondarised as our point of departure and focus. At the same time, we seek to experiment with interweaving the invisibilised accounts and experiences with ways of being-knowing and doing of agents such as government officials, non-governmental organisations, activists, corporations and church leaders, to name just a few. In so doing, we seek to avoid both romanticisation and demonisation of ways of being-knowing and doing and instead articulate what emerges in encounters of variously located actors, entities and agents.

"Political ontology … refers to a field of study that focuses on the interrelations amongst worlds, including the conflicts that ensue as different ontologies strive to sustain their own existence in their interaction with other worlds" (Escobar, 2017, p. 24). Inspired by Escobar (2017, pp. 243–244), in this endeavour, we attended to "power laden practices which make real/realize a world or ontology …" as well as making visible the 'ontological dimension of extractivist development models', particularly large-scale ones that sometimes involve large-scale displacements of populations.

By attending to ways of knowing-being and doing, our intention is to explore ways of worlding as enactments of particular ontologies, which are not part of and are not subsumed by the dominant and hegemonic 'one-world world' (Law, 2011), and open up the space for pluriversality.

The pluriverse

> For once we put the practices that do realities alongside one another and start to take those practices seriously, once we start to insist that differences are not just matters of perspective, once we start to insist that what the practices do is real in the context of those practices, once we do all of these things, then we start to discover multiple times, multiple spaces, multiple relations, multiple origins, and multiple modes of causality or noncausality. We find, in short, that reals start to multiply and the framing assumption that there is a single reality, a universe, starts to dissolve.
>
> *(Law & Lien, 2018, pp. 157–158)*

As Escobar (2012, p. xxvii) noted, "the world is a pluriverse, ceaselessly in movement, an ever-changing web of interrelations involving humans and non-humans". The pluriverse signals struggles for bringing about 'worlds and knowledges otherwise'—that is, "worlds and knowledges constructed on the basis of different ontological commitments, epistemic configurations and practices of being, knowing and making/doing" (p. 49). The notion of pluriverse originated from 'worlds usually labelled indigenous', particularly under the Zapatista dictum "The world we want is a world in which many worlds fit" (de la Cadena & Blaser, 2018, p. 1).

Academic studies of territorial and/or ontological struggles as well as resistance to extractivist development models and conservationist endeavours are mainly about South America, probably because the main theorists of political ontology (Mario Blaser, Arturo Escobar and Marisol de la Cadena) focus on that part of the world as their geographic area of study. However, as this chapter has argued, we need to visibilise and deploy ways of thinking and theorising about experiences of space, territory and temporalities drawing and deriving from Africa and African diaspora experiences and thought, hence the centrality of attending to ways of being-knowing and doing in Africa. That is not to say that one cannot build on the work from other parts of the world, including decolonial South American thinkers, which offer us crucial analytical and heuristic devices to understand beings and becomings.

Escobar (2017, p. 243) stressed that little is known about 'struggles for the defense of worlds' and ontological struggles in Africa. Taking Mozambique as a case in point, Wiegink (2022) studied the loss of a burial ground when people in Tete were resettled so a mine could be constructed as well as the resettled people's refusal of burying the dead in the new cemetery of the resettled village. She highlighted the reasons for the refusal and the importance of taking into account 'the socio-spiritual' dimensions of resettlement but stressed that her analysis "has not been predominantly one of a clash of different world-views" (Wiegink 2022, p. 102).

Conflicts and misunderstandings between the parties involved in the Tete case might be interpreted through what Viveiros de Castro (quoted in Blaser, 2009) called an 'uncontrolled equivocation', that is, a misunderstanding which occurs "not because there are different perspectives on the world but because there are different worlds and this is not recognized" (Blaser, 2009, p. 16). For instance, a cemetery, not to mention dead family members, might be a different thing for the resettled population than for state and corporate officials. Whilst the latter tend to emphasise the material loss, for the former, it is about relating to the dead, such as through ceremonies, as well as identity and belonging (also see Witter & Satterfield, 2014). Similar cases were also described for the Sahel in Chapter 1. Indeed, as stressed by Li (2014), the presence of spirits can be a mode of inscription, that is, it could be a way of claiming a particular territory. If we connect conceptions of life and death as a continuum, the circularity of time, relationships between the living and the dead (Honwana, 2002; Meneses, 2020) and the centrality of ancestral spirits as protectors of the living as Wiegink (2022) also did, the refusal to use the new cemetery in Tete could in fact be conceptualised in terms of a clash between worlds and thus constitutes an ontological claim.

Speculative openings

Instead of offering conclusions, we would like to leave the discussions open and would rather refer to 'speculative openings', as posited by de la Cadena and Blaser (2018, p. 17).

We propose openness as an analytical attitude—rather than closing in a normative manner what could be fruitful discussions—to allow innovative ideas and formulations to blossom and flourish. Why not think in terms of both being and becoming instead of being *or* becoming? What unexpected ways of being and becoming might emerge?

At the same time, by decentring and even demystifying Western assumptions about land, water and nature, which seem self-evident and hence not worth questioning or problematising, we aim to historicise these concepts and circumscribe them within a particular space and time so as to eventually provincialise them.

Research questions, as well as the questions one asks oneself and of research participants, are usually already tainted by and loaded with assumptions; thus, it is quite difficult to ask questions which are open (not in the sense of 'closed questions' that leave no room for a variety of answers or invite one particular type of answer) but open in the sense of not carrying one particular way of being-knowing and doing. Therefore, to research ways of being-knowing and doing, to borrow from Ndlovu-Gatsheni, one needs to think, unthink and rethink what one thinks one 'knows', whilst being aware that such a process is a transformative one also for the researcher or researchers.

To detach from the pretence of a single world/universe and open up to myriad ways of being and becoming human in relation to other beings, whether it be what one calls 'land' or 'water', probably implies taking seriously not only humans but other beings and entities as well, whilst attempting to leave hierarchies aside and not measure them according to a supposedly universal standard. Ways of being-knowing and doing are entangled and co-constitutive (Jasanoff, 2004) or, as Escobar described it, "… Knowledges within the inevitable coincidence of knowing, being and doing" (Escobar, 2018, p. 147).

Is the present planetary crisis a manifestation of the one-world world crisis (Escobar, 2017, p. 247)? That is, are the conjoined crises we are facing related to the logic and cosmovision associated with modern ontology and hence the one-world world? Or, in other words, are we dealing with the exacerbating outcome of the gigantic 'terraforming' process that was triggered by the conjunction of colonialism with modernisation, as suggested by Amitav Ghosh (2021)?

Amongst the discussions and dilemmas we faced when writing this joint paper, the extension of ontological deprivation to other humans beyond colonised subjectivities was paramount. Does sub-humanisation also imply the absence of reason in all cases? Can we say that, for instance, women, the poor and immigrants are also considered devoid of the capacity to reason as black persons and other colonised peoples were and are?

What are the full implications of challenging the one-world world and claiming the pluriverse? What are the implications in terms of the notions and vocabularies we use to think things through, such as the 'state'? Borders? 'Land and water rights'? When thinking in terms not of the universe but of the pluriverse, what forms of politics might emerge? What forms of politics are possible? What futurities could emerge?

What if we think through and of 'land' and 'water' not as resources to be used, exploited, managed and depleted but as conditions of possibility for life on earth?

This chapter and this book are an invitation to multiple ways of thinking of water and land as lenses for the pluriverse. It is a starting point for other ways of being-knowing and doing.

Notes

1 See the website of the Alternatives to Extractivism (A2E) research initiative: https://alternatives2extractivism.wordpress.com/.
2 Translation of "denotando simultaneamente, um estado particular do ser e um tornar-se" (Ramose, 2009, p. 169).
3 See Law (2011) on 'one-world world' (OWW).

References

Banik, D., & Mawdsley, E. (2023). South-South cooperation and global development in a multipolar world: China and India in Africa. *Journal of International Development*, *35*, 539–548. https://doi.org/10.1002/jid.3789
Blaser, M. (2009). The threat of the Yrmo: The political ontology of a sustainable hunting program. *American Anthropologist*, *111*(1), 10–20. https://doi.org/10.1111/j.1548-1433.2009.01073.x
Blaser, M. (2013). Notes towards a political ontology of "environmental" conflicts. In L. Green (Ed.), *Contested ecologies dialogues in the South on nature and knowledge* (p. 16). HSRC.
Bendix, D., Müller, F., & Ziai, A. (2019). Postdevelopment alternatives in the North. In E. Klein, & C. E. Morreo (Eds.), *Postdevelopment in practice: Alternatives, economies, ontologies*. Routledge. https://doi.org/10.4324/9780429492136
Chukwudi Eze, E. (2008). *On reason: Rationality in a world of cultural conflict and racism*. Duke University Press.
Coole, D., & Frost, S. (2010). Introducing the new materialisms. In D. Coole, & S. Frost (Eds.), *New materialisms: Ontology, agency and politics* (pp. 1–41). Duke University Press.
Daley, P., & Murrey, A. (2022). Defiant scholarship: Dismantling coloniality in contemporary African geographies. *Singapore Journal of Tropical Geography*, *43*, 159–176. https://doi.org/10.1111/sjtg.12422

Davis, A. (2016). *Freedom is a constant struggle: Ferguson, Palestine and the foundations of a movement*. Haymarket Books.

de la Cadena, M., & Blaser, M. (2018). *A world of many worlds*. Duke University Press. https://doi.org/10.1215/9781478004318

de Garine, I. (2003). Aspetti ecologici e simbolici dell'acqua tra le popolazioni del nord del Camerun. In V. Teti (Ed.), *Storia dell'acqua. Mondi materiali e universi simbolici* (pp. 357–371). Donzelli.

Dieckmann, U. (2021). *Mapping the unmappable? Cartographic explorations with indigenous peoples in Africa*. Transcript Verlag. https://doi.org/10.14361/9783839452417

Escobar, A. (2012). *Encountering development: The making and unmaking of the Third World*. Princeton University Press.

Escobar, A. (2017). Sustaining the pluriverse: The political ontology of territorial struggles in Latin America. In M. Brightman, & J. Lewis (Eds.), *The anthropology of sustainability: Beyond development and progress* (pp. 237–272) Palgrave McMillan. https://doi.org/10.1057/978-1-137-56636-2

Escobar, A. (2018). *Designs for the pluriverse: Radical interdependence, autonomy, and the making of worlds*. Duke University Press. https://doi.org/10.1215/9780822371816

Ferguson, J. (1994). *The anti-politics machine: Development, depoliticization, and bureaucratic power in Lesotho*. University of Minnesota Press.

Ginzburg, C. (1980). *The cheese and the worms: The cosmos of a sixteenth century miller*. Johns Hopkins University Press.

Ginzburg, C. (1991). *Ecstasies: Deciphering the witches' sabbath*. Pantheon Books.

Ghosh, A. (2021). *The nutmeg's curse. Parables for a planet in crisis*. Murray.

Goffman, E. (1961). *Asylums: Essays on the social situation of mental patients and other inmates*. Anchor Books.

Gonzalez-Vicente, R. (2019). Where is the South? Global, postcolonial and intersectional perspectives. In E. Mawdsley, E. Fourie, & W. Nauta (Eds.), *Researching South-South development cooperation: The politics of knowledge production*. Routledge. https://doi.org/10.4324/9780429459146

Grosfoguel, R. (2016). Del «extractivismo económico» al «extractivismo epistémico» y al «extractivismo ontológico»: una forma Destructiva de conocer, ser y estar en el mundo. *Tabula Rasa*, 24. https://doi.org/10.25058/20112742.60

Grovogu, S. (2011). A revolution nonetheless: The global South in international relations. *The Global South*, *5*(1), 175–190. https://doi.org/10.2979/globalsouth.5.1.175

Haraway, D. (1989). *Primate visions: Gender, race, and nature in the world of modern science*. Routledge.

Haraway, D. (1991). *Symians, cyborgs, and women: The reinvention of nature*. Routledge.

Haraway, D. (2016). *Staying with the trouble: Making kin in the Chthulucene*. Duke University Press.

Harding, S. (1987). *Feminism and methodology*. Indiana University Press.

Harding, S. (1998). *Is science multicultural?: Postcolonialisms, feminisms and epistemologies*. Indiana University Press.

Harrowell, E., Davies, T., & Disney, T. (2017). Making space for failure in geographic research. *The Professional Geographer*, *70*(2), 230–238. https://doi.org/10.1080/00330124.2017.1347799

Henig, D., & Knight, D. M. (2023). Polycrisis: Prompts for an emerging worldview. *Anthropology Today*, *38*(2), 3–6. https://doi.org/10.1111/1467-8322.12793

Holbraad, M., & Pedersen, M. (2017). *The ontological turn: An anthropological exposition*. Cambridge University Press. https://doi.org/10.1017/9781316218907

Honwana, A. (2002). *Espiritos vivos, tradições modernas: Possessão de espiritos e reintegração social pos-guerra no sul de Moçambique*. Promedia.

Hurrell, A. (2013). Narratives of emergence: Rising powers and the end of the Third World? *Brazilian Journal of Political Economy*, *33*(2), 203–221. https://doi.org/10.1590/S0101-31572013000200001

Icaza, R., & Vázquez, R. (2016). The coloniality of gender as a radical critique of developmentalism. In W. Harcourt (Ed.), *The Palgrave handbook on gender and development: Critical engagements in feminist theory and practice* (pp. 62–76). Palgrave. https://doi.org/10.1007/978-1-137-38273-3_5

Jasanoff, S. (2004). *States of knowledge: The co-production of science and social order*. Routledge.

Kapoor, I. (2004). Hyper-self-reflexive development? Spivak on representing the Third World 'other.' *Third World Quarterly*, *25*(4), 627–647. http://www.jstor.org/stable/3993737

Kohn, E. (2013). *How forests think: For an anthropology beyond the human*. University of California Press.

Kothari, A., de Maria, F., & Acosta, A. (2014). Buen Vivir, Degrowth and Ecological Swaraj: Alternatives to sustainable development and the green economy. *Development*, *57*, 3–4.

Kothari, A., Salleh, A., Escobar, A., Demaria, F., & Acosta (2019). *Pluriverse: A post-development dictionary*. Tulika Books.

Latour, B. (1993). *We have never been modern*. Harvard University Press.

Law, J. (2011). What's wrong with a one-world world. *Heterogeneities*, September 25, http://www.heterogeneities.net/publications/Law2011WhatsWrongWithAOneWorldWorld.pdf.

Law, J., & Lien, M. (2018). Denaturalizing nature. In M. de la Cadena & M. Blaser (Eds.), *A world of many worlds* (pp. 131–172). Duke University Press. https://doi.org/10.1215/9781478004318

Lebœuf, J.-P. (1973). Personne et système du Monde chez les Kotoko. In R. Bastide, & G. Dieterlen (Eds.), *La Notion de personne en Afrique noire*. Actes du colloque international du Centre national de la recherche scientifique, Paris 11-17 octobre 1971 (pp. 373–386). L'Harmattan.

Li, T. M. (2014). What is land? Assembling a resource for global investment. *Transactions of the Institute of British Geographers*, *39*(4), 589–602. https://doi.org/10.1111/tran.12065

Maldonado-Torres, N. (2007). On the coloniality of being. *Cultural Studies*, *21*, 2–3. https://doi.org/10.13169/statecrime.7.2.0349

Maldonado-Torres, N. (2008). Lewis Gordon: Philosopher of the human. *The CLR James Journal*, *14*(1), 103–137.

Matusse, A. (2019). Laws, parks, reserves, and local peoples: A brief historical analysis of conservation legislation in Mozambique. *Conservation & Society*, *17*(1), 15–25. https://www.jstor.org/stable/26554467

Mavhunga, C. (2018). *The mobile workshop: The tsetse fly and African knowledge production*. MIT Press.

Mawdsley, E., Fourie, E., & Nauta, W. (2019). *Researching South-South development cooperation: The politics of knowledge production*. Routledge. https://doi.org/10.4324/9780429459146

Mbembe, A. (2024). *Brutalism*. Duke University Press. https://doi.org/10.1215/9781478027720

Meneses, P. (2020). Desafios Ambientais a Sul: o Ubuntu como ética de ligação entre a comunidade e a natureza. In G. Santos, & S. Sampaio (Eds.), *Observatório da vida estudantil: interdisciplinaridade, vida estudantil e diálogo de saberes* (pp. 21–61). Edufba.

Mignolo, W. (2018). Decoloniality and phenomenology: The geopolitics of knowing and epistemic/ontological colonial differences. *The Journal of Speculative Philosophy*, *32*(3), 360–387.

Mignolo, W. (2015). Sylvia Wynter: What does it mean to be human? In K. McKittrick (Ed.), *Sylvia Wynnter: On being human as praxis* (pp. 106–124). Duke University Press.

Mignolo, W., & Walsh, C. (2018). *On decoloniality: Concepts, analytics, praxis*. Duke University Press. https://doi.org/10.1215/9780822371779

Mohan, G. (2021). Below the belt? Territory and development in China's international rise. *Development and Change*, *52*, 54–75. https://doi.org/10.1111/dech.12612

Mpofu, B., & Ndlovu-Gatsheni, S. J. (2019). *Rethinking and unthinking development: Perspectives on inequality and poverty in South Africa and Zimbabwe*. Berghan Books. https://doi.org/10.1515/9781789201772

Murrey, A. (2018). 'When spider webs unite they can tie up a lion': Anti-racism, decolonial options and theories from the South. In E. Fiddian-Qasmiyeh, & P. Daley (Eds.), *Routledge handbook of South-South relations* (pp. 59–75). Routledge. https://doi.org/10.4324/9781315624495

Ndlovu-Gatsheni, S. J. (2012). Coloniality of power in development studies and the impact of global imperial designs on Africa. *The Australasian Review of African Studies, 33*(2), 48–73. https://search.informit.org/doi/10.3316/informit.066503630435392

Ndlovu-Gatsheni, S. (2018). *Epistemic freedom in Africa: Deprovincialization and decolonization*. Routledge. https://doi.org/10.4324/9780429492204

Ndlovu-Gatsheni, S. (2020). *Decolonization, development and knowledge in Africa: Turning over a new leaf*. Routledge. https://doi.org/10.4324/9781003030423

Nyamnjoh, F. (2006). Re-thinking communication research and development in Africa. In P. Zeleza (Ed.), *The study of Africa: Disciplinary and interdisciplinary encounters* (Vol. I, pp. 393–416). CODESRIA.

Obeng-Odoom, F. (2020). *Property, institutions, and social stratification in Africa*. Cambridge University Press. https://doi.org/10.1017/9781108590372

Oyěwùmí, O. (1997). *The invention of women: Making an African sense of western gender discourses*. University of Minnesota Press.

Pase, A. (2024). Emptying and filling. Maps of inland Africa. In T. Rossetto, & L. Lo Presti (Eds.), *The Routledge handbook of cartographic humanities* (pp. 235–245). Routledge. https://doi.org/10.4324/9781003327578

Pereira da Silva, F. (2023). *En busca de la comunidad: caminos del pensamiento crítico en la periferia global*. Editorial Theseo.

Ramose, M. (2009). Globalização e Ubuntu. In B. Sousa Santos, & P. Meneses (Eds.), *Epistemologias do Sul* (pp. 135–177). Almedina.

Rao, R. (2020). *Out of time: The queer politics of postcoloniality*. Oxford University Press. https://doi.org/10.1093/oso/9780190865511.001.0001

Rodriguez, J. K., & Ridgway, M. (2023). Intersectional reflexivity: Fieldwork experiences of ethnic minority women researchers. *Gender, Work & Organization, 30*(4), 1273–1295. https://doi.org/10.1111/gwao.12977

Roque de Pinho, J. (2009). *"Staying together": People-wildlife relationships in a pastoral society in transition, Amboseli ecosystem, Southern Kenya* [Unpublished doctoral thesis,. Colorado State University].

Sarr, F. (2020). *Afrotopia*. University of Minnesota Press.

Scott, J. C. (1998). *Seeing like a state. How certain schemes to improve the human condition have failed*. Yale University Press.

Seignobos, C., & Iyébi-Mandjek, O. (Eds.). (2000). *Atlas de la province Extrême-Nord Cameroun*. IRD.

Sousa Santos, B. (2009). Para além do Pensamento Abissal: das linhas globais a uma ecologia de saberes. In B. Sousa Santos, & M. Meneses (Eds.), *Epistemologias do Sul* (pp. 23–73). Almedina.

Sousa Santos, B. (2018). *O Fim do Imperio Cognitivo: a afirmacao das epistemologias do Sul*. Almedina.

Sousa Santos, B., & Meneses, M. (2009). Introdução. In B. Sousa Santos, & M. Meneses (Eds.), *Epistemologias do Sul* (pp. 9–21). Almedina.

Tamale, S. (2020). *Decolonization and Afrofeminism*. Daraja Press.

Terblanché-Greeff, A. C., & Nel, P. (2023). Measuring context-specific collectivism: The Metzian Ubuntu Inventory. *South African Journal of Philosophy*. https://doi.org/10.1080/02580136.2023.2206343

Thiong'o, N. wa (1986). *Decolonizing the mind: The politics of language in African literature*. James Currey.

Tsing, A. L. (2015). *The mushroom at the end of the world: On the possibility of life in capitalist ruins*. Princeton University Press.

Waisbich, L. T., Roychoudhury, S., & Haug, S. (2021). Beyond the single story: 'Global South' polyphonies. *Third World Quarterly*, *42*(9), 2086–2095. https://doi.org/10.1080/01436597.2021.1948832

Witter, R., & Satterfield, T. (2014). Invisible losses and the logics of resettlement compensation. *Conservation Biology*, *28*(5), 1394–1402.

Wiegink, N. (2022). Dislocation of the dead: Land, burial and resettlement around coal mines in Mozambique. In C. Navarra, & C. Rodrigues (Eds.), *Transformations of rural spaces in Mozambique* (pp. 91–105). Zed Books.

Wynter, S. (2003). Unsettling the coloniality of being/power/truth/freedom: Towards the human, after man, its overrepresentation – An argument. *The New Centennial Review*, *3*(3), 257–337. https://doi.org/10.1353/ncr.2004.0015

Yusoff, K. (2018). *A billion black anthropocenes or none*. University of Minnesota Press.

Index

Note: Page references in *italics* denote figures, in **bold** tables and with "n" endnotes.

FSC
www.fsc.org
MIX
Papier | Fördert gute Waldnutzung
FSC® C083411

Kolibri 360 GmbH
Ferdinand-Jühlke-Straße 7
99095 Erfurt, Deutschland
produktsicherheit@kolibri360.de